Lowland Floodplain Rivers

British Geomorphological Research Group Symposia Series

Geomorphology in Environmental Planning

Edited by
J. M. Hooke

Floods
Hydrological, Sedimentological and
Geomorphological Implications

Edited by
Keith Beven and **Paul Carling**

Soil Erosion on Agricultural Land

Edited by
K. Boardman, J. A. Dearing and **I. D. L. Foster**

Vegetation and Erosion
Processes and Environments

Edited by
J. B. Thornes

Lowland Floodplain Rivers
Geomorphological Perspectives

Edited by
P. A. Carling and **G. E. Petts**

Lowland Floodplain Rivers

Geomorphological Perspectives

Edited by
P. A. Carling
Institute of Freshwater Ecology
Ambleside, UK

and

G. E. Petts
Department of Geography
University of Technology
Loughborough, UK

JOHN WILEY & SONS
Chichester · New York · Brisbane · Toronto · Singapore

Other Wiley Editorial Offices

John Wiley & Sons, Inc., 605 Third Avenue,
New York, NY 10158-0012, USA

Jacaranda Wiley Ltd, G.P.O. Box 859, Brisbane,
Queensland 4001, Australia

John Wiley & Sons (Canada) Ltd, 22 Worcester Road, Rexdale,
Ontario M9W 1L1, Canada

John Wiley & Sons (SEA) Pte Ltd, 37 Jalan Pemimpin   05-04,
Block B, Union Industrial Building, Singapore 2057

Library of Congress Cataloging-in-Publication Data:

Lowland floodplain rivers   :   geomorphological perspectives   /   edited by
    P. A. Carling and G. E. Petts.
        p.   cm. — (British Geomorphological Research Group symposia
    series)
      Includes bibliographical references and index.
      ISBN   0-471-93119-5 (ppc)
      1. Floodplains.   I. Carling, Paul.   II. Petts, Geoffrey E.
    III. Series.
      GB561.L69   1992   91–21774
      551.3′5—dc20   CIP

British Library of Congress Cataloguing in Publication Data:

A catalogue record for this book
is available from the British Library.

ISBN   0-471-93119-5

Typeset in 10/12 pt Times by Acorn Bookwork, Salisbury, Wiltshire
Printed and bound in Great Britain by Courier International Ltd, East Kilbride

Contents

vi   Contents

List of Contributors

K. Beven Centre for Research on Environmental Systems, Lancaster University, Bailrigg, Lancaster, LA1 4YQ, UK

M. A. Bickerton Freshwater Environments Group, International Centre of Landscape Ecology, Department of Geography, Loughborough University of Technology, Loughborough, Leicestershire, LE11 3TU, UK

A. G. Brown Department of Geography and School of Archaeological Studies, University of Leicester, Leicester, LE1 7RH, UK

A. Bullock Institute of Hydrology, Maclean Building, Crowmarsh Gifford, Wallingford, OX10 8BB, UK

P. A. Carling Institute of Freshwater Ecology, Far Sawrey, Ambleside, Cumbria, LA22 OLP, UK

N. J. Clifford School of Geography and Earth Resources, University of Hull, Hull, HU6 7RX, UK

M. F. Goodchild Department of Geography, University of California, Santa Barbara, California 93106, USA

M. T. Greenwood Freshwater Environments Group, International Centre of Landscape Ecology, Department of Geography, Loughborough University of Technology, Loughborough, Leicestershire, LE11 3TU, UK

A. Gustard Institute of Hydrology, Maclean Building, Crowmarsh Gifford, Wallingford, OX10 8BB, UK

S. K. Hamilton Department of Biological Sciences, University of California, Santa Barbara, California 93106, USA

G. L. Harris ADAS Field Drainage Experimental Unit, Trumpington, Cambridge, CB2 2LF, UK

Q. He Department of Geography, University of Exeter, Exeter, Devon, EX4 4RJ, UK

A. D. Howard Department of Environmental Sciences, University of Virginia, Charlottesville, VA 22903, USA

M. Keough Department of Geography, University of Leicester, Leicester, LE1 7RH, UK

K. Kern Institut für Wasserbau und Kulturtechnik, Universität Karlsruhe, Kaiserstrasse 12, DS-7500 Karlsruhe 1, Germany

A. R. G. Large   Freshwater Environments Group, International Centre of Land-scape Ecology, Department of Geography, Loughborough University of Technology, Loughborough, Leicestershire, LE11 3TU, UK

D. M. Lawler   School of Geography, University of Birmingham, Edgbaston, Birmingham, B15 2TT, UK

W. M. Lewis, Jr   Centre for Limnology, Department of Environmental, Population and Organismic Biology, University of Colorado, Boulder, Colorado 80309-0334, USA

J. M. Melack   Department of Biological Sciences, University of California, Santa Barbara, California 93106, USA

T. A. Quine   Department of Geography, University of Exeter, Exeter, Devon, EX4 4RJ, UK

T. Parish   Ecological Processes Section, NERC Institute of Terrestrial Ecology, Monks Wood Experimental Station, Huntingdon, PE17 2LS, UK

G. E. Petts   Freshwater Environments Group, International Centre of Landscape Ecology, Department of Geography, Loughborough University of Technology, Loughborough, Leicestershire, LE11 3TU, UK

K. S. Richards   Department of Geography, University of Cambridge, Downing Place, Cambridge, CB2 3EN, UK

M. C. Thoms   River Murray Laboratory, Department of Zoology, University of Adelaide, Adelaide, South Australia 5001, Australia

C. R. Thorne   Department of Geography, University of Nottingham, University Park, Nottingham, NG7 2RD, UK

K. F. Walker   River Murray Laboratory, Department of Zoology, University of Adelaide, Adelaide, South Australia 5001, Australia

D. E. Walling   Department of Geography, University of Exeter, Exeter, Devon, EX4 4RJ, UK

Series Preface

The British Geomorphological Research Group (BGRG) is a national multidisciplinary Society whose object is 'the advancement of research and education in gemorphology'. Today, the BGRG enjoys an international reputation and has a strong membership from both Britain and overseas. Indeed, the Group has been actively involved in stimulating the development of geomorphology and geomorphological societies in several countries. The BGRG was constituted in 1961 but its beginnings lie in a meeting held in Sheffield under the chairmanship of Professor D. L. Linton in 1958. Throughout its development the Group has sustained important links with both the Institute of British Geographers and the Geological Society of London.

Over the past three decades the BGRG has been highly successful and productive. This is reflected not least by BGRG publications. Following its launch in 1976 the Group's journal, *Earth Surface Processes* (since 1981 *Earth Surface Processes and Landforms*) has become acclaimed internationally as a leader in its field, and to a large extent the journal has been responsible for advancing the reputation of the BGRG. In addition to an impressive list of other publications on technical and educational issues, including our 30 *Technical Bulletins* and *Geomorphological Techniques*, edited by A. Goudie, BGRG symposia have led to the production of a number of important works. These have included *Nearshore Sediment Dynamics and Sedimentation* edited by J. R. Hails and A. P. Carr; *Geomorphology and Climate* edited by E. Derbyshire; *Geomorphology, Present Problems and Future Prospects*, edited by C. Embleton, D. Brunsden and D. K. C. Jones, *Mega-geomorphology* edited by R. Gardner and H. Scoging, *River Channel Changes* edited by K. J. Gregory, and *Timescales in Geomorphology* edited by R. Cullingford, D. Davidson and J. Lewin. This sequence of books culminated in 1987 with a publication of the *Proceedings of the First International Geomorphology Conference* edited by Vince Gardiner. This international meeting, arguably the most important in the history of geomorphology, provided the foundation for the development of geomorphology into the next century.

This open-ended BGRG Symposia Series has been founded and is now being fostered to help maintain the research momentum generated during the past three decades, as well as to further the widening of knowledge in component fields of gemorphological endeavour. The series consists of authoritative volumes based on the themes of BGRG meetings, incorporating, where appropriate, invited contributions to complement chapters selected from presentations at these meetings under the guidance and editorship of one or more suitable specialists. Whilst maintaining a strong emphasis on pure gemorphological research, BGRG meetings are diversifying, in a very positive way, to consider links between geomorphology *per se* and other disciplines such as ecology, agriculture, engineering and planning.

The first volume in the series was published in 1988. *Geomorphology in Environmental Planning*, edited by Janet Hooke, reflected the trend towards applied studies.

The second volume, edited by Keith Beven and Paul Carling, *Floods–Hydrological, Sedimentological and Geomorphological Implications*, focused on a traditional research theme. *Soil Erosion on Agricultural Land* reflected the international importance of the topic for researchers during the 1980s. This volume, edited by John Boardman, Ian Foster and John Dearing, formed the third in the series. The role of vegetation in geomorphology is a traditional research theme, recently revitalized with the move towards interdisciplinary studies. The fourth in the series, *Vegetation and Erosion—Processes and Environments*, edited by John Thornes, reflected this development in geomorphological endeavour, and raised several research issues for the next decade.

One of the recent trends in fluvial research concerns river channel adjustments, especially those consequent to engineering works and land use change. The present volume, *Lowland Floodplain Rivers—Geomorphological Perspectives*, edited by Paul Carling and Geoff Petts, provides a useful insight into such issues.

The BGRG Symposia Series will contribute to advancing geomorphological research and we look forward to the effective participation of geomorphologists and other scientists concerned with earth surface processes and landforms, their relation to Man, and their interaction with the other components of the Biosphere.

John Gerrard

May 1991   BGRG Publications

Preface

PERSPECTIVE ON LOWLAND RIVER – FLOODPLAIN SYSTEMS

One of the evident trends in fluvial research in recent years is an emphasis on the broader implications of natural river channel adjustments and the longer term geomorphological consequences of major engineering works and landuse change. Within the remit of planning authorities this concern now focuses on the concept of the 'river corridor'. Not only is the wetted section of the river itself included and considered in the concept, but also is the riparian zone and the floodplain, the ecosystems associated with these environments and the socio-economic infrastructure developed by communities throughout history. The river corridor is consequently a palimpsest of fragile and fragmentary units recording the history of interactions with the main river channel. Outwith this channel, the primary geomorphological component is the floodplain; a structural complex influenced to a greater or lesser extent by flood inundation and groundwater fluctuations as well as by channel migration. It is proper, therefore, to consider how river channels interact with floodplain environments. This collection of essays is designed to reflect the holistic nature of the river valley; to shift attention away from the river bed *per se* and to consider the data requirements needed to understand better the channel–floodplain relationships over a range of time-scales.

THE REACH SCALE

In order to achieve the above objectives, it is useful to review contemporary endeavour at modelling river channel evolution at the reach scale rather than at the scale of individual sections. For, not only is a recapitulation of the state-of-the-art necessary to advance theoretical considerations, but also, and just as valuably, it serves to emphasize where field and laboratory data, required for verification procedures, are lacking or deficient.

Howard provides such a review based on his own recent modelling exercises, and notes at the outset, that there is an abundance of information concerning historical changes in lowland rivers that can be used to model systems but major difficulties still exist. In part, this is because a wide range of experts are needed; such as surveyors, photogrammetric analysts, dendrochronologists, sedimentologists, geomorphologists, engineers, hydraulic modellers, and geologists amongst others, whose efforts historically have not always been well integrated. Although the driving components of channel migration consist basically of flow, topography and sediment transport, it is in the near-bank region that attention should be focused; a region traditionally avoided in hydraulic studies wherein rivers are often assumed to be infinitely wide with bed adjustments predominating over bank deposition or erosion. Howard uses a simple consideration of near-bank velocity or shear stress, flow depth and bank erodibility as a principle component of his model, but notes that there is much uncertainty concerning the appropriate form of the bank erosion relationship. In part, concern for the minuteae of flow structure close to banks may not be required for modelling, but Howard's review highlights the dearth of even the most simplistic bank erosion formulations.

Although Howard's model does not model floodplain stratigraphy and sediment composition implicitly, as the outer bank of a meander recedes, structured point-bar deposition occurs against the inner bank. Howard notes that pioneering models of the hydraulics of meander migration have been linked to depositional facies modelling, but idealized translation and enlargement of individual bends is assumed and the principles cannot be applied readily at the reach-length scale. The spatial variability in floodplain stratigraphy induced by meander migration is complicated further by the suite of deposits associated with chute cut-off, avulsion and splays, about which little is known. Finally, as a first approximation, Howard's model includes deposition of fines by diffusion during overbank flows and as a simple function of height, although he notes that more complex models for overbank flow and fine particle transport now exist and could be incorporated if required. Finally, such processes as floodplain stripping and coarse sediment deposition on the floodplain surface are other, probably minor, processes that Howard identifies as poorly parameterized. In summary, it seems that the present generation of models can reproduce, at least superficially, the time series of patterns of meanders observed in nature. However, the theoretical development has outstripped empirical validation for reach-length scales in contrast to studies found on the scale of the individual bend section. In particular, there is a rarity of reach-length studies of floodplain inundation and deposition at a fine enough resolution for validating models of overbank deposition.

CHANNEL DYNAMICS

Clifford and Richards emphasize that detailed and simultaneous measurement of velocity and shear stress across rivers over a wide range of flows is required to characterize complex hydraulic behaviour and the response of channel form. Locally high (transient) shear stresses can occur in pools that exceed those recorded over riffles, but these should not be used alone to infer that a shear stress reversal occurs during high flows. In fact an alternative mechanism is required, to account for the maintenance of the distinctive topography and planform of pool–riffle sequences, which considers local variability and spatially distributed form–process feedback. The complexity of flow structure over the reach scale is again addressed by Bevan and Carling. Although most hydraulic data is gathered at the scale of the individual cross-section, practical problems of transport of sediments and dispersion of contaminants in lowland channels occur at the reach scale and are poorly understood. This is in part owing to mixing processes induced by secondary currents associated with bends and pool-riffle topography, but also to strong bank interactions, retarding flow, especially where riparian vegetation is well developed. Eulerian and Lagrangian techniques are used and compared to reveal aspects of the flow structure relevant to the reach scale, which would be difficult to parameterize using section data alone.

The three-dimensional structure of river currents is intimately related to the near-bank flow processes that drive bank recession. Thorne argues that to explain bend geometry and planform evolution, account must be taken of variation in bank properties that influence both basal scour and lateral recession. Taking the Red River as an example, Thorne addresses the resistance afforded by banks of differing materials to erosion by currents in bends, the mechanics of bank failure and the control exerted by bank composition on the scour depth through the bend. In particular, different materials have different geotechnical properties and therefore have different characteristic geometries and failure modes. The implications for bank recession and meander migration are considered. Generally, scour depth increases as the composition of the outer banks show greater resistance, especially in the case of stalled meanders formed against clay plugs. A thorough knowledge of floodplain geology is consequently required to explain and predict channel evolution, and temporal variation in suspended sediment supply from bank failure. Variation in bank properties, in part induced by subaerial weathering and geographical location along a river course, is further considered by Lawler. Lawler presents details of an automatic monitoring system designed to obtain time-series of bank recession rates over time-scales commensurate with the passage of a hydrograph or seasonal variation in weathering intensity. Although not suitable to monitor basal clean-out or rapid recession by block-fall, the technique is best suited for measuring spalling and dry ravel over limited areas of river bank.

FLOODPLAIN DEVELOPMENT

As noted by Howard, floodplain topography is poorly parameterized and could be investigated profitably by using remote sensing techniques with fine resolution. The power of remote sensing for detailed analysis of low-relief floodplain topography is enormous, but little studied; allowing classification of lake form, distinction of topographic features and possibly the mapping of spatial concentration of suspended solids and flow lines over inundated areas. Hamilton *et al.* demonstrate this potential using radar imagery to map the myriad lakes on the Amazon and Orinoco floodplains. Their measurements indicate that the flooded topographic lows have statistical self-similarity with respect to area, implying that relief on the floodplains displays fractal characteristics. Such a result, if of general application, would have implications for modelling spatial variation in floodplain deposition. Walling, Quine and He address the problem of measuring the spatial variability in deposition rates over floodplain surfaces using bomb- and Chernobyl-derived [137]Cs. This is a logical development to its use in dating lake sedimentation, and although still somewhat problematic, offers the possibility of indicating approximate depths of deposition over given time periods. Although it is unlikely that the technique would find wide utility, it offers the important possibility of mapping in detail, and over large areas, spatial patterns of sedimentation with respect to topographic variation.

Overbank deposition couples with deposition from channel processes to provide a three-dimensional sedimentary body. Brown and Keough use meticulous study of exposures to argue that the floodplain of the River Nene is of polygenetic origin, including pedogenic processes, flood deposition, and localized stripping but inconsequential channel erosion throughout wide areas. In fact, large areas of the floodplain have not been reworked in the last 2000 years and the investigation also provided valuable information on the temporal sequence of events. Two major periods of channel change and lateral deposition were identified, as well as a period dominated by channel stability and vertical deposition. Associated with this depositional history are changes in the water table that have biased the nature of the biotic record. It is studies such as these that are vital to provide the time-frame and spatial and vertical co-ordinates for simulation exercises.

CHANNELS, FLOODPLAINS AND MAN

Fluctuations in water tables, whether induced by man or natural, influence the groundwater flux of nutrients and pollutants, and the local floodplain ecology. Coincidental with land drainage are the associated changes in landuse that often accompany conversion of floodplain marsh or water-meadow to arable fields. Increased use of fertilizers and tillage alters the soil structure and the composition of natural species present. In recent years conversion of floodplains to arable land has been cited as the cause of increased nitrate in arterial river systems, but the hydrological processes in floodplains are poorly known. Harris and Parish examine how changes in agricultural practice influence soil processes and nitrate leaching, with highest concentrations recorded in former floodplain areas. Lowering of the water table and the mineralization of organic floodplain soils is shown to result in high nitrate losses in drainage water, which is likely to persist for several decades. Concentrations exceeding 175 mg l^{-1} were recorded in pipe-drain outfalls from recently drained floodlands.

Lowered groundwater levels, reduced frequency and duration of inundation, and changed patterns of erosion and sedimentation can have marked effects on the mosaic of ecological patches that characterize the floodplain ecotone. Petts et al. examine the influence of hydrology and geomorphology on floodplain ecology, focusing on ecological successions within abandoned channels. An approach is presented to construct successional models and to assess the conservation value of ecological units. Using (Staphylinidae) Coleoptera as functional describers, the riparian zone is shown to provide an important but very different habitat from the wetland and wet woodland successional units of cut-off channels in the floodplain. A diversity of species are shown to depend on these seasonally inundated floodplain habitats, including a number of nationally rare species. In the light of the changing agricultural policy, it is argued that creation of new backwaters and restoration of floodplain cut-off units could make a major contribution to conservation development.

River regulation also has influenced instream habitats significantly. Thoms and Walker argue that lowland rivers in Australia, distinguished by low gradients and long-term sediment storage in the floodplains, have been studied inadequately with respect to the long-term implications of adjustments to engineering. Taking the case of the River Murray from 1906 to present, the paper outlines the morphological adjustment of the river to multiple weir construction in 1922–1935. The weirs trap 73% of the sediments, but regulated competent flows still peak at the channel capacity. The fact that the planform is relatively stable, is attributed to cohesive banks and an overall low streampower, and yet some adjustments still are not complete, reflecting the long relaxation times that may be associated with channel changes.

Establishing deterministic linkages between fluvial geomorphology and ecology remains elusive, but considerable progress has been made in developing management tools. In the USA the instream flow incremental methodology (IFM) has become a standard approach for assessing ecological (especially fisheries) impacts of flow regulation.

Bullock and Gustard explain how a programme of research is being developed to implement the IFM for British lowland rivers. Habitat versus flow relationships are developed for various species at different life stages and may be used to set minimum flows, discharge consents and compensation flow levels for example. The work clearly is at the development stage, but also should prove a useful tool in those cases where restoration of habitat is contemplated and design criteria sought at the planning stage.

PROSPECT FOR RESTORATION

Perhaps it is appropriate to end this compendium with a consideration of our ability to restore lowland rivers (which have been constrained from meandering) to a semi-natural state that serves a multitude of purposes and users. This preferred balance between controlled rivers vital for the economy and more natural rivers important for ecological diversity, aesthetic and recreational needs is addressed by Kern. The nub of research effort, as exemplified by the contributions in this volume, must lie in successful management of river and floodplain systems. The success or otherwise of attempts to predict what a river system will do when given several degrees of freedom is a most appropriate test. The problem from the German perspective is that all lowland rivers were controlled during the nineteenth century and little is known of the geomorphology and hydraulic character of pre-regulation rivers. Nevertheless, public opinion requires that consideration be given to restoring water courses to 'near-natural' conditions.

Kern outlines the detailed planning principles used to arrive at the most practical solution, starting with the 'Leitbild' concept, which details desirable stream properties in the absence of human pressure, and leads on to optimal and feasible solutions. This process is demonstrated with a fascinating case study of the Upper Danube, where meander sequences have been reimposed on a straightened reach so that high flows still pass down the straightened channel. One prime question was how to design a natural meander in soils that may cut-back 100 m in 30 years? Here then we have turned full circle, returning to the requirements of the modeller and the necessity of detailed field observation. In the event, the meanders were designed by physical hydraulic modelling, but as Kern would admit, no one is really sure how the system will evolve with time!

Paul Carling
March 1991   Geoff Petts

1 Modeling Channel Migration and Floodplain Sedimentation in Meandering Streams

ALAN D. HOWARD
University of Virginia

INTRODUCTION

Meandering streams are one of the few geomorphological systems for which an abundant historical record exists of changes in channel pattern and associated floodplain erosion and deposition. Despite the evidence from surveys, aerial photographs, topographic mapping, process measurements, dendrochronology and floodplain stratigraphy, geomorphologists and sedimentologists are just beginning to construct realistic process models of meandering stream evolution. The model discussed here combines simulated bank erosion and channel migration with a simple model of floodplain sedimentation. Such simulation modeling has both practical and theoretical utility for prediction of channel and floodplain changes, validation of theoretical process models, and increased understanding of the sedimentological structure of fluvial deposits, with implications for petroleum geology and groundwater flow.

The model discussed here has three major components. The first is a model of flow, bed topography, and sediment transport in meandering streams. This component has been the primary stumbling block in developing simulation models of stream meandering and sedimentation, because appropriate theoretical models have become available during the last decade only. The second component is a relationship between near-bank velocity and depth and corresponding rates of bank erosion and lateral migration. The final ingredient is a process model of floodplain sedimentation. The marriage of a realistic model of meandering with floodplain sedimentation is the novel contribution of this paper.

In the first section the basic structure of the model is presented. Results of some simulations are presented in the second section to illustrate the essential features of the model. The present version of the model is preliminary; the discussion presents possible enhancements and extensions of the model, and research needed to validate and improve such simulation models.

Lowland Floodplain Rivers: Geomorphological Perspectives. Edited by P. A. Carling and G. E. Petts
© 1992 John Wiley & Sons Ltd

FLOW AND SEDIMENT TRANSPORT MODEL

Since the bank erosion rate is assumed to be related to near-bank flow velocity and depth, an essential element is a mathematical model to predict flow and bed configuration within a sinuous channel. Several such models are available, ranging from linearized one-dimensional (along-stream) models with implicit representation of cross-stream variations in flow, topography, and sediment transport (Johannesson and Parker, 1989; Odgaard, 1989a,b; Parker and Johannesson, 1989) through two-dimensional (downstream and cross-stream) solutions (e.g. Nelson and Smith, 1989a,b). All of these models incorporate simplifying assumptions of the governing equations in order to make numerical solutions feasible. The Johannesson and Parker (1989) model (abbreviated JP) is adopted here, because it captures the essential features of flow, bed topography, and transport in meandering streams, and it is easy to implement and computationally efficient. The JP model is a descendent of the pioneering paper by Ikeda, Parker and Sawai (1981), and it provides good predictions of the bed topography and the flow characteristics in experimental meandering channels with narrow width, vertical banks, and mobile sediment beds. Furthermore, the JP model, when combined with the assumption that bank erosion rates are proportional to near-bank flow velocity, gives accurate estimates of bank erosion rates in natural channels (see later discussion). The Odgaard (1989a,b) model is very similar, and could be investigated as an alternative. In the model, local depth (h) and downstream vertically averaged velocity (u) are resolved into a section mean (H and U) and a dimensionless perturbation (h_1 and u_1) (Figure 1.1):

$$u = U(1 + u_1) \tag{1.1}$$

$$h = H(1 + h_1) \tag{1.2}$$

Generally, we are interested in near-bank values, indicated as u_{1b} and h_{1b}. At the channel centerline u_1 and h_1 are assumed to equal zero. Several simplifing assumptions are incorporated, among which are a spatially and temporally constant channel width and a linear cross-stream variation in the vertically averaged downstream velocity, with negligible sidewall effects on near-bank flows (Figure 1.1a). In other words, the effect of sidewalls on the downstream velocity field is assumed to be important in a narrow zone at the bank only, and the velocity distribution in this zone is not modeled explicitly. The bed and water surface are assumed to be sloping linearly in the cross-stream direction (Figure 1.1b), although the magnitude and direction of the slope vary downstream. Thus the model provides a crude representation of the point bar as a uniformly sloping bed. Furthermore, the energy gradient and average channel depth are assumed to be uniform in the downstream direction. The model predicts the steady-state values of flow, bed topography and sediment transport; therefore, transient bedforms, such as ripples, dunes and migrating transverse bars, are not modeled and their presence is assumed not to introduce systematic effects on local time-averaged depth and velocity. Exposition of the JP model in this paper will be limited to identification of the important model input parameters and local variables and presentation of the governing differential equations. The parameters that are input to the simulation to describe channel and sediment characteristics (Table 1.1) are assumed to be constant areally. Several additional parameters are

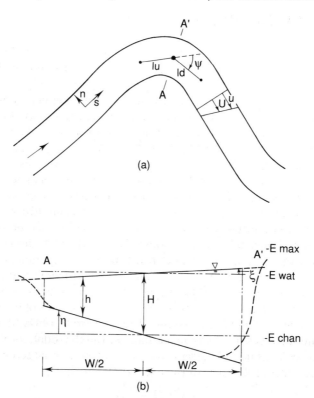

Figure 1.1 Planform (a) and cross-section (b) of a meandering stream showing definitions: (a) cross-stream and downstream coordinate system, downstream vertically
averaged velocity (equation (1.1)), and measurement of channel planform curvature
using centerline nodes to represent the stream (equation 1.16)); and (b) channel depth
(equation (1.2)), bed and water surface elevations (equations (1.A8a) and (1.A8b)), and
reference levels for the depositional model (equations (1.17) and (1.18)). Reference levels
for bed- and water-surface elevations are the average elevations of the bed and water
surface, respectively. The inner and outer near-bank locations shown by vertical long-
short dashed lines. The assumed bed cross-stream profile is shown by the heavy solid
line, and possible actual banks are shown by dashed lines

Table 1.1 Flow and transport input parameters

Parameter	Description
β	Cross-stream slope effect on cross-stream bedload transport $(1.5)^a$
M	Exponent relating velocity to bedload transport rate (5.0)
F_0	Froude Number: $F_0 = U_0/(g\,H_0)^{1/2}$, where U_0 and H_0 are reach-averaged velocity and depth for a straight channel with gradient equal to the valley gradient, and g is the gravitational acceleration (0.5)
γ_0	Channel width/depth ratio: $\gamma = W/H_0$, where W is channel width (20.0)
C_f	Coefficient of friction: $C_f = (u_*/U)^2$, where u_* is the shear velocity (0.01)

[a]Values in parentheses are those assumed for simulations reported here.

Table 1.2   Local variables

Variable	Description
$\mathscr{C}_w = W\mathscr{C}$	Channel planform curvature normalized by width
$\mathscr{C}_{ws} = W\mathscr{C}_s$	Normalized secondary current cell strength
$S_W = S/W$	Downstream distance normalized by width
u_{1b}	Dimensionless near-bank velocity perturbation
h_{1b}	Dimensionless near-bank depth perturbation

derived from the input parameters (Appendix A). Several local variables are calculated by the model for each location along the stream (Table 1.2). The governing differential equations in the JP model and the solution method for calculating the depth and velocity perturbations are presented in Appendix A. In these equations all distances and the channel curvature are non-dimensionalized by the average channel width, W, so that the distance unit is *width-equivalents*.

Discussion of Flow and Transport Model

Flow, sediment transport and bank erosion in natural stream channels occurs over a spectrum of time and spatial scales. Fluvial sediment transport and its associated bedforms can be ordered into a sequence of increasing time-scale of development: (1) motion of individual particles; (2) ripples; (3) dunes; (4) alternate transverse bars; (5) point bars associated with channel curvature and meander development. An increasing spatial scale is generally associated with this sequence. Although not always warranted, mathematical modeling of any given bedform type usually relies on an averaged representation of the effects of the smaller, more transitory bedforms and their associated flow features and sediment transport phenomena. For example, a model of aeolian dune development utilizes sediment transport formulae that predict the integrated particle flux rather than motion of individual particles (Howard *et al.*, 1978). Furthermore, superimposed ripples are incorporated only through their averaged effects on velocity profiles and sediment transport. The JP model predicts the time-independent average values of bed topography and flow in meandering streams forced by channel curvature only, and thus does not treat migrating dunes and alternate bars (although a time-dependent version of the JP model can be used to predict properties of alternate bars in straight channels—Parker and Johannesson, 1989). In particular, the time-independence assumption implies that a completely straight channel of uniform width, gradient and average depth will have a level, planar bed.

The work of Ikeda, Parker and Sawai (1981) showed that flow asymmetries set up by channel curvature imply development of a regular meandering pattern if bank erosion rates correlate with near-bank velocity perturbations; simulations by Howard and Knutson (1984) and those in the present paper (Figure 1.2) demonstrate that the Ikeda model and its descendants imply development of meanders from a channel that is straight except for small perturbations normal to the flow direction. Thus the curvature-forced perturbations of velocity and depth are a sufficient mechanism to cause development of meanders.

However, other models of stream meandering (Parker, 1976; Callander, 1978; Fredsoe, 1978) incorporate the assumption that development of stream meanders are caused by periodic flow asymmetries associated with alternate bars. Observations of meander development in an initially straight gravel channel apparently forced by alternate bars (Lewin, 1976) offer support for this model, at least for wide, gravel-bed streams. However, models of meander evolution relying on alternate bars have two deficiencies as a universal explanation for meandering: (1) they predict a non-

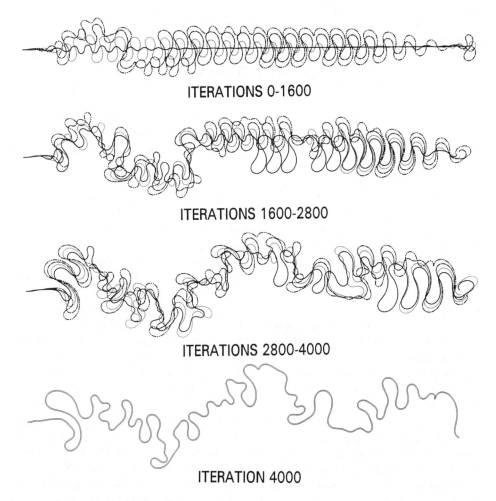

ITERATIONS 0-1600

ITERATIONS 1600-2800

ITERATIONS 2800-4000

ITERATION 4000

Figure 1.2 Evolution of the channel centerline for 4000 iterations using the flow and bed topography model of Johannesson and Parker (1989) coupled with the assumption of bank erosion proportional to the near-bank velocity perturbation, u_{1b} (equation (1.6)), and input parameters from Table 1.1. The simulation starts from a straight stream with small, random normal perturbations, which is not shown but would be an essentially straight horizontal line. Flow is from left to right. In the three top panels the channel centerline is shown at 400 iteration intervals with the sequence solid, dotted, long-dot-dot, and long-short-short lines. In the two central panels the final centerline from the preceding panel is shown as a solid line. The final panel shows the banks of the final channel position

meandering planform for channels too narrow ($\gamma < 10$) for development of alternate bars (the curvature-based model allows meandering under such conditions); (2) although alternate bars can remain fixed in position for specific combinations of flow conditions and cross-sectional geometry, under a range of conditions alternate bars migrate down-channel at a time-scale more rapid than bank erosion rates, at least for streams with cohesive banks.

Therefore, both curvature-forced variations in velocity and depth and alternate bars may control development of meanders. The natural wavelengths of meandering associated with the curvature forcing and alternate bar forcing may not be the same, leading to the possibility of multiple wavelength scales. In many cases migrating alternate bars occur in meandering channels (e.g. Kinoshita, 1961; Fukuoka, 1989; H. Ikeda, 1989; Whiting and Dietrich, 1989; Tubino and Seminara, 1990). Although the present model incorporates the assumption that migrating bars do not affect average bank erosion rates in a systematic manner, such interactions may occur (Whiting, 1990). For certain combinations of width/depth ratio and flow parameters, alternate bars become stationary, and if their natural wavelength is the same as the meander wavelength, a 'resonance' occurs, under which conditions the linearized models, such as JP, predict very large amplitude bars (Blondeaux and Seminara, 1985; Colombini, Seminara and Tubino, 1987; Parker and Johannesson, 1989; Seminara and Tubino, 1989; Tubino and Seminara, 1990). Whether such high-amplitude resonance occurs in natural channels is uncertain at present; non-linear effects may dampen and modify such resonance (G. Parker and W. Dietrich, pers. comm.). Another, and possibly related observation is that alternate bars migrate freely in low amplitude sinuous channels but can become suppressed in high-amplitude sinuous channels, possibly reforming in very high-amplitude meanders (Kinoshita, 1961; Fukuoka, 1989; Tubino and Seminara, 1990; Whiting, 1990). Such locking and suppression may induce systematic variations in flow and bed topography that is not accounted for by the linearized models such as JP, and which could affect bank migration rates (Seminara and Tubino, 1989; Whiting and Dietrich, 1989). This possibility is addressed further in the Discussion section through statistical comparison of the morphometry of simulated and natural meanders. Finally, the JP model also is clearly inadequate in the case where the width/depth ratio is great enough ($\gamma \geq 40$) for braiding to become important.

In conclusion, the present flow and bed topography model is best suited to channels in which resonant conditions do not occur and where alternate bars, if present, migrate rapidly through the channel in comparison with bank erosion rates. In fact, the JP model appears to be unable to provide a numerically stable solution to flow and bed topography for channels of arbitrary meander planform under conditions close to resonance. The natural conditions most likely to match these restrictions are low width/depth ratio, relatively cohesive banks, and a high suspended-load to bed-load ratio.

BANK EROSION RATE LAWS

Any of four constraints (or processes) may limit the rate of bank erosion. These constraints are, or may be, sequentially linked, so that the slowest among them controls the overall rate.

(1) The rate of deposition of the point bar.
(2) The ability of the stream to remove the bedload component of the sediment eroded from the bank deposits via a net transport flux divergence.
(3) The ability of the stream to entrain sediment from *in situ* or mass-wasted bank deposits.
(4) The rate with which weathering acts to diminish bank sediment cohesion to the point that particles may be entrained by the flow or bank slumping may occur.

Constraint (1) would be limiting for the case where deposition of the point bar were to lag behind bank erosion, so that flow velocities diminish as the channel widens and possibly shallows, with a corresponding decrease in bank erosion until bar deposition 'catches up'. Alternatively, rapid deposition on the point bar might narrow the channel and increase velocities and corresponding bank erosion rates. Neill (1984) related bank erosion to bedload transport rates, which, in part, determines point-bar deposition rates. Observations of rapid bank erosion below cut-offs (Kondrat'yev, 1968; Kulemina, 1973; Brice, 1974b; Bridge *et al.*, 1986) have been suggested to result from efflux of sediment to the next bend (Nanson and Hickin, 1983). However, rapid erosion can also occur owing to high near-bank velocities resulting from the steeper gradients through the cut-off, large curvatures at the cut-offs and changes in bend flow pattern (Howard and Knutson, 1984; Bridge *et al.*, 1986).

Constraint (1) is probably not the limiting factor in most meandering streams. Banks, generally, are more cohesive than the bed so that processes of bank erosion are limiting. Exceptions could occur if the banks are composed of non-cohesive sediments finer in gain size than the channel bed. However, in this case the channel width/depth ratio is likely to be large enough that a braided stream pattern will develop.

The distinctions between constraints (2) through to (4) are subtle but important. Constraint (2) will be the rate limiting factor where the banks are non-cohesive or easily disaggregated and the resulting accumulation of bedload-sized sediment at the foot of banks inhibits further bank erosion until it is removed. The overall rate of bank erosion will thus be related to near-bank flow, sediment transport, and bank height and composition (Hasegawa, 1989a,b). If the banks and slumped bank material are slightly cohesive, the rate of bank erosion will be determined by the detachment capability of the flow (constraint 3), and overall bank erosion rates will be less than if constraint 2 were limiting. If the bank material is strong (e.g. indurated alluvium or rock walls), then erosion by particle entrainment or mass-wasting can be limited by processes of bank disaggregation, such as frost action or chemical weathering, that may or may not be related directly to flow characteristics (constraint 4). It seems likely that all four cases may occur and vary in importance among streams, from place to place along a given stream, and through time at a given location. A variety of processes and material factors that may control bank erosion have been observed, including slumping and toppling (Laury, 1971; Thorne and Lewin, 1979; Thorne and Tovey, 1981; Thorne, 1982; Pizzuto, 1984; Ullrich, Hagerty and Holmberg, 1986; Osman and Thorne, 1988; Thorne and Osman, 1988), freeze-thaw (Wolman, 1959; Lawler, 1986a,b), removal of sediment from the base of the cut bank (Nanson and Hickin, 1986; Hasegawa, 1989a,b), vegetation type and density (Brice, 1964; Pizzuto, 1984; Odgaard, 1987; Hasegawa, 1989a,b), and soil type (Grissinger, 1966, 1982; Turnbull, Krinitsky and Weaver, 1966; Goss, 1973; Murray, 1977). None the less, fairly simple models relating flow characteristics to bank erosion rates are successful in many stream systems. Overall, the regularity of form and migration pattern of most meandering streams also suggests that fairly simple relationships can be used to predict long-term rates of bank erosion.

One such relationship expresses the bank erosion rate, ζ, as a function of the difference between the near-bank shear stress, τ_b, and the average boundary shear stress, τ:

$$\zeta = \mathscr{E}\,(\tau_b - \tau)/\tau \qquad (1.3)$$

where $\mathscr{E}$ is bank erodibility (units of length per unit time), which may depend upon bank sediment characteristics, flow properties and channel planform shape. This equation can be re-expressed in terms of flow velocities using the definition of shear velocity and the assumed constancy of the coefficient of friction (Table 1.1):

$$(\tau_b - \tau)/\tau = (u_b^2 - U^2)/U^2 \qquad (1.4)$$

By definition (1.1)

$$u_b^2 = U^2\,(1 + u_{1b})^2 \simeq U^2\,(1 + 2u_{1b}) \qquad (1.5)$$

where in the right-hand side the squared (higher order) term in the velocity perturbation has been dropped. This results in a linear relationship between bank erosion rate and the velocity perturbation

$$\zeta \simeq 2\,\mathscr{E}\,u_{1b} \qquad (1.6)$$

This linear relationship between bank erosion and velocity perturbation has been assumed in the models of Ikeda, Parker and Sawai (1981), Beck (1984), Beck, Mefli and Yalamenchili (1984), Parker (1984), Howard and Knutson (1984), Parker and Andrews (1986), Hasegawa (1989a,b), and Odgaard (1987). Hasegawa (1989a,b) shows a strong correlation between observed bank erosion rates and u_{1b} in short stretches of natural meanders. Pizzuto and Meckelnburg (1989) and Odgaard (1987) have made similar observations. Parker (1984) and Furbish (1988) show that the observations of Hickin and Nanson (1975,1984) and Nanson and Hickin (1983,1986) of a direct relationship between mean bend curvature and bank erosion rate in gentle bends is consonant with a proportionality between erosion rate and the velocity perturbation.

Since u_{1b} lags significantly the downstream changes in curvature (Figure 1.3), meanders both migrate downstream and enlarge in amplitude, with eventual neck cut-offs (Parker and Andrews, 1986; Howard and Knutson, 1984; Parker, 1984). However, as individual meander loops increase in amplitude, these models also predict that inflection points tend to become fixed and little downstream migration occurs; this is illustrated in Figure 1.2, in which successive positions of the simple asymmetric loops at the right side of panels 2 and 3 intersect at nearly fixed positions near the inflection points.

Equation (1.6) is probably most relevant to erosion rate being limited by detachment of either *in situ* or slumped cohesive bank sediment (constraint 3). Erosion rates of cohesive sediments are commonly found to correlate with the applied fluid shear force (Parthenaides, 1965; Parthenaides and Paaswell, 1970; Akky and Shen, 1973; Parchure and Mehta, 1985; Ariathurai and Arulandan, 1986; Kuijper, Cornelisse and

Winterwerp, 1989), and Howard and Kerby (1983) found that channel erosion rates in badlands on mudstones and shales were related linearly to bed shear.

However, there is likely to be a critical near-bank shear stress, τ_t, below which bank erosion ceases. This suggests that equation (1.6) should be rewritten as

$$\zeta = \mathscr{E}\,(\tau_b - \tau_t)/\tau \simeq \mathscr{E}\,(2u_{1b} - \pi) \tag{1.7}$$

where

$$\pi = \left\{ \begin{array}{c} 0 \\[2mm] \dfrac{\tau_t - \tau}{\tau} \end{array} \right. \quad \begin{array}{l} \text{for } \tau_t \leqslant \tau \\[4mm] \text{for } \tau_t > \tau \end{array}$$

and erosion occurs for positive values of $(2u_{1b} - \pi)$ only. Since overall channel width is presumably determined by a dominant discharge near bankfull, the value of τ_t would be the average boundary shear associated with that stage. Note that equation (1.7) implies that localized bank erosion would occur at low stages and more generalized erosion at high stages, and that (1.7) is equivalent to equation (1.6) if the dominant discharge is such that $\tau \geqslant \tau_t$.

Odgaard (1989a,b) has suggested that bank erosion is related to the depth perturbation rather than the velocity perturbation. By analogy to equation (1.6)

$$\zeta \simeq 2\,\mathscr{E}h_{1b} \tag{1.8}$$

Odgaard (1989b) presents data that suggests that the depth perturbation is better at predicting the location of first outer bank erosion along bends of the Nishnabotna River, Iowa. Odgaard presents little justification for his erosion model, but mentions the analyses of Osman and Thorne (1988) and Thorne and Osman (1988), which indicate that bank stability decreases with bank height. Nanson and Hickin (1986) found a good correlation between bank erosion rates and the grain size of sediment exposed in the deepest scour holes in meander bends of Canadian rivers, with the implication that deeper bank sediments (representing bedload rather than overbank deposition) are less cohesive and therefore entrained more readily. Lapointe and Carson (1986) feel that bank erosion near the beginning of bends is related more to great depth rather than high velocity.

There is an important consequence to patterns of meander evolution if erosion rates are related to h_{1b} rather than u_{1b}. For flow and sediment characteristics that are typical of natural stream channels, the depth perturbation is nearly in phase with, or may even lead the curvature (Figure 1.3). This implies that meanders would tend to grow in amplitude, with negligible downstream migration.

Hasegawa (1989a,b) provides an analysis of the factors controlling bank erosion if transport of eroded bank sediment (constraint 2) is the limiting factor. Six terms related to transport rate, sediment characteristics, flow properties and bank geometry emerge from his analysis (Hasegawa, 1989a, his equation 8). Three of the terms are of second-order importance only, and the remaining three can be summarized as follows for conditions where sediment transport rate is well above threshold conditions:

$$\zeta = \mathscr{E}(u_{1b} - h_{1b}/2 - H_b/3H) \tag{1.9}$$

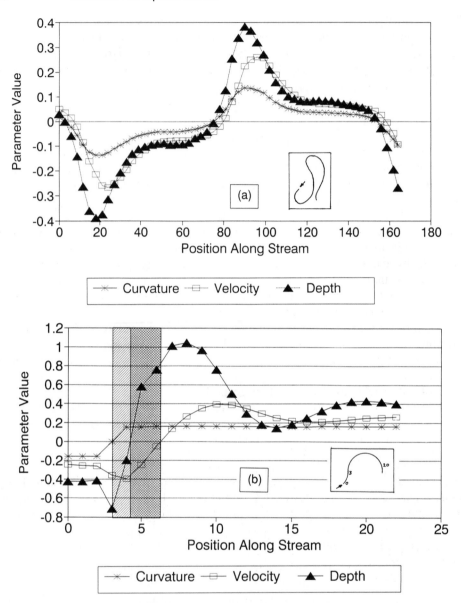

Figure 1.3 Local channel curvature and near-bank velocity and depth perturbations in meanders as predicted by the Johannesson and Parker model (1989) using parameters from Table 1.1. X-axis is position in bend measured downstream in width-equivalent units. Curvature is width-normalized; (a) a large asymmetric meander such as occurs in iterations 1600–2800 near the right-hand side of Figure 1.2; the channel planform is shown in inset; (b) a meander with an abrupt change of sign of curvature, with planform shown in inset. In (b) the initial velocity and depth perturbations are close to equilibrium values for constant curvature. Note the overshooting effects in the near-bank depth and velocity responses to curvature change. Ruled and cross-ruled areas delineate the zone along the stream in which the curvature and velocity perturbations have opposite sign and bank erosion is directed towards the inner, convex bank if it is proportional to the near-bank velocity perturbation u_{1b}. The cross-ruled area shows the zone in which the depth and velocity perturbations have opposite sign

where ζ is the rate of bank erosion, $\mathscr{E}$ depends upon bank sediment characteristics and transport parameters, and H_b is the height of the bank above water level. Hasegawa suggests that the third term will be of smaller magnitude than the other two and can be neglected. However, where a stream impinges on a tall bank of non-cohesive sediment (e.g. a terrace) the emergent bank height would become important. Hasegawa (1989a) also suggests that depth perturbation (second term) 'does not directly arrest erosion, but rather works only to decrease the erosion rate' and 'can be left out of consideration' (p. 226). However, this reasoning is counter to his analysis, which suggests that h_{1b}, the magnitude of which may exceed u_{1b} in tight bends, is of direct importance. Inclusion of the depth perturbation term may indeed have important effects on patterns of bank migration, because the depth and velocity perturbations are out of phase (Figure 1.3). The negative weighting of depth in Hasegawa's relationship (opposite to the positive weighting in Odgaard's model) is a result of the greater amount of sediment contributed from higher banks, and its effect would be to shift the locus of maximum erosion downstream from the locus of maximum near-bank velocity, thereby increasing the ratio of rates of downstream meander migration to meander enlargement.

Where erosion rates are limited by disaggregation processes (constraint 4) acting on the channel banks (such as frost action, wetting and drying, or progressive bank failure), erosion rates may have an upper limit that is independent of the local flow perturbation.

Thus there is considerable uncertainty concerning an appropriate form for the bank erosion relationship. A general relationship is suggested here that includes weighted values of both the depth and velocity perturbation

$$\zeta \simeq 2\,\mathscr{E}\,(a u_{1b} + \varepsilon h_{1b}) \tag{1.10}$$

where the weight a is probably positive, and ε may be positive, negative, or zero. If very high emergent banks occur locally, an additional term may be included.

Bank erodibility $\mathscr{E}$ may depend upon a number of factors, including bank sediment characteristics, processes determining the rate of bank disaggregation and bank height. Hasegawa's (1989a,b) analysis for transport-limited bank erosion indicates a dependency on sediment density, friction angle, and porosity as well as transport stage. Hickin and Nanson (1984) and Nanson and Hickin (1986) relate bank erodibility to median grain size, d, of sediment at the base of cut banks and the ratio of stream power to bank height. This suggests that

$$\mathscr{E} = \frac{\tau\,\gamma\,U}{\mathscr{F}(d)} \tag{1.11}$$

where $\mathscr{F}(d)$ is an empirical bank resistance function (units of shear stress) that has a form similar to the classic critical tractive force diagrams. Nanson and Hickin (1986) feel that their results are consistent with transport of eroded sediments (constraint 2) being the rate-limiting process. Hasegawa (1989a,b) finds an inverse relationship between measured penetration resistance of banks and bank erodibility. Bank erodibility also may be a function of type and density of bank vegetation.

The present simulation model utilizes equation (1.10) to predict bank erosion rates, with $a = 1$ and $\varepsilon = 0$, in accord with most previous models of bank erosion.

SIMULATION PROCEDURES FOR CHANNEL MIGRATION

The simulation procedures for the bank erosion and channel migration component of the model are similar to those adopted by Howard and Knutson (1984). A number of simplifying assumptions are made, including (1) constant bank erodibility, $\mathscr{E}$, (2) uniform width-averaged sediment load, (3) slow enough bank migration so that the erosion by individual flow events can be represented by a continuous process, (4) a single thread channel with spatially and temporally constant width, and (5) constant input of water and sediment from upstream and a constant downstream base level, so that the stream is not aggrading or downcutting. The *valley* gradient is assumed to be constant, so that the average *channel* gradient is inversely proportional to sinuosity, $\mathscr{T}$. The input parameters β, M, C_f and W are assumed to be constant temporally, but the depth, width/depth ratio, mean velocity and Froude number must be corrected for changing sinuosity:

$$H = H_0 \, \mathscr{T}^{1/3} \tag{1.12}$$

$$\gamma = \gamma_0 \, \mathscr{T}^{-1/3} \tag{1.13}$$

$$U = U_0 \, \mathscr{T}^{-1/3} \tag{1.14}$$

$$F = F_0 \, \mathscr{T}^{-1/2} \tag{1.15}$$

where H_0, γ_0, and F_0 are the values for a straight channel with a gradient equal to the valley gradient, S_0 (the above follow from the relationships $\tau = \varrho g H S = \varrho u_*^2$, $C_f = (u_*/U)^2$, and $\mathscr{T} = S_0/S$).

The simulation proceeds by repeated iterations, each iteration proceeding downstream through the individual points, or nodes, that represent the channel centerline. Individual nodes have a nominal downstream spacing of one width-equivalent. At each node the local near-bank velocity and depth are calculated by the procedures outlined in Appendix A. Local dimensionless curvatures $\mathscr{C}_w$ (Table 2) used in these procedures are calculated by

$$\mathscr{C}_w = 2 \, W \, \Psi/(l_u + l_d) \tag{1.16}$$

where Ψ is the angular change in direction (positive for clockwise downstream turning) at the node and l_u and l_d are the distances to the adjacent upstream and downstream nodes. As the stream migrates, the distance between individual nodes may increase or decrease, necessitating addition or removal of nodes, as discussed by Howard (1984) and Howard and Knutson (1984).

Each point is moved, corresponding to bank erosion and channel migration, by an amount proportional to ζ. This erosion is directed normal to the stream centerline (in the $\bar{n}$ direction, Figure 1.1), moving the centerline to the left (facing downstream) if ζ is positive, and to the right if ζ is negative (in the JP model the near-bank depth and velocity perturbations are equal in magnitude and opposite in sign on opposing banks). Owing to the weighting of upstream curvatures implied by the governing equations, this erosion may be contrary, locally, to the direction of the local curvature.

When separate portions of the channel centerline approach closer than a predetermined distance, a neck cut-off occurs by deleting the points representing the abandoned channel. The program checks for potential neck cut-offs each 50 iterations only, so that the critical distance is set to 1.2 widths to minimize the occurrence of channel overlaps. Chute cut-offs and avulsions are not incorporated in the present model.

DEPOSITION MODEL

Major physiographic features in meandering streams include point bars, natural levees, crevasse splays, back swamps, overbank channels, and abandoned channel segments. In the present model, levees, point bars, back swamps and channel fills are modeled as an additive combination of two processes, point bar deposition and overbank sediment diffusion.

Bridge (1975), Jackson (1976), Allen (1977), and Willis (1989) have pioneered models coupling meander migration with depositional facies modeling. These studies have been concerned primarily with stratigraphy and sedimentology of point bars, and have relied on simple idealizations of translation and enlargement in single bends. Here, a more general model of meander migration is used and long-term evolution of floodplain deposits is considered. However, no attempt is made here to model sedimentary facies of the floodplain sediments. Leeder (1978) and Bridge and Leeder (1979) have modeled sedimentary facies deposited by streams in depositional basins, including the effects of avulsions. However, these models do not attempt detailed reconstruction of topography or sedimentary facies within meander belts. The present modeling thus falls in temporal and spatial scales in between the detailed point bar models of Bridge (1975) and others and the basin modeling of Leeder (1978) and Bridge and Leeder (1979). Enhancements of the present approach would be suitable for examination of the sedimentological structure of meander belts.

In accord with observations and theory (Kesel et al., 1974; Pizzuto, 1987), deposition of the coarse fraction of suspended load is modeled as a processes of diffusion from the main channel, with rates decreasing with distance from the channel. However, fine sediment deposition is modeled as slow settling from quiescent flow that is assumed to be independent of location. Several studies also have shown that floodplain deposition rates in meandering streams decrease with floodplain age (Wolman and Leopold, 1957; Everitt, 1968; Nanson, 1980), presumably because older floodplain locations are higher (and thus less frequently and less deeply flooded) and generally farther from the stream channel. Accordingly, deposition rate, Φ, is modeled as a function of relative floodplain height, the rates of fine sediment deposition, and a characteristic diffusion length scale:

$$\Phi = (E_{\max} - E_{\mathrm{act}}) \left[\nu + \mu \exp(- D/\lambda) \right] \qquad (1.17)$$

where $E_{\max}$ is a maximum floodplain height, E_{act} is the local floodplain height, ν is the position-independent deposition rate of fine sediment, μ is the deposition rate of coarser sediment by overbank diffusion, λ is a characteristic diffusion length scale, and D is the distance to nearest channel (both measured in channel-width equivalent units). This model is assumed to provide a crude representation of both deposition very close to the channel (banks and levees) as well as more distant overbank sedimentation.

Deposition of the point bar by the migrating channel is accounted for by making the initial floodplain elevation prior to overbank deposition equal to the near-bank channel-bed elevation, η_b (Figure 1.1). Specifically, when migration results in a channel migrating into a floodplain cell, the elevation is reset to the mean channel bed elevation E_{chan}. However, when the channel subsequently migrates past the flood-plain cell, the elevation E_{act} is initially set equal to

$$E_{act} = E_{chan} + \eta_{1b}\, H \qquad (1.18)$$

where E_{wat} is an assumed mean water surface elevation and η_{1b} is the near-bank perturbations of depth below E_{wat} (see Appendix A) for the bank opposite the direction of migration. Note that H equals $(E_{wat} - E_{chan})$. The three elevations $E_{chan} < E_{wat} < E_{max}$ are parameters input to the model (Figure 1.1). Elevations are measured relative to the local E_{chan} and do not account for the valley gradient. Note that equation (1.18) is applied to the newly vacated cell prior to calculation of sediment deposition (equation (1.17)).

Floodplain stratigraphy and sediment composition are not modeled explicitly. Floodplain elevations and ages are stored in a matrix that overlies the meander belt. In the present simulations, each matrix cell corresponds to a square area with sides equal to one width-equivalent.

In summary, the depositional model incorporates both a crude model of point bar sedimentation expressed as a variable advancing bank initial elevation (equation (1.18)) and a bank and overbank depositional component (equation (1.17)).

SIMULATION RESULTS

Figure 1.2 shows the planform evolution of the centreline of a meandering channel simulated with the present model, starting from a stream that is straight except for small, random normal perturbations. Model input parameters are given in Table 1.1. The length of the simulated valley section (and the initial stream length) is 512 widths. The resulting pattern of channel evolution and cut-off development is similar to that obtained by Howard and Knutson (1984) using the earlier flow model of Ikeda, Parker and Sawai (1981).

The initial pattern of migration is very regular and develops the classic 'Kinoshita' loop shape, which is skewed upstream and increases to considerable amplitude prior to cut-off. This is a shape that is characteristic of the solutions to the governing equations (Parker, 1984; Parker and Andrews, 1986) and common in natural streams (Carson and Lapointe, 1983). There are local differences in rapidity of initial growth that depend upon the random perturbations of the initial input stream. However, after cut-offs begin, the stream pattern becomes much more varied in form of meanders owing to the disturbances that propagate throughout the meander pattern as a result of cut-offs. At these advanced stages, the pattern becomes much more similar to natural meandering streams, with their commonly complex loop shapes. As a result of chance occurrence of two or more cut-offs on the same side of the valley, the overall meander belt can develop a wandering path, as noted by Howard and Knutson (1984). The sharp bends that result from cut-offs are very rapidly converted to more gentle bends, commonly by reverse migration caused by maximum flow velocities occurring on the inside of very sharp bends. The development of varied meander forms from an initially regular pattern indicates that the combination of meander growth, the occurrence of cut-offs, and the influence of complicated initial and boundary conditions (including variations in bank resistance that are held constant) implies a 'sensitivity to initial conditions' in the meandering process. That is, small differences in initial geometry or boundary conditions between two otherwise identical streams will cause different meander patterns. Also, predictability of future meander pattern decreases with time and past river patterns become increasingly uncertain with elapsed time (unless topographic or stratigraphic evidence is available).

The deposition model is illustrated in Figure 1.4, which portrays the evolution of a square region (dimensions of 100×100 width-equivalents) extracted from the middle of a simulated stream approximately five times longer than the square region. Although simulation parameters are the same as those for Figure 1.2, slightly different initial conditions resulted in a different pattern of meander evolution. The simulation starts with an existing meandering stream and shows bank erosion and sedimentation occurring during the course of 2100 iterations. Contours of floodplain age (measured in iterations) are shown in Figure 1.4a. The resulting patterns of meander-loop growth afford examples of most of the types discussed by Brice (1974b) and Hickin (1974). Some portions of the floodplain have not been occupied by the meandering channel during the simulation (outside the dashed lines). Floodplain elevations for two values of the parameter μ, controlling the rate of overbank deposition by sediment diffusion, are shown in Figure 1.4b and 4c. The assumed value is -10 for E_{chan}, 11 for E_{wat} and 20 for E_{max} (arbitrary units).

Similarly, Figure 1.5 shows meander and floodplain evolution for confined meanders developed between non-erodible valley walls. Note that the meanders develop a characteristic asymmetric pattern, with gentle bends terminating abruptly in sharp bends at valley walls, similar to natural confined meanders (Lewin, 1976; Lewin and Brindle, 1977; Allen, 1982; Howard and Knutson, 1984).

Figure 1.6 shows the average relationship between floodplain elevation and floodplain age for the simulations shown in Figure 1.4. As would be expected from the model assumptions, deposition rates decrease with increasing age of the floodplain (or alternatively, with increasing elevation of the floodplain).

These simulations exhibit many of the essential features of natural meandering streams, including overbank deposits gradually increasing in elevation away from the channel, rapid isolation of abandoned channels by filling near the main channel (modeled here as resulting from sediment diffusion from the main channel, but in natural channels advectional transport through the abandoned channel also would occur), and slower infilling of oxbow lakes primarily by deposition from suspension. Note that two neck cut-offs have occurred at the left edge of Figure 1.4 just prior to the end of the simulation, so that the abandoned channels have not been closed by sedimentation, as has occurred for the loops abandoned earlier in the simulation on the right side. Also, the sharp change of curvature at the site of the cut-offs has not yet been smoothed out by rapid meander growth at the cut-off site. There is considerable variability of channel migration rates from bend to bend, and the slope of floodplain surface in the interior of bends is generally steeper the slower the migration (compare Figures 1.4a and 1.4b). For higher rates of overbank deposition most of the floodplain

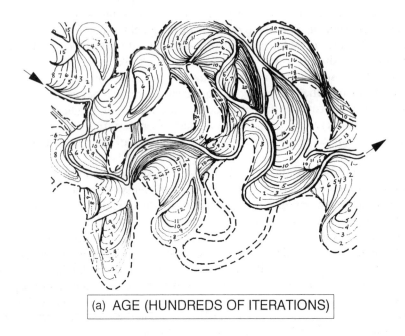

(a) AGE (HUNDREDS OF ITERATIONS)

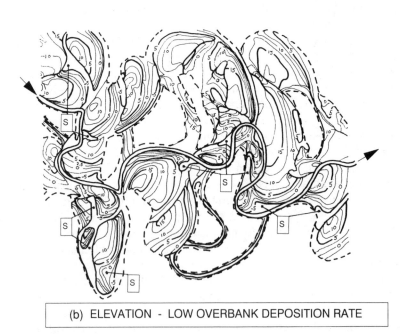

(b) ELEVATION - LOW OVERBANK DEPOSITION RATE

Figure 1.4 Simulations of floodplain evolution in a freely meandering stream for 2100 iterations; (a) contours of floodplain age, in hundreds of iterations. Location of present stream is shown by arrows and lines delineating its banks. Floodplain areas older than 2100 iterations are bordered by dashed lines and are uncontoured; (b) contours of

rapidly reaches values close to E_{max} and cut-offs are rapidly isolated into oxbows (Figures 1.4c and 1.5c). Both simulations have low values for floodplain sedimentation, v, so that oxbow lakes are filled very slowly.

The simulations with a low rate of overbank sedimentation (Figures 1.4b and 1.5b) exhibit a depositional feature that is a consequence of out-of-phase relationships between near-bank velocity and bed elevation perturbations. Where curvature changes abruptly downstream the depth adjusts quite rapidly on the new outer bank, and generally overshoots its value for constant curvature, but velocity responds more slowly (Figure 1.3b). In this figure the curvature changes at position 3 from negative to a constant positive value. In the zones indicated by ruling and cross-hatching the velocity perturbation is opposite in sign to the curvature, indicating that the highest velocity is directed towards the inside of the bend, where, from equation (1.10), bank erosion will occur. In addition, in the cross-hatched zone, the depth perturbation is positive, so that the depth is greatest on the outside bank. This means that, from equation (1.18), deposition on newly created floodplain on the outside bank must start from very low relative elevations (a scour hole). This zone of very low initial elevations is short (about 2.5 width-equivalents). Just downstream from the cross-hatched zone the velocity perturbation is positive, indicating that the more normal pattern of erosion is directed towards the outer bank. In this zone the depth perturbation is positive and large in magnitude, so that the point-bar elevation on the inner bank is large; therefore, from equation (1.18), floodplain deposition starts from

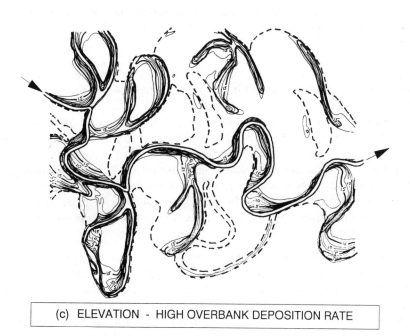

(c) ELEVATION - HIGH OVERBANK DEPOSITION RATE

Figure 1.4 *(cont.)* floodplain elevation for depositional parameters having the values $\lambda = 3$ width-equivalents, $v = 0.0003$ vertical units per iteration, and $\mu = 5v$. Location of low-elevation sloughs indicated by lines from 'S' boxes. (c) Contours of elevation for $\lambda = 3$, $v = 0.0003$, and $\mu = 50v$

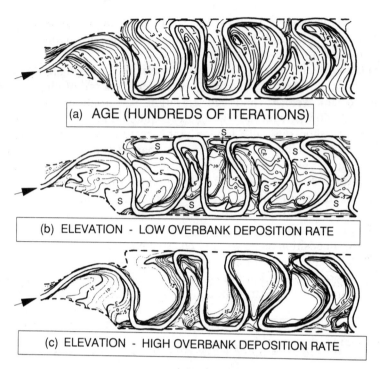

Figure 1.5 Simulations of floodplain evolution of meanders confined within a valley 24 width-equivalents across. Low elevation sloughs are indicated in (b) by 'S'. See Figure 1.4 for further information

a high relative level. Similar effects can occur in narrow zones where curvature either increases or decreases abruptly, but it occurs most strongly where curvature changes sign abruptly.

The simulation modeling indicates that these short zones with lower than average initial floodplain elevations are located in consistent positions relative to bends as the channel migrates, in places leaving behind depressions, or sloughs, in the floodplain deposits. These sloughs are most commonly located in the axial position of sharp meander bends, and are best developed on the downstream end of the point bar near the curvature inflection leading to the next bend. Several of these depressions are labeled with 'S' in Figures 1.4b and 1.5b. These floodplain depressions exhibit several consistent patterns.

(1) The depressions are best developed at strong changes in curvature, which in both simulated and natural streams occurs generally at short, abrupt bends. Such abrupt bends are best developed near the site of recent neck cut-offs (Figure 1.4) or where meander migration is confined by valley walls (Figure 1.5). Long meander bends of the classic Kinoshita form have little lag between velocity and depth perturbation (relative to bend length) and negligible zones in which the sign of these perturbations are opposite and large in magnitude (Figure 1.3a).
(2) Natural and experimental channels abound in similar features, which may result from a mechanism similar to that incorporated in the model. Figure 1.7 shows

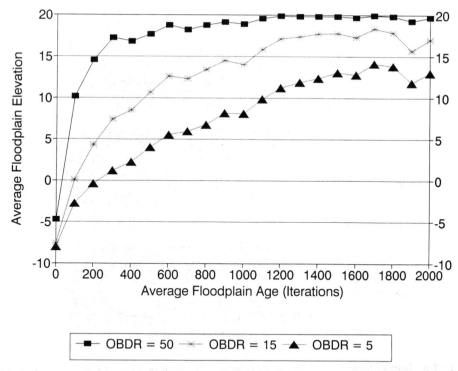

Figure 1.6 Average relationship between floodplain age and floodplain elevation for simulations with three values of the model paramter, μ (overbank deposition rate, OBDR, expressed in multiples of the parameter ν). The curves for OBDR = 50 and OBDR = 5 correspond to the simulations shown in Figures 1.4b and 1.4c, respectively

depth contours for two of the experimental runs of Friedkin (1945). Figures 1.8a and 1.8b show natural meandering channels with such sloughs or floodplain depressions. The present modeling suggests that they are best explained simply as sites of retarded deposition owing to the low initial near-bank bar elevations, as noted by Wolman and Brush (1961). Fine sediments accumulating in such sloughs and floodplain lows have been called concave bank benches (Cary, 1969; Taylor and Woodyer, 1978; Woodyer, 1975; Hickin, 1979; Nanson and Page, 1983). Lewin (1978) attributes floodplain sloughs extending upstream from the outside bank of confined and unconfined sharp bends to formation as residual depressions from migrating deep scour pools—essentially the same mechanism as occurs in the simulations (Figures 1.4b and 1.5b).

(3) In the type of sharp bends associated with slough development, large depth perturbations on the inside bank tend to occur in initial portions of the bend, leading to high bars. Since the depth perturbation diminishes through the bend, a typical bar form emerges that is high on the upstream end, diminishing in height downstream and toward the inside of the bend, and terminating in the slough. The sloughs are commonly deepest at their downstream end and shallow or disappear upstream. See the laboratory channels in Figure 1.7 and the natural channel in Figure 1.8b. In some cases in both the simulations and natural channels these

20

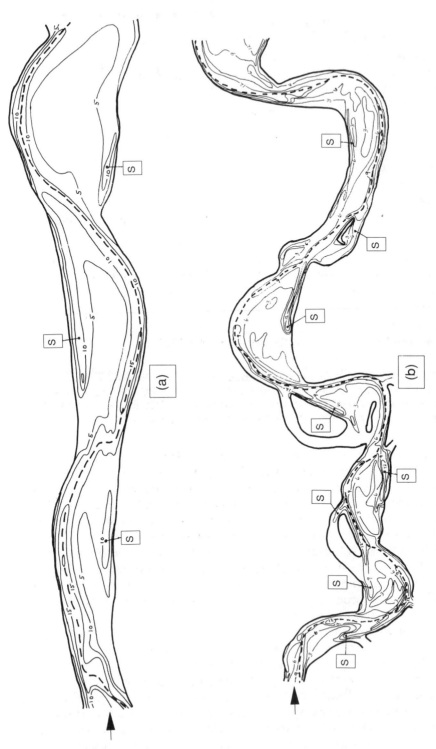

Figure 1.7 Channel and bank elevations for two laboratory meandering channels (Friedkin, 1945). Channel thalweg shown in dashed lines and sloughs are indicated by 'S'. Flow is from left to right; (a) bank stabilization test 5 with relative depths below banks in hundredths of feet. Note prominent near-bank sloughs opening downstream; (b) bank stabilization test 6 with relative depths below banks in hundredths of feet. This run models bank erosion of a portion of the Mississippi River. Note sloughs and mid-channel islands

sloughs may extend completely through the meander axis, creating a second channel and a mid-channel bar (Figure 1.8 (a and c)). The depositional mechanisms modeled here may thus offer an explanation for these commonly occurring bars (Leopold and Wolman, 1957; Hooke, 1986; Bridge et al., 1986). However, natural bars in meandering channels are influenced by flow through the slough, which is not modeled in the present simulations.

(4) The model suggests that these sloughs and bar forms are best developed in streams where depth perturbations are large (high sinuosity and small values of the parameter β governing the slope of point-bars) and where suspended sediment deposition rates are modest (Nanson and Page, 1983). If this is not the case, then bank and floodplain deposition is more uniform (Figures 1.4c and 1.5c).

(5) The simulations suggest that these bar and low floodplain features of wide meandering streams cannot be understood solely in terms of adjustments of bar form to contemporary flow and sediment transport, but are in addition intimately related to the kinematics of bank migration.

(6) The sloughs should be preferred locations for the development of chute cut-offs. Since the sloughs are fixed in location, as the bend sinuosity increased during migration of the main channel the sloughs become more advantageous as a flow route. This is supported by observations by Lewis and Lewin (1983) that chute cut-offs are most common in tight bends and at axial locations (where the sloughs are best developed).

The present explanation for these floodplain depression, sloughs and mid-channel bars may conflict with the dominant current interpretation that these forms result from migration of and deposition from alternate bars coupled with channel migration (Lewin, 1976; Bridge, 1985; Bridge et al., 1986). Alternate bars typically have their highest point located away from the bank with a corresponding depressed zone immediately adjacent to the bank. Such a form is observed in natural channels (e.g. Kinoshita, 1961; Bluck, 1976; Lewin, 1976), flume experiments (Wolman and Brush, 1961; Whiting, 1990), and in the simulations of Nelson and Smith (1989a,b) and Shimizu and Itakura (1989). Bridge (1985) and Bridge et al. (1986) suggest that point and mid-channel bars, as well as associated depressions resulting in sloughs, are created from portions of migrating alternate bars that become fixed (or trapped) as the channel shifts owing to bank erosion. However, the JP model does account for sediment continuity and thus predicts time-averaged depositional effects of migrating alternate bars during channel shifting. The question that remains to be resolved is whether the development of fixed point and mid-channel bars and associated depressions are primarily related to transitory effects of migrating bars interacting with channel shifting (and thus not represented in the present model) or are adequately represented by the shifting steady-state bed topography of the JP model. In conclusion, bar sedimentation simulated using the JP model provides a sufficient, if not necessarily accurate, explanation for development of sloughs and mid-channel bars.

The simulations shown in Figures 1.4 and 1.5 show little tendency towards formation of natural levees. A number of other simulations with widely varying values of the deposition parameters in equation (1.17) likewise exhibited no obvious natural levees. One explanation may be the inclusion in the model of decreasing sedimentation rate

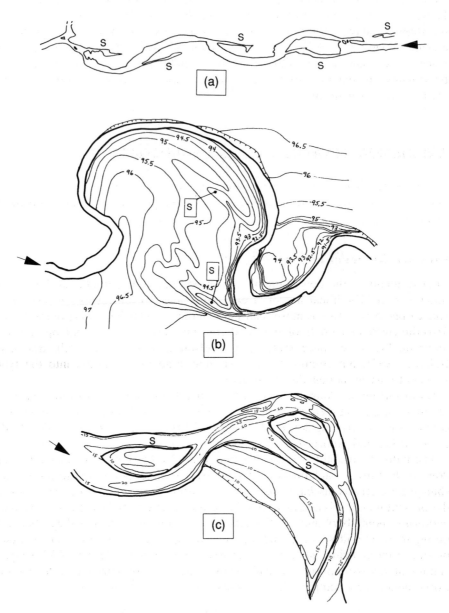

Figure 1.8 Natural channels with sloughs and mid-channel bars, with sloughs and floodplain depressions marked with 'S'; (a) channel banks of a partially confined river, showing prominent sloughs and mid-channel island (modified by permission from Lewin, 1976); (b) floodplain elevations (in feet) for two bends of Watts Branch, Maryland, showing linear depressions extending upstream across point bars from sharp bends (reproduced by permission from Wolman and Leopold, 1957); (c) relative depths (in feet) and banks of two bends of the Calamus River, with thalweg indicated by dashed line (reproduced by permission from Bridge *et al.*, 1986). Note mid-channel bars at bend apices

with increase of surface elevation, including the upper limiting elevation E_{max}. However, simulations with a very high distance-dependent deposition rate, μ, small diffusion distance, λ, and low ν (equation (1.17)), which give average floodplain elevations well below E_{max}, fail to produce levees. Alternatively, pronounced natural levees may occur only in situations where channel beds aggrade relative to their floodplains, so that the difference in near-channel and remote sedimentation rates accumulates through time.

DISCUSSION: PROPOSED MODEL ENHANCEMENTS

The present model could be revised in several ways to improve the fidelity to natural processes and increase the range of environments and features included in the simulation.

Flow and Sediment Transport Model

Improvements in the flow and bed topography model might address the limitations noted earlier. The linear model of Odgaard (1989a,b) represents the point bar by a convex bed profile that is more realistic than the straight-line profile of the JP model. Two-dimensional models of flow and transport are available that operate on the alternate bar time–space scales (e.g. Nelson and Smith, 1989a,b; Shimizu and Itakura, 1989), but computational costs may limit incorporation into the types of sedimentological models discussed here.

Local sorting of sediment in curved channel flows could be incorporated in future sedimentological modeling. Some attempts to address this have been undertaken in connection with existing flow models (Allen, 1970; Bridge, 1976,1977,1984a; S. Ikeda, 1989; Parker and Andrews, 1985).

The natural variability of flow has important implications with respect to channel flow and bedforms, bank erosion, and overbank sedimentation. Bedforms in channels change geometry as discharge changes, but with a lag in some cases (Allen, 1974). In the present model it is assumed that the response time of the large curvature-induced bedforms (point bars and related features) is slow enough that they change little during individual flow events and that the averaged response is the same as would be produced by a steady, *dominant* discharge, presumably a near-bankfull stage. The validity of this assumption can be tested by field study, appropriately scaled laboratory experiments or by computer experiments.

Bank Erosion Processes and Cut-offs

Although bank erosion rates are here assumed to be proportional to the near-bank velocity perturbation (equation (1.6)), the model incorporates the more general rate law incorporating a possible positive or negative contribution of the depth perturbation (equation (1.10)). As previously discussed, a variety of other assumptions might be appropriate.

The model can be extended to provide for both random and systematic spatial variations in bank erodibility, such as decreased erodibility of valley walls and clay plugs deposited in cut-offs (Fisk, 1947). At present the model does provide for a finite valley width with walls of zero erodibility (Figure 1.5).

Bank erosion rates also are assumed to respond to a dominant discharge. Discharge variation affects not only the magnitude of the velocity and depth perturbations, but also their distribution around the bend (lags in response to curvature changes become greater at higher discharges). Variable discharge following a log-normal frequency distribution has been incorporated into an earlier version of the channel migration model. Preliminary experiments suggested that patterns of bank erosion and channel migration were not changed significantly by inclusion of variable discharge.

Neck cut-offs presently are assumed to occur when centerlines approach closer than a critical distance. However, stochastic modeling would be more realistic for neck cut-offs, with probabilities decreasing with greater neck width and increasing with higher discharges (Bridge, 1975).

The development of high sinuosity in meandering channels is restricted primarily by chute cut-offs. Chute cut-offs generally occur across recently deposited point bar and low floodplain deposits and the probability of their occurrence presumably increases the greater the decrease in bend length provided by the cut-off and the lower the elevation across the cut-off site. The probability also should depend upon main channel flow velocity and depth near the chute and the angle with which the upstream end of the chute intersects the main stream at its upstream end, decreasing in likelihood as the angle approaches 90° and proportionally less flow is diverted. Cut-offs are initiated primarily during high flows, although complete diversion is a slow process generally (Fisk, 1947; Bridge *et al.*, 1986).

As discussed above, chute cut-offs are more common in streams with low rates of overbank sedimentation and a high width/depth ratio, and commonly occur along sloughs. Since the model simulates development of these sloughs, such cut-offs could be incorporated into the model realistically.

Avulsions are abandonment of long sections of an existing meander belt in favor of a more direct route across the existing floodplain. Avulsions are common only where the streams are aggrading relative to their floodplain (Allen, 1978; Leeder, 1978; Bridge and Leeder, 1979; Bridge, 1984b; Alexander and Leeder, 1987; Brizga and Findlayson, 1990), becoming the dominant mechanism of channel shifting on alluvial fans and river deltas. However, since such situations are required for preservation of thick meandering stream deposits, incorporation of avulsions into the model is desirable for sedimentological studies. Modeling of avulsions requires book-keeping of stream elevation changes relative to the floodplain, which could be done in an *ad hoc* manner (e.g. assuming a constant rate of rise) or by modeling the long-valley routing of bedload and regional elevation changes resulting from tectonic, sea-level, or consolidation processes. A simple implementation of avulsion probability would incorporate relative channel-floodplain elevations, bank and natural levee height, and possibly the magnitude of the sinuosity change and individual flood heights.

Deposition Modeling

The deposition rate laws incorporated equations (1.17) and (1.18) admittedly are crude, but they incorporate most of the features noted in empirical studies, including the role of channel bars and deposition rate decreasing with floodplain height and distance from the main channel. Unfortunately, little published data exists for testing and refining of this model.

An obvious extension would be to include grain size and stratigraphic information by modeling deposit thickness, bedform, and grain size. Grain size and bedforms of point-bar deposits can be related to within-channel sorting and flow parameters much in the manner adopted by Allen (1970), Bridge (1975,1978,1984a) and Bridge and Jarvis (1977,1982). In overbank sedimentation each grain size range might be treated using a relationship such as equation (1.17) with varying parameters. For example, fine sizes would have large diffusion length, λ, and low intrinsic rate, μ, eliminating the need for a separate deposition rate of fine sediment, ν. This would yield the decreasing grain size with distance noted in overbank deposition (Fisk, 1947; Kesel *et al.*, 1974; Pizzuto, 1987).

Overbank sedimentation also may be different depending upon location inside versus outside of the nearest bend. This could be added by incorporating a multiplicative term $(1 + \sigma \eta_{1b})$ in equation (1.17), where σ is a scale factor to account for secondary current effects on sediment diffusion from the main channel. Fisk (1947) suggests that natural levees are more common on the outside of meander bends, suggesting a positive value for σ. This could result from the radially outward near-surface secondary flows within bends. However, the higher initial elevations of the older, eroding banks may enhance levee development relative to point bars. The higher near-bank bed elevations at point bars results from near-bed, inward secondary flows, which also enhance inward suspended load transport, suggesting the opposite sign for σ. It may be that σ changes sign from coarser to finer suspended load sizes.

Variations in discharge and associated flow depths are an essential control on overbank sedimentation. In the present model an averaged response to many overbank flow events is assumed, which may be satisfactory for prediction of long-term deposition rates and resultant floodplain topography, but it is clearly inadequate for simulation of floodplain stratigraphy, which generally exhibits stratification resulting from individual flood events. Incorporation of overbank events would be fairly straightforward, with deeper flows corresponding to larger E_{max} (the water level), μ and λ in equation (1.17) for a given grain size range. For example, Pizzuto (1986) has modeled channel bank height as a function of the frequency distribution of flood depths and sediment loading, using an approach pioneered by Wolman and Leopold (1957).

Crevasse-splay deposits also could be incorporated as a stochastic model component, with initiation probability presumably a function of bank height, flood depth and possibly position within a bend. Similarly, scroll bar topography and deposits might be included in high-resolution floodplain modeling in an empirical fashion.

Deposition rates and grain sizes in the present model are parameterized by elevation and distance from the river, based upon a simple diffusional model of overbank sediment transport. However, as Pizzuto (1987) points out, advectional flow transport can lead to patterns of deposition rates and grain sizes not describable by the above parameters. This is particularly important for flows in chutes and sloughs, where both suspended load advection and bedload transport may occur. It may be possible to improve the model by empirical corrections, or overbank flow patterns and associated depositional processes might be included as an additional component if computational costs are not excessive. Techniques for modeling of overbank flows have been developed (Knight and Demetriou, 1983; Yen and Yen, 1984; Ervine and Ellis, 1987; Knight, 1989; Gee, Anderson and Baird, 1990; Miller, 1990).

Very large floods may cause erosional and depositional features that clearly are outside the range of the present model framework. One such effect is widening of the main channel, which can be very dramatic where banks are readily eroded (Schumm and Lichty, 1963; Burkham, 1972; Osterkamp and Costa, 1987), sometimes resulting in a change of channel pattern from meandering to braided. Other effects include development of new chutes or re-excavation of old chutes, stripping of the floodplain surface, or deposition of a veneer of coarse gravel (Graf, 1983; Nanson, 1986; Ritter and Blakely, 1986; Baker, 1988; Kochel, 1988). However, such effects are most important in mountain or confined valleys and do not appear to be common in the classic meandering of the lowland rivers, which are also those that are most commonly represented in the stratigraphic record.

DISCUSSION: MODEL VALIDATION

The present model appears to replicate the major features of natural meandering streams. In addition, suggestions have been made to improve the fidelity of model representation of natural processes and deposits. However, despite many years of observations, relevant data for model validation, calibration and revision remains fragmentary and inconclusive. Further development of the model should proceed only hand-in-hand with field, laboratory, map, and theoretical work.

Field studies generally have involved investigation of one or, at most, a few meander bends, including studies of flow and sediment transport (Bathurst, 1979; Dietrich, Smith and Dunn, 1979; Bridge and Jarvis, 1982; Dietrich and Smith, 1983, 1984; Dietrich, 1987; Dietrich and Whiting, 1989), bank erosion processes (Hughes, 1977; Lawler, 1986a,b), channel and overbank sedimentation processes, and floodplain stratigraphy. Such studies have proven very useful in validation of theoretical models and elucidation of the types and rates of processes occurring in natural channels.

Laboratory studies (e.g. Fisk, 1947; Wolman and Brush, 1961; Hickin, 1969; R. Hooke, 1975; Desheng and Schumm, 1987; Odgaard and Bergs, 1988) offer controlled conditions useful for unraveling complicated process interactions and validation of theoretical models. However, time and costs limit the range of experiments, and certain processes, especially overbank sedimentation and effects of cohesive bank sediments, are difficult to scale to laboratory dimensions.

Theoretical models underpin quantitative prediction, simulation and interpretation of natural phenomena, but development of theory requires field or laboratory observations for validation, and theory may be limited in applicability owing to model shortcomings or computational costs.

Further development of simulation models requires quantitative studies of morphology, migration and deposition rates, as well as sedimentological and stratigraphic relationships, over spatial scales extending through several to dozens of meander bends and over temporal scales incorporating extensive channel shifting and associated deposition and cut-offs. Several crucial needs for studies at such *reach-length* spatio-temporal scales are discussed below.

Reach-Length Studies

Validation of Migration Model

The spatio-temporal pattern of channel migration depends both on the flow and transport model as well as the model for bank erosion. The combined assumptions used in the simulations clearly are sufficient to produce meander patterns that are visually similar to many meandering streams of high sinuosity. However, theoretical model development clearly has outpaced empirical validation. Two approaches that can be used to test theoretical models are (1) static comparison of simulated and natural meander planform geometry and (2) kinematic comparison of predicted and simulated channel migration.

Static Comparisons

Howard and Hemberger (in press) have developed a suite of 40 statistical variables to characterize meander morphometry. These include *ensemble* statistics based on measures of sinuosity, spectral characteristics, fitting of autoregressive models, and moments of channel curvature. In addition, *half-meander* statistics are based upon breaking the channel into individual half-meanders, or half-loops at inflection points and statistically characterizing their sinuosity, length, shape and asymmetry. These variables were measured on 57 long reaches of freely meandering channels from 33 rivers. In addition, these statistics also were measured on planforms generated by two theoretical models, the first being the disturbed periodic model (DPM) of Ferguson (1976,1977) (which stochastically generates meander planforms but does not simulate meander kinematics) and the second being the simulation model of Howard and Knutson (1984) (HKM) based upon the theoretical model of Ikeda, Parker and Sawai (1981).

Discriminant analysis is used to compare the natural streams with the two theoretical models (Figure 1.9). Two discriminant variables that are linear combinations of the suite of morphometric variables clearly are able to separate natural streams from the two theoretical models despite their visual similarity. The DPM and HKM simulated streams have less variability of centerline curvature for a comparable sinuosity and a narrower range of half-meander sizes than the natural streams. In addition, the HKM streams have more sinuous half-meanders, greater overall sinuosity, and more strongly asymmetrical meanders (upstream skewing) than natural streams. Although the statistical analysis clearly points out deficiencies in the theoretical models, the suite of variables are very sensitive to small systematic differences in morphometry, and the HKM model remains a good first approximation to natural meandering for highly sinuous streams.

There are several possible reasons for the morphometric differences between HKM simulated streams and natural streams:

(1) Random variations in bank erodibility were incorporated into some simulations with the HKM model, with little improvement in planform similarity to natural streams. However, systematic variations of bank erodibility occur in natural streams in that bank erosion may be hindered or stopped by natural levees and

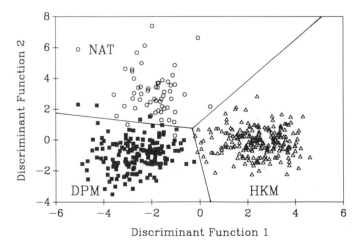

Figure 1.9 Scores for two discriminant functions of meander morphology for natural streams (NAT, open circles), streams simulated with the disturbed periodic model of Ferguson (1976; DPM, filled squares), and streams simulated with the kinematic model of Howard and Knutson (1984; HKM, open triangles). Discriminant boundary fences between the three classes of meanders are shown as the intersecting lines. Note that the two linear combinations of meander morphology indices produce almost complete discrimination between the three classes

exposure of clay plugs from old oxbow lakes (Fisk, 1947; H. Ikeda, 1989; Thorne, this volume). Such effects could be investigated in future revision of the model.

(2) Some natural meandering streams appear to have short meanders superimposed upon larger meanders, often generating the compound or cumuliform forms noted by Brice (1974a,b) and Hickin (1974). One cause may be temporal change in dominant wavelength, such as would result from a long-term decrease in flood peaks, so that short, new meanders develop on older, larger ones. Although this may occur on a few natural streams, a more universal cause may be the simultaneous operation of two distinct processes affecting meandering in streams. The first is the secondary circulation caused by stream curvature that is incorporated in the theory of Ikeda, Parker and Sawai (1981) and the HKM model. The second is the formation of alternate bars owing to sediment transport – flow interactions. Curvature-forced alternate bars are incorporated into the JP model used in the present simulations, but were not included in the HKM simulations used for the morphometric comparison. The inclusion of the curvature-forced bars allows for resonant interactions and overdeepening effects (Johannesson and Parker, 1989; Parker and Johannesson, 1989) that may have significant effects on meander morphometry. In addition, there may be additional systematic interactions between alternate bars and meander planform not accounted for in the JP model (including the locking of migrating bars in tight bends discussed above), as suggested by experiments (Whiting and Dietrich, 1989; Whiting, 1990).

(3) Bank erosion rates in the HKM model are assumed to be proportional to the velocity perturbation. As discussed above, a variety of other functional forms may occur. A few variations in the functional dependence of bank erosion rates with

the velocity and depth perturbations were incorporated in the HKM simulations used in the discriminant analysis, with no dramatic improvement of meander morphometrics compared with natural streams.

Kinematic Comparisons. Analysis of the kinematic pattern of channel migration of natural and simulated channels potentially is much more sensitive than the static comparisons and can offer important clues as to specific sources of model deficiencies. Unfortunately, although a rich data base of historical meander change exists, analyses of meander kinematics to date have been qualitative primarily or have yielded only summary statistics, such as average bank erosion rates (e.g. Brice, 1974a,b; Dort, 1978; Hooke, 1977,1979,1980,1984). Studies of the systematic variation in erosion rates with channel curvature and sediment properties have utilized isolated bends (Hickin and Nanson, 1975,1984; Nanson and Hickin, 1983, 1986), and have not accounted fully for upstream control of local flow and bed characteristics implied in theoretical models (such as the JP model used here) (Parker, 1984; Furbish, 1988). Howard and Knutson (1984) have used an earlier version of the flow model to simulate several decades of channel shifting on the White River of Indiana, with generally encouraging results. Short-term predictions of channel shifting also have been made by Parker (1982), Beck (1984), and Beck, Mefli and Yalamanchili (1984). The flow and transport model has been compared with flume studies of meandering channels (Johannesson and Parker, 1989). Pizzuto and Meckelnburg (1989), Hasegawa (1989a,b) and Furbish (1988) have compared observed bank erosion for individual bends or short reaches with model predictions. Although these comparisons generally are encouraging, they are too few and too rudimentary to comprise a thorough testing of the flow and erosion model. What is needed is systematic analysis of meander kinematics on a number of long reaches of natural and simulated channels and a comparison with model predictions.

Reach-length historical records of change of natural meandering channels can be used to test (or develop) predictive models of channel migration, involving both forward and inverse techniques.

Forward Analysis. The simplest forward method is to use theoretical models of flow and bend topography (such as the JP model) together with assumptions regarding an appropriate bank erosion rate law. A historical pattern of the river is used for initial conditions, and the simulation model predicts future migration, which can be compared with actual channel shifting. An appropriate criterion for degree of correspondence might be least-squares difference between the actual and predicted channel pattern. Cut-offs have to be treated specially, since either the predicted occurrence of a cut-off that did not occur or the reverse would lead automatically to very large least-squares discrepancies. Appropriate model parameters (e.g. F, β, γ, $\mathscr{E}$) can be estimated from field data. The disadvantage of this approach is that it is laborious to experiment with model parameters and bank erosion rate law assumptions to find the best fit to the observed changes.

A more flexible forward approach is to use the natural channel shift data to construct a spatial series of erosion rates together with the corresponding spatial series of channel curvatures. Because the channel is shifting through time, an intermediate natural channel half-way between the initial and final locations must be constructed. These data can be used in several ways. The combined flow and bank erosion models can be used, with appropriate parameter assumptions, to estimate bank erosion rates for the intermediate channel planform. This approach gives the flexibility of examining the bank erosion model separately *assuming* that the flow and bed topography model is correct.

Inverse Analysis. Spatial series of bank erosion rates and curvature from natural streams, as discussed above, can be used in a variety of ways to fit 'time series' models to the observed data statistically. A variety of techniques have been developed to develop such models, the simplest of which are linear (Box and Jenkins, 1976). Transfer modeling techniques can be used to develop autoregressive and/or moving average (ARMA) models relating the time series of curvature and observed erosion rates. These linear models imply a governing differential equation whose parameters can be inferred from the fitted model. For example, the linear model of flow velocity and bank erosion of Ikeda, Parker and Sawai (1981) utilizes a single governing differential equation relating the curvature series to bank erosion rates. However, the multiple equation model of JP (Appendix A) could not be fit readily by ARMA transfer model techniques. However, an intermediate approach is again available if the JP model is used in a forward manner to predict velocity and depth perturbations, and transfer function techniques are used to derive an appropriate bank erosion model, including possible indentification of systematic spatial variations in bank erodibility, such as clay plugs.

Spatio-temporal information of neck and chute cut-offs and relevant information on point-bar heights, presence of sloughs, etc. also can be related to the spatial series of curvature and erosion rates in order to develop predictive models.

Validation of Deposition Model

The deposition model is the most difficult of the components to validate, because relevant data are difficult to obtain. The simplest type of quantitative comparisons between simulated and natural floodplains are statistics relating floodplain age, elevation and distance from stream channels. The analysis also could include as factors the position on inside or outside of bend and distance to abandoned channels. A few local studies have related floodplain elevation to age (Everitt, 1968; Kesel *et al.*, 1974; Nanson, 1980). However, reach-length measurements of floodplain elevations at sufficient vertical and horizontal resolution to be useful in model validation are rare. A few large rivers, notably the Mississippi, have been mapped at 5–10 ft contour intervals, but few reaches have slight enough influence of man (e.g. levee build-up, bank revetments, and artificial cut-offs) to form a useful data base. However, low-level stereo aerial photography is available for many rivers, and could be used to make detailed elevation measurements. Floodplain ages can be determined by dendrochronology (Everitt, 1968; Hickin and Nanson, 1975), although many rapidly meandering streams have sufficient historical record from maps and aerial photographs to be used to construct age maps.

Depositional features, such as point- and mid-channel bars, sloughs and floodplain, can be characterized by their morphometric or sedimentological characteristics (elevation, size, bedforms, grain size, etc.) and position along the channel, and related to the spatial series of curvature and erosion rates by time series analysis.

Information on floodplain sedimentology and stratigraphy is more difficult to obtain. Cores and trenches are the obvious but tedious method, supplemented with archaeological and age-dating techniques to provide rate information (e.g. Brakenridge, 1984,1985; Brown, 1987,1990; Walling and Bradley, 1989). Judicious use of sections exposed in cut banks also can be useful. Ancient deposits in the sedimentary record serve as comparisons. Relatively short duration studies of rates and spatial patterns of deposition and erosion are quite practical for reach-length studies, using surveying, coring and use of markers such as sand or gypsum.

CONCLUSIONS

The present model combines a model for flow and bed topography in meandering streams (Johannesson and Parker, 1989) with the assumption that bank erosion rates are related to the near-bank perturbations of downstream velocity and channel depth. This model provides realistic migration of simulated channels, although the simulated channels tend to be somewhat more asymmetric, sinuous and regular than natural channels.

The floodplain deposition model, which assumes that deposition rates decrease with distance from the closest channel and with increasing floodplain elevation, produces simulated topography that resembles that of natural floodplains, including point bars and oxbow lakes. Bank sedimentation is assumed to be initiated from the near-bank depths predicted by the flow-bed topography model. This produces linear depressions or sloughs at the downstream, inside margins of point-bar complexes in locations of sharp bends. Similar sloughs or mid-channel bars are found in natural channels at sharp bends, particularly at locations of confined meandering and recent cut-offs.

Both the meandering and depositional models can be modified in a number of ways to increase the range of features that are simulated (such as floodplain stratigraphy) or to improve the fidelity to natural processes. However, both existing model assumptions and suggested modifications will require validation through studies of natural meandering processes, particularly over reaches of several bends or more.

ACKNOWLEDGEMENTS

The author wishes to thank Paul Carling and Geoffrey Petts for providing the opportunity to present this paper at the Lowland Rivers Conference, and John Bridge, William Dietrich, and James Pizzuto for insightful comments on an earlier draft. Gary Parker has patiently explained to me the details of his model and its underlying assumptions, but errors of presentation are mine.

LIST OF SYMBOLS

A	coefficient of transverse bed slope	Equations (1.A5) and (1.A16)
C_f	coefficient of friction	Table 1.1
$\mathscr{C}$	Channel planform curvature (see Figure 1.1) (L^{-1})	Table 1.2
D	distance to nearest channel (L)	Equation (1.17)
E	elevation (L)	Equations (1.17) and (1.18)
$\mathscr{E}$	bank erodibility (LT^{-1})	Equation (1.3)
F	Froude number	Table 1.1
$\mathscr{F}(d)$	bank shear resistance, a function of median grain size, d, of sediment at base of cut bank $(ML^{-1}T^{-2})$	Equation (1.11)
g	gravitational acceleration (LT^{-2})	Table 1.1
h	local channel depth (see Figure 1.1) (L)	Equation (1.2)
H	section average channel depth (L)	Equation (1.2)
l	distance between nodes defining channel centerline (L)	Equation (1.16)
M	exponent relating sediment transport rate to velocity	Table 1.1
n	cross-stream distance from centerline (L)	Figure 1.1
s	downstream distance along centerline (L)	Figure 1.1
$\mathscr{T}$	sinuosity of channel reach	Equation (1.12)
u	local vertically-averaged velocity (LT^{-1})	Equation (1.1)
U	section average velocity (LT^{-1})	Equation (1.1)
W	channel width (L)	Table 1.1
Z	weighting coefficients for differentials	Equation (1.A19)
α	weighting coefficient for velocity for bank erosion	Equation (1.10)
β	transverse bed slope effect on transverse bedload transport	Table 1.1
γ	channel width/depth ratio	Table 1.1
Γ	parameter governing alternate bar wavelength	Equation (1.A7)
δ	parameter governing phase shift of secondary flow	Equation (1.A4)
ε	weighting coefficient of depth for bank erosion	Equation (1.10)
ζ	bank erosion rate (LT^{-1})	Equation (1.3)
η	bed elevation (L)	Figure 1.1
λ	characteristic diffusion scale length (L)	Equation (1.17)

μ	weighting coefficient for coarse-grained sediment deposition (T^{-1})	Equation (1.17)
ν	weighting coefficient for fine-grained sediment deposition (T^{-1})	Equation (1.17)
ζ	local water surface elevation (L)	Figure 1.1
π	dimensionless excess shear stress	Equation (1.7)
ϱ	fluid density (ML^{-3})	Table 1.1
τ	average bed shear stress ($ML^{-1}T^{-2}$)	Equation (1.3)
Φ	floodplain deposition rate (LT^{-1})	Equation (1.17)
χ, χ_1, χ_{20}	velocity profile shape parameters	Equations (1.A1)–(1.A3)
Ψ	angle, in radians, between successive centerline nodes (see Figure 1.1)	Equation (1.16)

L, length; M, mass; T, time.

Subscripts

0	value for straight (non-sinuous) channel
1	dimensionless perturbation (except for χ_1)
b	near-bank value at $n = W/2$
c	value derived from curvature-forced solution with tractive force balance
f	value correcting curvature-forced solution for sediment transport continuity (except for C_f)
d	downstream value
i	enumeration index
s	variable expressing effect of secondary flow strength and phase shift
t	value at threshold of erosion
u	upstream value
w	value non-dimensionalized by multiplying or dividing by channel width

APPENDIX A

The Johannesson–Parker Flow and Sediment Transport Model

The following parameters are defined in terms of the input parameters (Table 1.1) and used in the differential equations presented below:

$$\chi_1 = 0.077/C_f^{1/2} \tag{1.A1}$$

$$\chi = \chi_1 - 1/3 \tag{1.A2}$$

$$\chi_{20} = (\chi^3 + \chi^2 + 2\chi/5 + 2/35)/\chi_1^3 \tag{1.A3}$$

$$\delta = \chi_1^2 \, (\chi + 1/4)/(\chi^2/12 + 11\chi/360 + 1/504) \tag{1.A4}$$

$$A = 2(\chi + 2/7)/(0.267\beta \, (\chi + 1/3)) \tag{1.A5}$$

$$A_s = 724 \, (2\chi^2 + 4\chi/5 + 1/15)/(\gamma^2 \, \chi_1) \tag{1.A6}$$

$$\Gamma = 4\beta/(\gamma^2 \, C_f) \tag{1.A7}$$

These terms are explained more fully in the List of Symbols.

As with channel depth and velocity, dimensionless perturbations of bed and water-surface elevations are defined:

$$\eta_1 = \eta/H \qquad (1.A8a)$$

$$\zeta_1 = \zeta/H \qquad (1.A8b)$$

These differ from the depth and velocity perturbation definitions (equations (1.1) and (1.2)) because the mean values of η and ζ are zero.

Depth, velocity and bed-elevation perturbations are resolved into a component resulting from curvature effects (subscript c) and a correction accounting for sediment transport continuity (subscript f):

$$u_{1b} = u_{1cb} + u_{1fb} \qquad (1.A9)$$

$$h_{1b} = h_{1cb} + h_{1fb} \qquad (1.A10)$$

$$\eta_{1b} = \eta_{1cb} + \eta_{1fb} \qquad (1.A11)$$

The basic differential equations that must be solved are presented below. The equations are equivalent to the JP equations, but are normalized by channel width rather than half-width, maximum curvature and wavenumber as in JP. The three equations given below must be solved sequentially in order to determine the velocity perturbations:

$$\frac{d\,\mathscr{C}_{ws}}{dS_w} + \delta\gamma\, C_f\, \mathscr{C}_{ws} = \delta\gamma\, C_f\, \mathscr{C}_w \qquad (1.A12)$$

$$\frac{\partial u_{1cb}}{\partial S_w} + 2\gamma\, C_f\, u_{1cb} = 0.5 \left\{ -\chi_{20}\, \frac{\partial\,\mathscr{C}_w}{\partial S_w} + \gamma\, C_f\, \left[(F^2\, \chi_{20} - 1)\, \mathscr{C}_w \right.\right.$$

$$\left.\left. + (A + A_s)\, \mathscr{C}_{ws} \right] \right\} \qquad (1.A13)$$

$$\frac{\partial^2 u_{1fb}}{\partial S_w^2} + \gamma\, C_f\, [3 - M + (\pi/2)^2\, \Gamma]\, \frac{\partial u_{1fb}}{\partial S_w} + 2\, [\gamma\, C_f\, (\pi/2)]^2\, \Gamma\, u_{1fb} =$$

$$\gamma\, C_f\, (M - 1)\, \frac{\partial u_{1cb}}{\partial S_w} - 0.5\gamma\, C_f\, \left[F^2\, \chi_{20}\, \frac{\partial\,\mathscr{C}_w}{\partial S_w} + A\, \frac{\partial\,\mathscr{C}_{ws}}{\partial S_w} \right] \qquad (1.A14)$$

Having solved for velocity using equations (1.A12)–(1.A14), the following equations give the depth perturbations:

$$\eta_{1fb} = - [1/(\gamma\, C_f)]\, \frac{\partial u_{1fb}}{\partial S_w} - 2\, u_{1fb} \qquad (1.A15)$$

$$\eta_{1cb} = -0.5\, A\, \mathscr{C}_{ws} \qquad (1.A16)$$

$$\zeta_{1b} = 0.5\, F^2\, \chi_{20}\, \mathscr{C}_w \qquad (1.A17)$$

$$h_{1b} = \zeta_{1b} - \eta_{1b} \qquad (1.A18)$$

The solution for these equations marches downstream. First and second derivatives are expressed as a weighted sum of the local ($i = 0$) and upstream ($i > 0$) values of the differentiated variable, using formulae for asymmetric differentials. For example, $d\,\mathscr{C}_{ws}/dS_w$ becomes

$$d\,\mathscr{C}_{ws}/dS_w = \sum_{i=0}^{n} \mathscr{C}_{ws_i}\, Z_i, \qquad (1.A19)$$

where Z_i are functions of the upstream distances S_{wi} to the stream nodes.

The Z_i are found by the method of undetermined coefficients (Gerald and Wheatley, 1989, their Appendix B) through solution of simultaneous equations (the Z_i are different for each location and iteration since the S_{w_i} vary downstream and temporally). The resulting difference equation is solved for the unknown local value ($\mathscr{C}_{ws_0}$ in this case). The number of polynomial terms, n, is specified (typically four or five). This autoregressive approach has been suggested by Furbish (1988,1989) and is equivalent functionally to the convolution approach used by Howard and Knutson (1984) and Johannesson and Parker (1989).

For the first few stream locations (when there are less than n upstream points) the derivatives are set to zero. This means that a lead-in section of stream is required in order to obtain good estimates of the variables, ideally, one or more meander wavelengths long.

REFERENCES

Akky, M. R. and C. K. Shen (1973). Erodibility of a cement-stabilized sandy soil. In *Soil Erosion: Causes and Mechanisms*, US Highway Research Board Special Report 135, pp. 30–41.

Alexander, J. and M. R. Leeder (1987). Active tectonic control on alluvial architecture. In F. G. Ethridge, R. M. Flores and M. D. Harvey (eds), *Recent Developments in Fluvial Sedimentology*, Society of Economic Paleontologists and Mineralogists, Special Publication No. 39, pp. 243–252.

Allen, J. R. L. (1970). A quantitative model of grain size and sedimentary structures in lateral deposits. *Geological Journal*, **7**, 129–146.

Allen, J. R. L. (1974). Reaction, relaxation, and lag in natural sedimentary systems: general principles, examples and lessons. *Earth Science Reviews*, **10**, 263–342.

Allen, J. R. L. (1977). Changeable rivers: some aspects of their mechanics and sedimentation. In K. J. Gregory, (ed.), *River Channel Changes*, Wiley, pp. 15–45.

Allen, J. R. L. (1978). Studies in fluviatile sedimentation: an exploratory quantitative model for the architecture of avulsion-controlled alluvial suites. *Sedimentary Geology*, **21**, 129–147.

Allen, J. R. L. (1982). Free meandering channels and lateral deposits. In *Sedimentary Structures: Their Character and Physical Basis*, Vol. 2, Elsevier, pp. 53–100.

Ariathurai, R. and K. Arulandan (1986). Erosion rates of cohesive soils. *Journal of the Hydraulics Division, American Society of Civil Engineers*, **104**, 279–298.

Baker, V. R. (1988). Flood erosion. In V. R. Baker, R. C. Kochel and P. C. Patton (eds), *Flood Geomorphology*, Wiley, pp. 123–137.

Bathurst, J. C. (1979). Distribution of boundary shear stress in rivers. In D. D. Rhodes and G. P. Williams (eds), *Adjustments of the Fluvial System*, Kendall/Hunt, pp. 95–116.

Beck, S. (1984). Mathematical modeling of meander interaction. In C. M. Elliott (ed.), *River Meandering*, American Society of Civil Engineers, pp. 932–941.

Beck, S., D. A. Mefli and K. Yalamanchili (1984). Lateral migration of the Genessee River, New York. In C. M. Elliott (ed.), *River Meandering*, American Society of Civil Engineers, pp. 510–517.

Blondeaux, P. and G. Seminara (1985). A unified bar-bend theory of river meanders. *Journal of Fluid Mechanics*, **157**, 449–470.

Bluck, B. J. (1976). Sedimentation in some Scottish rivers of low sinuosity. *Transactions, Royal Society of Edinburgh*, **69**, 425–456.

Box, G. E. P. and G. M. Jenkins (1976). *Time Series Analysis: Forecasting and Control*, Holden-Day, 575 pp.

Brakenridge, G. R. (1984). Alluvial stratigraphy and radiocarbon dating along the Duck River, Tennessee: implications regarding flood-plain origin. *Geological Society of America Bulletin*, **95**, 9–25.

Brakenridge, G. R. (1985). Rate estimates for lateral bedrock erosion based on radiocarbon ages, Duck River, Tennessee. *Geology*, **13**, 111–114.

Brice, J. C. (1964). Channel patterns and terraces of the Loup Rivers in Nebraska. *U. S. Geological Survey Professional Paper*, **422-D**, 41 pp.

Brice, J. C. (1974a). Meandering pattern of the White River in Indiana – an analysis. In M. Morisawa (ed.), *Fluvial Geomorphology*, State University of New York, pp. 178–200.

Brice, J. C. (1974b). Evolution of meander loops. *Geological Society of America Bulletin*, **85**, 581–586.

Bridge, J. S. (1975). Computer simulation of sedimentation in meandering streams. *Sedimentology*, **22**, 3–43.

Bridge, J. S. (1976). Bed topography and grain size in open channel bends. *Sedimentology*, **23**, 407–414.

Bridge, J. S. (1977). Flow, bed topography, grain size and sedimentary structure in open channel bends: a three-dimensional approach. *Earth Surface Processes*, **2**, 401–416.

Bridge, J. S. (1978). Palaeohydraulic interpretation using mathematical models of contemporary flow and sedimentation in meandering channels. In A. D. Miall (ed.), *Fluvial Sedimentology*, Canadian Society of Petroleum Geologists, pp. 723–742.

Bridge, J. S. (1984a). Flow and sedimentary processes in river bends: comparison of field observations and theory. In C. M. Elliott (ed.), *River Meandering*, American Society of Civil Engineers, pp. 857–872.

Bridge, J. S. (1984b). Large-scale facies sequences in alluvial overbank environments. *Journal of Sedimentary Petrology*, **54**, 583–588.

Bridge, J. S. (1985). Paleochannel patterns inferred from alluvial deposits: a critical evaluation. *Journal of Sedimentary Petrology*, **55**, 579–589.

Bridge, J. S. and J. Jarvis (1977). Velocity profiles and bed shear stress over various bed configurations in a river bend. *Earth Surface Processes*, **2**, 281–294.

Bridge, J. S. and J. Jarvis (1982). The dynamics of a river bend: a study of flow and sedimentary processes. *Sedimentology*, **29**, 499–541.

Bridge, J. S. and M. R. Leeder (1979). A simulation model of alluvial stratigraphy. *Sedimentology*, **26**, 617–644.

Bridge, J. S., N. D. Smith, F. Trent, S. L. Gabel and P. Bernstein (1986). Sedimentology and morphology of a low-sinuosity river: Calamus River, Nebraska Sand Hills. *Sedimentology*, **33**, 851–870.

Brizga, S. O. and B. L. Findlayson (1990). Channel avulsion and river metamorphosis: the case of the Thompson River, Victoria, Australia. *Earth Surface Processes and Landforms*, **15**, 391–404.

Brown, A. G. (1987). Holocene floodplain sedimentation and channel response of the lower River Severn, United Kingdom. *Zeitschrift fur Geomorphologie*, **31**, 293–310.

Brown, A. G. (1990). Holocene floodplain diachronism and inherited downstream variations in fluvial processes: a study of the river Perry, Shropshire, England. *Journal of Quaternary Science*, **5**, 39–51.

Burkham, D. E. (1972). Channel changes of the Gila River in Safford Valley, Arizona 1846–1970. *U. S. Geological Survey Professional Paper*, **655-G**, 24 pp.

Callander, R. A. (1978). River meandering. *Annual Review of Fluid Mechanics*, **10**, 129–158.

Carey, W. C. (1969). Formation of flood plain lands. *Journal of the Hydraulics Division, Proceedings of the American Society of Civil Engineers*, **95**, 981–994.

Carson, M. A. and M. F. Lapointe (1983). The inherent asymmetry of river meanders. *Journal of Geology*, **91**, 41–55.

Colombini, M., G. Seminara and M. Tubino (1987). Finite-amplitude alternate bars. *Journal of Fluid Mechanics*, **181**, 213–232.

Desheng, J. and S. A. Schumm (1987). A new technique for modeling river morphology. In V. Gardiner (ed.), *International Geomorphology 1986*, Wiley, pp. 681–690.

Dietrich, W. E. (1987). Mechanics of flow and sediment transport in river bends. In K. S. Richards (ed.), *River Channels: Environment and Process*, Basil Blackwell, pp. 179–227.

Dietrich, W. E. and J. D. Smith (1983). Influence of the point bar on flow through curved channels. *Water Resources Research*, **19**, 1173–1192.

Dietrich, W. E. and J. D. Smith (1984). Bedload transport in a river meander. *Water Resources Research*, **20**, 1355–1380.

Dietrich, W. E. and P. Whiting (1989). Boundary shear stress and sediment transport in river meanders of sand and gravel. In S. Ikeda and G. Parker (eds), *River Meandering*, Water Resources Monograph 12, American Geophysical Union, pp. 1–50.

Dietrich,, W. E., J. D. Smith and T. Dunne (1979). Flow and sediment transport in a sand bedded meander. *Journal of Geology*, **87**, 305–314.

Dort, W. Jr. (1978). *Channel Migration Investigation: Historic Channel Change Maps, Kansas City District*, U. S. Army Corps of Engineers, 50 pp.

Ervine, D. A. and J. Ellis (1987). Experimental and computational aspects of overbank floodplain flow. *Transactions of the Royal Society of Edinburgh: Earth Sciences*, **78**, 315–325.

Everitt, B. L. (1968). Use of the cottonwood in an investigation of the recent history of a flood plain. *American Journal of Science*, **266**, 417–439.

Ferguson, R. I. (1976). Disturbed periodic model for river meanders. *Earth Surface Processes*, **1**, 337–347.

Ferguson, R. I. (1977). Meander migration: equilibrium and change. In K. J. Gregory (ed.), *River Channel Changes*, Wiley, pp. 235–263.

Fisk, H. N. (1947). *Fine-Grained Alluvial Deposits and their Effect on Mississippi River Activity*, Waterways Experiment Station, U. S. Army Corps of Engineers, 82 pp.

Fredsoe, J. (1978). Meandering and braiding of rivers. *Journal of Fluid Mechanics*, **84**, 609–624.

Friedkin, J. F. (1945). *A Laboratory Study of the Meandering of Alluvial Rivers*, Mississippi River Commission, Waterways Experiment Station, U. S. Army Corps of Engineers, 40 pp.

Fukuoka, S. (1989). Finite amplitude development of alternate bars. In S. Ikeda and G. Parker (eds), *River Meandering*, Water Resources Monograph 12, American Geophysical Union, pp. 237–266.

Furbish, D. J. (1988). River-bed curvature and migration: how are they related? *Geology*, **16**, 752–755.

Furbish, D. J. (1989). The relation of thalweg position to channel curvature along the Ohio River. *Southeastern Geology*, **29**, 143–154.

Gee, D. M., M. G. Anderson and L. Baird (1990). Large-scale floodplain modeling. *Earth Surface Processes and Landforms*, **15**, 513–523.

Gerald, C. F. and P. O. Wheatley (1989). *Applied Numerical Analysis*, Addison-Wesley, 679 pp.

Goss, D. W. (1973). Relation of physical and mineralogical properties to streambank stability. *Water Resources Bulletin*, **9**, 140–144.

Graf, W. L. (1983). Flood-related channel change in an arid-region river. *Earth Surface Processes and Landforms*, **8**, 125–139.

Grissinger, E. H. (1966). Resistance of selected clay systems to erosion by water. *Water Resources Research*, **2**, 131–138.

Grissinger, E. H. (1982). Bank erosion of cohesive materials. In R. D. Hey, J. C. Bathurst and C. R. Thorne (eds), *Gravel-bed Rivers*, Wiley, pp. 273–287.

Hasegawa, K. (1989a). Studies on qualitative and quantitative prediction of meander channel

38   Lowland Floodplain Rivers

shift. In S. Ikeda and G. Parker (eds), *River Meandering*, Water Resources Monograph 12, American Geophysical Union, pp. 215–236.

Hasegawa, K. (1989b). Universal bank erosion coefficient for meandering rivers. *Journal of Hydraulic Engineering*, **115**, 744–765.

Hickin, E. J. (1969). A newly-identified process of point bar formation in natural streams. *American Journal of Science*, **267**, 999–1010.

Hickin, E. J. (1974). Development of meanders in natural river-channels. *American Journal of Science*, **274**, 414–442.

Hickin, E. J. (1979). Concave-bank benches in the Squamish River, British Columbia, Canada. *Canadian Journal of Earth Science*, **16**, 200–203.

Hickin, E. J. and G. C. Nanson (1975). Character of channel migration on the Beatton River, northwest British Columbia, Canada. *Geological Society of America Bulletin*, **86**, 487–494.

Hickin, E. J. and G. C. Nanson (1984). Lateral migration rates of river bends. *Journal of Hydraulic Engineering*, **110**, 1557–1567.

Hooke, J. M. (1977). The distribution and nature of changes in river channel patterns: the example of Devon. In K. J. Gregory (ed.), *River Channel Changes*, Wiley, pp. 265–280.

Hooke, J. M. (1979). An analysis of the processes of river bank erosion. *Journal of Hydrology*, **42**, 39–62.

Hooke, J. M. (1980). Magnitude and distribution of rates of river bank erosion. *Earth Surface Processes*, **5**, 143–157.

Hooke, J. M. (1984). Meander behavior in relation to slope characteristics. In C. M. Elliott (ed.), *River Meandering*, American Society of Civil Engineers, pp. 67–76.

Hooke, J. M. (1986). The significance of mid-channel bars in an active meandering river. *Sedimentology*, **33**, 839–850.

Hooke, R. Le B (1975). Distribution of sediment transport and shear stress in a meander bend. *Journal of Geology*, **83**, 543–565.

Howard, A. D. (1984). Simulation model of meandering. In C. M. Elliott (ed.), *River Meandering*, American Society of Civil Engineers, pp. 952–963.

Howard, A. D. and Hemberger, A. T. (in press). Multivariate characterization of meandering. *Geomorphology*.

Howard, A. D. and G. Kerby (1983). Channel changes in badlands. *Geological Society of America Bulletin*, **94**, 739–752.

Howard, A. D. and T. R. Knutson (1984). Sufficient conditions for river meandering: a simulation approach. *Water Resources Research*, **20**, 1659–1667.

Howard, A. D., J. B. Morton, M. Gad-el-Hak and D. B. Pierce (1978). Sand transport model of barchan dune equilibrium. *Sedimentology*, **25**, 307–338.

Hughes, D. J. (1977). Rates of erosion on meander arcs. In K. J. Gregory (ed.), *River Channel Changes*, Wiley, pp. 193–205.

Ikeda, H. (1989). Sedimentary controls on channel migration and origin of point bars in sand-bedded meandering rivers. In S. Ikeda and G. Parker (eds), *River Meandering*, Water Resources Monograph 12, American Geophysical Union, pp. 51–68.

Ikeda, S. (1989). Sediment transport and sorting in bends. In S. Ikeda and G. Parker (eds), *River Meandering*, Water Resources Monograph 12, American Geophysical Union, pp. 103–126.

Ikeda, S., G. Parker and K. Sawai (1981). Bend theory of river meanders, 1, linear development. *Journal of Fluid Mechanics*, **112**, 363–377.

Jackson, R. G., II (1976). Depositional model of point bars in the lower Wabash River. *Journal of Sedimentary Petrology*, **46**, 579–594.

Johannesson, J. and Parker, G. (1989). Linear theory of river meanders. In S. Ikeda and G. Parker (eds), *River Meandering*, Water Resources Monograph 12, American Geophysical Union, pp. 181–214.

Kesel, R. H., K. C. Dunne, R. C. McDonald, K. R. Allison and B. E. Spicer (1974). Lateral erosion and overbank deposition on the Mississippi River in Louisiana caused by 1973 flooding. *Geology*, **2**, 461–464.

Kinoshita, R. (1961). *Investigation of channel deformation in Ishikari River*, Report of Bureau of Resources, Department of Science and Technology, Japan (in Japanese), 174 pp.

Knight, D. W. (1989). Hydraulics of flood channels. In K. Beven and P. Carling (eds), *Floods: Hydrological, Sedimentological and Geomorphological Implications*, Wiley, pp. 83–105.

Knight, D. W. and J. D. Demetriou (1983). Flood plain and main channel flow interaction. *Journal of the Hydraulics Division, Proceedings of the American Society of Civil Engineers*, **109**, 1073–1082.

Kochel, R. C. (1988). Geomorphic impact of large floods: review and new perspectives on magnitude and frequency. In V. R. Baker, R. C. Kochel and P. C. Patton (eds), *Flood Geomorphology*, Wiley, pp. 169–187.

Kondrat'yev, N. Ye (1968). Hydromorphological principles of computations of free meandering. 1. Signs and indices of free meandering. *Soviet Hydrology: Selected Papers*, **4**, 309–335.

Kuijper, C., J. M. Cornelisse and J. C. Winterwerp (1989). Research on erosive properties of cohesive sediments. *Journal of Geophysical Research*, **94**, 14341–14350.

Kulemina, N. M. (1973). Some characteristics of the process of incomplete meandering of the channel of the upper Ob' River. *Soviet Hydrology: Selected Papers*, **6**, 518–534.

Lapointe, M. F. and M. A. Carson (1986). Migration patterns of an asymmetric meandering river: the Rouge River, Quebec. *Water Resources Research*, **22**, 731–743.

Laury, R. L. (1971). Stream bank failure and rotational slumping: preservation and significance in the geologic record. *Geological Society of American Bulletin*, **82**, 1251–1266.

Lawler, D. M. (1986a). Bank erosion and frost action: an example from South Wales. In V. Gardiner (ed.), *International Geomorphology 1986*, Wiley, pp. 575–590.

Lawler, D. M. (1986b). River bank erosion and the influence of frost: a statistical examination. *Transactions of the Institute of British Geographers, N. S.* **11**, 227–242.

Leeder, M. R. (1978). A quantitative stratigraphic model for alluvium, with special reference to channel deposit density and interconnectedness. In A. D. Miall (ed.), *Fluvial Sedimentology*, Canadian Society of Petroleum Geologists, pp. 587–596.

Leopold, L. B. and M. G. Wolman (1957). River channel patterns—braided, meandering, and straight. *U. S. Geological Survey Professional Paper*, **282-B**, 85 pp.

Lewin, J. (1976). Initiation of bed forms and meanders in coarse-grained sediment. *Geological Society of America Bulletin*, **87**, 281–285.

Lewin, J. (1978). Meander development and floodplain sedimentation: a case study from mid-Wales. *Geological Journal*, **13**, 25–36.

Lewin, J. and B. J. Brindle (1977). Confined meanders. In K. J. Gregory (ed.), *River Channel Changes*, Wiley, pp. 221–233.

Lewis, G. W. and J. Lewin (1983). Alluvial cut-offs in Wales and the Borderlands. In *Modern and Ancient Fluvial Systems*, Blackwell Scientific Publications, pp. 145–154.

Miller, A. J., and M. G. Wolman (1990) 2D simulation of flood flow patterns in mountain valleys. *EOS (Transactions of the American Geophysical Union)*, **71**, 510.

Murray, W. A. (1977). Erodibility of coarse and sand-clayey silt mixtures. *Journal of the Hydraulics Division, Proceedings of the American Society of Civil Engineers*, **103**, 1222–1227.

Nanson, G. C. (1980). Point bar and floodplain formation of the meandering Beatton River, northeastern British Columbia, Canada. *Sedimentology*, **27**, 3–29.

Nanson, G. C. (1986). Episodes of vertical accretion and catastrophic stripping: a model of disequilibrium flood plain development. *Geological Society of America Bulletin*, **97**, 1467–1475.

Nanson, G. C. and E. J. Hickin (1983). Channel migration and incision on the Beatton River. *Journal of Hydraulic Engineering*, **109**, 327–337.

Nanson, G. C. and E. J. Hickin (1986). A statistical analysis of bank erosion and channel migration in western Canada. *Geological Society of America Bulletin*, **97**, 497–504.

Nanson, G. and K. Page (1983). Lateral accretion of fine-grained concave benches on meandering rivers. In *Modern and Ancient Fluvial Systems*, Blackwell Scientific Publications, pp. 133–143.

Neill, C. R. (1984). Bank erosion vs. bedload transport in a gravel river. In C. M. Elliott (ed.), *River Meandering*, American Society of Civil Engineers, pp. 204–211.

Nelson, J. M. and J. D. Smith (1989a). Flow in meandering channels with natural topography. In S. Ikeda and G. Parker (eds), *River Meandering*, Water Resources Monograph 12, American Geophysical Union, pp. 69–102.

Nelson, J. M. and J. D. Smith (1989b). Evolution and stability of erodible channel beds. In S. Ikeda and G. Parker (eds), *River Meandering*, Water Resources Monograph 12, American Geophysical Union, pp. 321–378.

Odgaard, A. J. (1987). Streambank erosion along two rivers in Iowa. *Water Resources Research*, **23**, 1225–1236.

Odgaard, A. J. (1989a). River-meander model. I: development. *Journal of Hydraulic Engineering*, **115**, 1433–1450.

Odgaard, A. J. (1989b). River-meander model. II: applications. *Journal of Hydraulic Engineering*, **115**, 1451–1464.

Odgaard, A. J. and M. A. Bergs (1988). Flow processes in a curved alluvial channel. *Water Resources Research*, **24**, 45–56.

Osman, A. M. and C. R. Thorne (1988). Riverbank stability analysis. I: development. *Journal of Hydraulic Engineering*, **114**, 134–150.

Osterkamp, W. R. and J. E. Costa (1987). Changes accompanying an extra-ordinary flood on a sand-bed stream. In L. Mayer and D. Nash (eds), *Catastrophic Flooding*, Allen & Unwin, pp. 201–224.

Parchure, T. M. and A. J. Mehta (1985). Erosion of soft cohesive sediment deposits. *Journal of the Hydraulics Division, American Society of Civil Engineers*, **111**, 1308–1326.

Parker, G. (1976). On the cause and characteristic scales of meandering and braiding in rivers. *Journal of Fluid Mechanics*, **76**, 457–480.

Parker, G. (1982). *Stability of the Channel of the Minnesota River Near State Bridge No. 93, Minnesota*, Project Report No. 205, St. Anthony Falls Hydraulic Laboratory, University of Minnesota, 33 pp.

Parker, G. (1984). Theory of meander bend deformation. In C. M. Elliott (ed.), *River Meandering*, American Society of Civil Engineers, pp. 722–732.

Parker, G. and E. D. Andrews (1985). Sorting of bed load sediment by flow in meander bends. *Water Resources Research*, **21**, 1361–1373.

Parker, G. and E. D. Andrews (1986). On the time development of meander bends. *Journal of Fluid Mechanics*, **162**, 139–156.

Parker, G. and H. Johannesson (1989). Observations on several recent theories of resonance and overdeepening in meandering channels. In S. Ikeda and G. Parker (eds), *River Meandering*, Water Resources Monograph 12, American Geophysical Union, pp. 379–416.

Parthenaides, E. (1965). Erosion and deposition of cohesive soils. *Journal of the Hydraulics Division, American Society of Civil Engineers*, **91**, 105–139.

Parthenaides, E. and R. R. Paaswell (1970). Erodibility of channels with cohesive boundaries. *Journal of the Hydraulics Division, American Society of Civil Engineers*, **96**, 755–771.

Pizzuto, J. E. (1984). Bank erodibility of sand-bed streams. *Earth Surface Processes and Landforms*, **9**, 113–124.

Pizzuto, J. E. (1986). Flow variability and bankfull depth of sand-bed rivers of the American Midwest. *Earth Surface Processes and Landforms*, **11**, 441–450.

Pizzuto, J. E. (1987). Sediment diffusion during overbank flows. *Sedimentology*, **34**, 301–317.

Pizzuto, J. E. and T. S. Meckelnburg (1989). Evaluation of a linear bank erosion equation. *Water Resources Research*, **25**, 1005–1013.

Ritter, D. F. and D. S. Blakely (1986). Localized catastrohic disruption of the Gasconade River flood plain during the December 1982 flood, southeast Missouri. *Geology*, **14**, 472–476.

Rohrer, W. L. (1984). Effects of flow and bank material on meander migration in alluvial rivers. In C. M. Elliott (ed.), *River Meandering*, American Society of Civil Engineers, pp. 770–782.

Schumm, S. A. and R. W. Lichty (1963). Channel widening and floodplain constriction along Cimarron River in southwestern Kansas. *U. S. Geological Survey Professional Paper*, **352-D**, 71–88.

Seminara, G. and M. Tubino (1989). Alternate bars and meandering: free, forced and mixed interactions. In S. Ikeda and G. Parker (eds), *River Meandering*, Water Resources Monograph 12, American Geophysical Union, pp. 267–320.

Shimizu, Y. and T. Itakura (1989). Calculation of bed variation in alluvial channels. *Journal of Hydraulic Engineering*, **115**, 367–384.

Taylor, G. and K. D. Woodyer (1978). Bank deposition in suspended-load streams. In A. D. Miall (ed.), *Fluvial Sedimentology*, Canadian Society of Petroleum Geologists, pp. 257–275.

Thorne, C. R. (1982). Processes and mechanisms of river bank erosion. In R. D. Hey, J. C. Bathurst, and C. E. Thorne (eds), *Gravel-bed Rivers*, Wiley, pp. 227–272.

Thorne, C. R. and J. Lewin (1979). Bank processes, bed material movement, and planform development in a meandering river. In D. D. Rhodes and G. P. Williams (eds), *Adjustments of the Fluvial System*, Kendall/Hunt, pp. 117–137.

Thorne, C. R. and Osman, A. M. (1988). Riverbank stability analysis. II: applications. *Journal of Hydraulic Engineering*, **114**, 151–172.

Thorne, C. R. and N. K. Tovey (1981). Stability of composite river banks. *Earth Surface Processes and Landforms*, **6**, 469–484.

Tubino, M. and G. Seminara (1990). Free-forced interactions in developing meanders and suppression of free bars. *Journal of Fluid Mechanics*, **214**, 131–159.

Turbull, W. J., E. L. Krinitsky and F. J. Weaver (1966). Bank erosion in the soils of the Lower Mississippi River. *Journal of Soil Mechanics and Foundation Engineering*, **92**, 121–137.

Ullrich, C. R., D. J. Hagerty and R. W. Holmberg (1986). Surficial failures of alluvial stream banks. *Canadian Geotechnical Journal*, **23**, 304–316.

Walling, D. E. and S. B. Bradley (1989). Rates and patterns of contemporary floodplain sedimentation: a case study of the river Culm, Devon, UK. *Geojournal*, **19.1**, 53–62.

Whiting, P. J. (1990). *Bar development and channel morphology*, Unpublished PhD dissertation, University of California, 196 pp.

Whiting, P. J. and W. E. Dietrich (1989). Multiple bars in highly sinuous flume bends: implications for bank erosion. *EOS (Transactions of the American Geophysical Union)*, **70**, 329.

Willis, B. J. (1989). Palaeochannel reconstructions from point bar deposits: a three-dimensional perspective. *Sedimentology*, **36**, 757–766.

Wolman, M. G. (1959). Factors influencing erosion of a cohesive river bank. *American Journal of Science*, **257**, 204–216.

Wolman, M. G. and L. B. Leopold (1957). River flood plains: some observations on their formation. *U. S. Geological Survey Professional Paper*, **282-C**, 87–109.

Wolman, M. G. and L. M. Brush, Jr. (1961). Factors controlling the size and shape of stream channels in coarse noncohesive sands. *U. S. Geological Survey Professional Paper*, **282-G**, 183–210.

Woodyer, K. D. (1975). Concave-bank benches on Barwon River, N. S. W. *Australian Geographer*, **13**, 36–40.

Yen, B. C. and C.-L. Yen (1984). Flood flow over meandering channels. In C. M. Elliott (ed.), *River Meandering*, American Society of Civil Engineers, pp. 554–561.

2 The Reversal Hypothesis and the Maintenance of Riffle–Pool Sequences: A Review and Field Appraisal

N. J. CLIFFORD
School of Geography and Earth Resources, University of Hull
and
K. S. RICHARDS
University of Cambridge

INTRODUCTION

Riffle–pool sequences are the characteristic reach-scale bedforms of low-gradient rivers with mixed load and gravel beds. Riffles are high points in the bed topography, associated over most of the flow range with fast, divergent flows, steep water surface slopes and coarser bed material. Pools are low points in the topography, with slow, convergent flows, slighter water surface slopes and finer bed material. Riffle–pool sequences have been the subject of geomorphological study as important features in the dynamic adjustment between form and process in alluvial channels, but a knowledge of their sedimentology and dynamics is also important with respect to water planning, the maintenance of stream ecology and the assessment and management of the recreational potential of rivers. The stability of riffles makes them ideal natural gauging sites, and since riffles function as natural weirs (Richards, 1978a), investigation of their regulatory role with respect to low flow hydraulics is important to the assessment of critical levels of water pollution, aquatic habitat and instream flow needs (Miller and Wenzel, 1985). There is strong evidence that riffle–pool sequences are the principal determinants of the longitudinal arrangement of stream ecosystems, which figures in quantitative river channel inventories directed to water resource planning (Mosley, 1985). Ecologically, reach-scale contrasts in flow and bed sedimentology that are also stable between successive floods are necessary for the maintenance of viable populations of several fish species and a range of invertebrate fauna (e.g. Stuart, 1953; Rutherford and Mackay, 1985), with catastrophic reduction in stocks following river channelization (Graesser, 1979; Cowx, Wheatley and Mosley, 1986). For these and for aesthetic reasons, the creation of riffle–pool sequences is advocated in 'environmentally sensitive' river management designs (e.g. Keller, 1978; Hey, 1986; Brookes, 1988), thus stimulating continued interest in their flow–sediment dynamics.

Lowland Floodplain Rivers: Geomorphological Perspectives. Edited by P. A. Carling and G. E. Petts
© 1992 John Wiley & Sons Ltd

RIFFLE–POOL MAINTENANCE AND REVERSAL HYPOTHESES

The most widely accepted account of riffle–pool maintenance is provided by the hydraulic reversal hypothesis, which describes a reversal in the spatial hierarchy of flow strength between riffle and pool as discharge rises. Of 12 investigations concerned with the hydraulics of riffle–pool sequences, six accept a reversal in one or more indices of flow strength (Keller, 1971; Andrews, 1979; Lisle, 1979; Teisseyre, 1984; O'Connor, Webb and Baker, 1986; Ashworth, 1987; Petit, 1987), three review the reversal hypothesis as a starting point for further investigation (Richards, 1976a; Jackson and Beschta, 1981; Carling, 1991), while three reject outright any sort of reversal (Teleki, 1972; Bathurst, 1982a; Bhowmik and Demissie, 1982).

Origins

The reversal hypothesis was formally presented as an account of the maintenance of riffle–pool sequences by Keller (1971), although Gilbert's (1914) observations of contrasting surface agitation in streams between high and low flows are credited by Keller (1972) as the source of the reversal idea. It is clear however, that the role of channel geometry in providing a spatially varied and stage-dependent control on the hierarchy of flow strength had been noted earlier by Seddon (1900), and Lane and Borland (1953) provide the first detailed account of a reversal in mean velocity to explain the maintenance of crossings and deeps in large, lowland, sand-bedded channels.

Outline

Figure 2.1 illustrates the nature and detail of the reversal hypothesis as it might apply to a 'typical' riffle–pool unit, composed of a downstream riffle and an upstream pool. In the region towards the origin of the graph, flow strength and therefore sediment transport competence are low. Riffle flow strength (indexed by measures of velocity or shear stress) exceeds that through the pool, and sediment movement occurs as a winnowing of fines from the riffle to the next downstream pool, which is the locus of lower stage deposition. As discharge rises, flow strength through riffle and pool converge, and more generalized movement of finer material occurs throughout the riffle–pool sequence. Eventually, at about 50–90% of the bankfull discharge, the plots describing flow strength at riffle and pool cross. From that time on until discharge declines, the low flow hierarchy in competence is 'reversed'. At and beyond the reversal discharge, it is assumed that competence through the pool is sufficient to move coarse fractions of the bed material from the pool to the next downstream riffle. The pool is scoured, while the riffle becomes an area of stable deposition for the coarser grain sizes.

Evaluation

The strength of the model of riffle–pool dynamics that the reversal hypothesis provides rests in its ability to resolve what might otherwise appear a paradox: that the stability of upstanding areas of apparently coarser sediment is maintained, despite

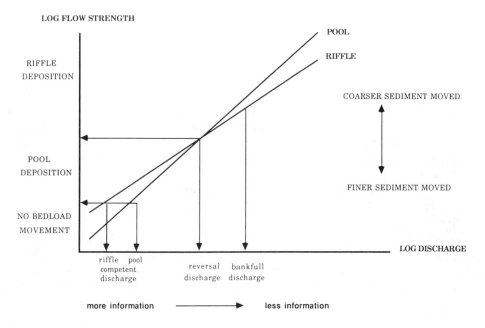

Figure 2.1 Schematic representation of flow–sediment interactions in a riffle–pool sequence, and the importance of a 'reversal' phenomenon

their coincidence at low flow with zones of increased flow strength and water surface slope. The model also has a number of weaknesses, however, which can be illustrated by considering again Figure 2.1.

The reversal hypothesis relies on the following assumptions: (1) that power law regularities (or close approximations to them) hold in both riffle and pool over the entire discharge range (this explains the logarithmic axes used in Figure 2.1 and the straight-line nature of the flow strength relationships); (2) that these relations are similar for riffles and pools in straight and sinuous channels; and (3) that the reversal occurs for all paired cross-sections designated as riffle and pool within a single riffle–pool unit. Four related issues arise from this that require examination before the reversal hypothesis can be accepted.

First, the hypothesis is formulated from many observations at or near low flow conditions, with few high flow counterparts. Keller's own data, for example, include no observations at high flow, and the plot of mean bottom velocity against discharge on which his argument was based illustrates a convergence, rather than a reversal. Belief in a reversal becomes an exercise in ratiocination: the high stage reversal is assumed, without corroboration, to account for the pattern of areal sorting of bed material observed at low stage (Keller, 1972, p. 918). Most natural river sections display more complex behaviour than conventional hydraulic geometry suggests, and there is no a priori reason why power functions should best describe variations in flow geometry with discharge (Richards, 1982, pp. 154–155). There is a general tendency in gravel-bedded rivers for the slope of the trend in mean velocity with respect to discharge to decrease as discharge increases, which is explained by a slackening in the rate at which flow resistance falls as discharge rises (Richards, 1973). Thus, while discriminant functions employing hydraulic geometry exponents provide dimensionless criteria for differentiating between riffles and pools (Richards, 1976a, 1982), the discriminant functions are independent of flow stage only in the range for which measurements are available and for which power functions exist. Recent observations (Carling, 1991; Keller, pers. comm., 1986) confirm that riffles and pools may possess hydraulic geometries that are significantly non-linear in logarithmic plots, which would not be revealed unless data over the entire discharge range were available. Log–log plots also must be interpreted carefully, because it is possible for convergent logarithmic plots to mask *absolute* increases in the difference between two quantities plotted in this way.

Second, perhaps the most striking feature of the literature dealing with the reversal hypothesis is the diversity of reversal parameters used to substantiate, reject or formulate alternatives to it. In addition to Keller's (1971) use of mean bottom velocity, mean section velocity (Andrews, 1979; Teissyre, 1984), mean boundary shear stress (Lisle, 1979), section-average shear velocity (Carling, 1991), stream power (O'Connor, Webb and Baker, 1986), and a variety of point measures of velocity and shear stress (Jackson and Beschta, 1981; Ashworth, 1987; Petit, 1987) have all been cited in accounts that offer support to the hypothesis, while Bhowmik and Demissie (1982), in rejecting velocity measures and the reversal hypothesis, use the Froude number as a criterion controlling size–sorting relationships through pools and riffles. Only Teissyre (1984) and Carling (1991) examine the simultaneous variation in several parameters, but Teissyre's data are limited to three observations, while Carling examines only section-average parameters. Both studies therefore provide a restricted basis for comparison. Although successive authors have assumed that the behaviour of different indices of flow strength are directly equivalent measures of changing competence, the assumption holds only where flow is strictly uniform. Since the riffle–pool sequence by definition involves non-uniform channel geometry and, under the predictions of the reversal hypothesis, its maintenance is explained in terms of a discharge-dependent interaction between an irregular channel boundary and the flow behaviour, patterns in the behaviour of section-averaged and point measurements at riffles and pools are not necessarily similar, and their *physical* equivalence as measures of flow competence may also be suspect. Not only does this cast doubt on the validity of individual explanations of riffle–pool maintenance, especially those that rely on section-averaged measures, but it implies also that the numerous references to reversals using *differing* reversal parameters cannot be interpreted necessarily as evidence that the reversal hypothesis has been well tested.

This difficulty is compounded since reliance on hydraulic parameters to infer likely patterns of sediment transport is reliable only in conditions where bed material movement is transport-limited. In supply-limited conditions, even were a reversal in competence to occur, this would not be reflected in a simple pattern of scour at the pool and fill at the riffle. Particular difficulties arise with inferences based upon mean section velocity and mean boundary shear stress. In the former case, unless there is significant filling of the pool at high discharge, a reversal in mean velocity requires systematic channel geometry or flow resistance differences, which are not present between all riffle and pool sections. In the latter case, mean boundary shear calculated from the reach-scale depth–slope product is unlikely to index sediment transport closely, since in the presence of an undulating topography, it is unlikely that the near bed (grain) component of total stress will be in balance with the downstream propulsive force of gravity at pool locations. Moreover, Clifford (1990), following an earlier suggestion by Lisle (1979) has shown that systematic differences in the occurrence of microtopographic bedforms occurs between riffles and pools, which is confirmed by Sear (in press). Riffles possess significantly greater microtopographic development than pools. Robert (1988) has argued that small-scale gravel bedforms contribute significant additional *form* drag, and there is, therefore, reason to suspect that at both riffle and pool, the stress available to transport sediment is only a fraction of the total flow resistance. In so far as the structuring of particles influences sediment supply by changing the threshold conditions, the existence of spatial variations in gravel bedform frequency results in variation in both sediment availability and the shear available for transport between riffle and pool locations. Consequently, the nature of the flow–sediment interaction in riffle–pool sequences may be more complicated than the reversal hypothesis implies.

Fourth, there has been almost complete neglect of the need for close spatial control in previous evaluations of form–flow interaction through riffle–pool sequences. This is important at a variety of scales. First, within individual *cross-sections*, it is possible that flow distribution is non-uniform, and that point measures of flow strength may not be integrated appropriately into a meaningful cross-section average. For example, Carling (1991) used section-averaged shear velocities, which is clearly only acceptable after rigorous analysis of velocity profiles from adjacent verticals to ensue that uniformity exists—as was the case in the River Severn (Carling, pers. comm.). Second, at the *reach scale*, the most basic consideration is that comparisons should be restricted to contiguous riffle–pool units. Nevertheless, several important studies (Andrews, 1979; Lisle, 1979; Jackson and Beschta, 1981) must be criticized for their attempts to make comparisons between widely separated sections in different riffle–pool units. Jackson and Beschta (1981) for example, compare a single *down*stream pool with two *up*stream riffles separated by an intervening pool and a 'small log step' (p. 521). In Lisle's (1979) study, the riffle and pool sections compared are 1 km apart, while in Andrew's (1979) study, there are no contiguous riffle and pool sections that demonstrate the mean velocity reversal he supports. Also, Keller (1972) noted that the demonstration of a reversal may depend upon those sections arbitrarily defined with respect to low flow topography as representative of riffle or pool as discharge rises, and the low flow hydraulic control of the riffle is drowned. Most generally, there is the implication that because riffle–pool sequences occur irrespective of *channel planform*, the reversal hypothesis also must be valid equally in meandering, braided and straight river sections. Differences in hydraulic behaviour seem to occur between curved and straight pool–riffle sequences: Carling (1991) notes that results from meandering sections are much more variable than from adjacent straight sections. 'Straight' riffle friction factors may be twice as large as those at meander bend inflection sites, possibly reflecting the permanent drowning-out of riffle skin friction in the backwater curve from an 'overdeepened' meander bend pool (Richards, 1976a), and observations of unit stream power have been made that demonstrate a reversal above bankfull for a curved reach, which is absent for a straight reach of the same river (Keller, pers. comm., 1986). With respect to observed behaviour of mean boundary stress and stream power indexed by water surface slope contrasts, the spacing of riffles in relation to overall channel gradient appears to be crucial (Carling, 1991).

There are, then, sufficient uncertainties surrounding acceptance of the reversal hypothesis as a general account of riffle–pool maintenance to warrant further study. These uncertainties persist in spite of the foundation of the hypothesis early in 'modern' studies of fluvial geomorphology, and despite its popularity in more recent accounts of the flow and sediment dynamics through riffle–pool sequences. There is an evident need for closer attention to the spatial variability of flow–topography–sediment interaction, both at the reach scale between riffle–pool units, as well as within individual riffle–pool units. The purpose of this paper, therefore, is to adopt a spatially disaggregated sampling design, and thereby: (1) to examine more closely the *simultaneous* behaviour of the *different* reversal parameters; (2) where variations occur between the measures, to account for the differences; (3) to assess the implications that these have for the reversal hypothesis as a general explanation of the maintenance of riffle–pool sequences.

METHODS

Measurements were collected over two field seasons, in November 1986 and November 1987, at a single riffle–pool unit in a straight reach of the River Quarme (grid reference SS920350). Location of the study reach within a straight section of the channel enabled curvature effects to be neglected, while concentration of effort at a single site enabled simultaneous monitoring of a variety of flow strength indices over a wide flow range with close spatial control.

Five cross-sections were located approximately perpendicular to the channel centre-line and monumented for all subsequent work according to the following scheme (Figure 2.2): four cross-sections, named in upstream sequence, riffle midpoint (RM), riffle crest (RC), pool tail (PT) and pool midpoint (PM); and a gauging section, chosen for its uniformity of bed and banks, absence of undercutting and apparently stable bed material. Gauging was undertaken according to standard 'mean-section' procedures, and discharge calculated using a BSI streamflow measurement program (Midgeley, Petts and Walker, 1986) implemented on a BBC microcomputer.

The riffle–pool unit was identified with respect to the low flow bed and flow geometry, with the riffle crest cross-section defined at the position of the break of water surface slope, perpendicular to the channel centreline. The pool midpoint was defined at half the inter-riffle crest distance using observations of the next upstream riffle, and the pool tail at half the distance between the pool midpoint and the riffle crest. These locations are entirely arbitrary, although reference to the local bed profile (Figure 2.3) shows that the riffle and pool midpoint cross-sections are located at the points of minimum and maximum centreline depth. They should, therefore, be physically 'representative' of hydraulic behaviour that depends on relative elevation of the channel bed. Designation of the riffle midpoint was most difficult. A distance of one-sixth of the inter-riffle crest distance downstream from the riffle crest was eventually chosen, to accommodate the notional distinction between this and a 'pool entry' point at one-sixth of the inter-crest distance further downstream.

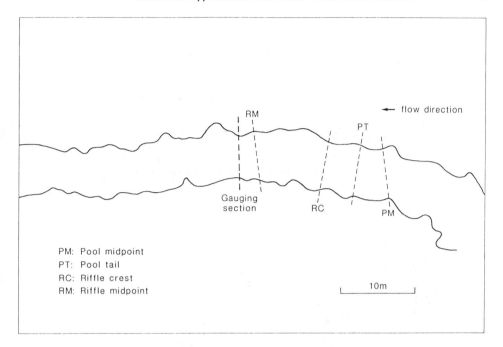

Figure 2.2 Plan of the study reach, River Quarme, showing location of pool and riffle cross-sections and the gauging section

Each of the riffle and pool cross-sections was subdivided into six segments by establishing a network of control points on which all subsequent measurements were based. 'Verticals' were located from a reference point at the midpoint of the section bed-width ('point C') with two verticals at equal fractions of the bed-width on either side (points A and B and D and E, respectively). Cross-section profiles of each section, together with the position of the verticals, are shown in Figure 2.4. Nylon cords were pegged tight across each section, and the position of each vertical was marked on these in permanent ink. In this way, the position of the verticals could be located accurately at all flow stages.

Extent of the Data Set

Direct monitoring and calculation of hydraulic variables was possible at seven discharge values, over the range 0.57–4.47 m^3 s^{-1} (approximately base flow to bankfull flow). These were:

(1) Local variables measured at all sections and verticals: local depth; 'bed velocity' ($v_{0.03}$), measured with an Ott C-3 current meter set at 0.03 m above the local bed defined with respect to its rod base-plate; and 'profile average velocity' ($v_{0.6d}$), measured as above at a height of 0.6 times the local depth measured from the water surface. At discharges of 0.70 m^3 s^{-1}, 0.89 m^3 s^{-1}, 1.81 m^3 s^{-1} and 2.61 m^3 s^{-1}, further velocity measurements were made at 0.05 m above the bed, to obtain

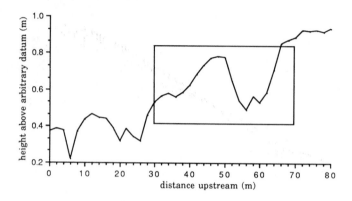

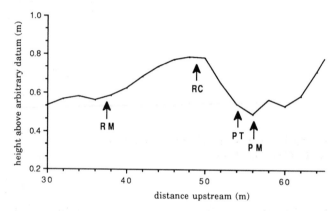

Figure 2.3 Bed profile of the River Quarme, showing location of principal measuring sections within the study reach

local bed shear stress (τ_1) from:

$$\tau_1 = \varrho \left(\frac{(v_2 - v_2)}{5.75 \log_{10}(y_2/y_1)} \right)^2 \tag{2.1}$$

where ϱ is the fluid density (kgm^{-3}), v is the velocity (ms^{-1}) at height y (m) above the bed.

(2) Mean variables. These consisted of section mean velocity and mean boundary shear stress. Section mean velocity was calculated by dividing discharge at the gauging section by cross-section area. Water surface slopes (S) were surveyed at four discharges in the intermediate and higher range (1.27–4.47 m^3 s^{-1}), and these were used in conjunction with the estimated hydraulic radius (R) at each section to calculate mean boundary shear stress from:

$$\tau_0 = \varrho g R S \tag{2.2}$$

At discharges below 1.27 m^3 s^{-1}, the water surface slope through the pool was too slight to record, whereas the flow over the riffle was too shallow and distorted by particle protrusion to allow measurements to be obtained.

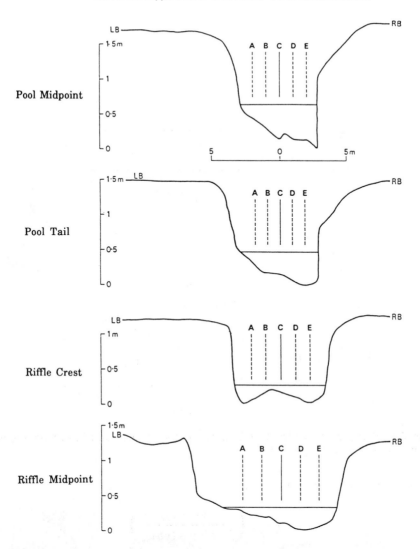

Figure 2.4 Cross-section profiles of pool and riffle sections in the River Quarme study reach. Locations of measurement control verticals also are shown, marked A–E. Left bank (LB) and right bank (RB) refer to view looking downstream

RESULTS AND DISCUSSION

Local Measures

Direct Observations of Cross-Section Velocity Distribution

Figure 2.5 shows near-bed velocities obtained at the grid of measuring points for three discharges. The influence of topography on flow geometry is clear in each case. At low flow (Figure 2.5a), channel topography is mirrored in longitudinal and cross-sectional

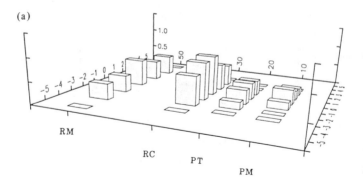

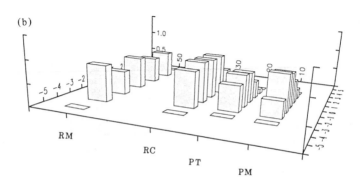

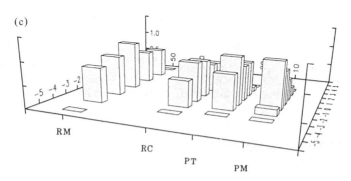

Figure 2.5 Graphical representation of near-bed velocities measured at control verticals through the riffle–pool study reach, River Quarme: (a) at a stage of 0.23 m; (b) at a stage of 0.47 m; and (c) at a stage of 0.60 m. Control vertical 'A' is in the foreground of each graph, and flow is from right to left. Unit on the vertical axis is velocity in ms^{-1}. Values of zero velocity are assumed at each bank

velocity contrasts: pool velocities are lower than riffle velocities, but measurable flow is confined to a restricted portion of the channel at all cross-sections. A more even distribution of velocity at riffle sections than at pool sections reflects a topographic influence via convergence and divergence. At an intermediate discharge (Figure 2.5b) there is reduced longitudinal variation, with similar maximum values of bed velocity at all sections. Cross-sectional flow distribution is still clearly different, however, with more even distributions at the riffle cross-sections. At the highest discharge (Figure 2.5c) bed velocities at the channel margins remain slight in all cases, but the pool midpoint stands out as the cross-section with the greatest non-uniformity of flow across the channel. Significantly, the plots also reveal that the location of maximum flow velocity shifts with changing discharge, indicating that consistent measurement of velocity at a fixed point may yield unsatisfactory data for testing the reversal concept.

The variability associated with an irregular cross-section velocity distribution can be expressed quantitatively by calculating the velocity distribution coefficient, used in the Bernoulli equation. This is particularly sensitive to lateral variation in velocity (Hulsing, Smith and Cobb, 1966, p. C13), and is computed by comparing the summation of velocities calculated for individual segments of a given cross-section to the mean velocity of the section according to:

$$\alpha = \frac{\sum v^3 \, \Delta A}{V^3 \, A} \tag{2.3}$$

where ΔA is an elementary area of the total cross-section A, v is the velocity of the water flowing through ΔA, and V is the cross-section mean velocity (Chow, 1981, p. 29). A value of unity indicates a uniform velocity distribution, with higher values indicating progressive non-uniformity. In applying equation (2.3), six elementary areas were defined by the area between control verticals A–E, and between the outer control verticals and the channel margins. The value at bankfull for the pool midpoint is missing because velocities for verticals B and D are missing owing to hazardous observation conditions.

Figure 2.6 confirms the visual impression that the velocity distribution of the pool sections is less uniform than that through the riffle, with the exception of the bankfull pool tail value. Gross tendencies in the behaviour of all sections are similar, showing a decline in non-uniformity as discharge rises. The hierarchy is generally: pool midpoint (least uniform); pool tail; riffle crest; and riffle midpoint (most uniform). At bankfull, the riffle midpoint has a value of unity. The range of values is considerably greater than the variation between 1.15 and 1.50 assumed in engineering analyses of open channels. In the vicinity of weirs, values greater than 2.00 have been recorded (Chow, 1981, p. 28), and measurements from shallow sections of high-gradient streams of approximately trapezoidal section reveal a range of coefficients between 1.00 and 2.00 (Jarrett, 1984; Bathurst, 1985). Both of these environments have greater similarity to the riffle sections measured here than to the pool, and offer some support for the veracity of the calculated riffle velocity distribution coefficients. The most extensive data from natural channels is that of Hulsing, Smith and Cobb (1966), where a range between 1.09 and 2.90 is quoted for natural, regular trapezoidal channels, and a range of 1.06 to 4.70 is reported for natural channels with flow obstructions, such as bridge piers. The values obtained for the pool sections here, then, are not without counter-

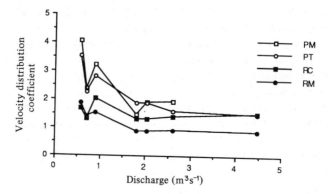

Figure 2.6 Velocity distribution coefficients at pool and riffle cross-sections, River Quarme

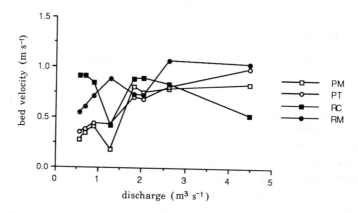

Figure 2.7 Near-bed velocity at channel centreline, at pool and riffle cross-sections, River Quarme

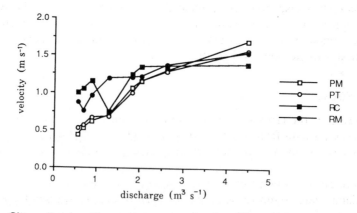

Figure 2.8 Channel centreline velocity at 60% depth, at pool and riffle cross-sections, River Quarme

parts in fluvial environments, and Hulsing, Smith and Cobb (1966, p. C19) raise the possibility of abrupt changes in rivers at sections immediately upstream of riffles. It is interesting too, in the light of observations of velocity distribution coefficients close to a value of 4.00 obtained at the outlets of closed conduits (Chow, 1981, p. 29), to recollect Seddon's belief (1900) in a stage-dependent transition from flow controlled by a series of weirs, to flow controlled by a series of orifices, and Keller's (1971) original formulation of a reversal in terms of a high velocity jet, or core, rather than a section-average phenomenon. The significance of Figures 2.5 and 2.6 lies in the doubt they cast on those reversals inferred from comparisons of velocity or shear stress measured at single points within riffle and pool cross-sections, such as those shown in Figures 2.7 and 2.8.

Near-bed and Profile-average Velocity

Figure 2.7 shows the behaviour of near-bed velocity obtained at control vertical C, the channel centreline. At low discharge, there is the anticipated hierarchy of lower values at pool sections, and a reversal apparently takes place at high discharge between both pool sections and the riffle crest, although not for the riffle midpoint. A similar plot of the profile average velocity at the channel centreline (Figure 2.8) also shows a reversal, which occurs at *ca.* 70% bankfull, this time between both pool sections and both riffle sections. However, the hierarchy at bankfull and other discharges differs between the two point-specific measures, so that different measures are not equivalent indicators of possible reversals. There is some suggestion from a comparison of Figures 2.7 and 2.8 that the existence or otherwise of a reversal depends, in part, on which pre-defined riffle or pool cross-sections are compared. A similar conclusion is reached by Keller (1972), but its importance has been neglected in subsequent accounts, apart from that of Petit (1987). Both plots also contain erratic behaviour, with 'transient reversals' at low and intermediate discharges. On the basis of Figure 2.7 for example, two 'transient reversals'—between the pool tail and riffle crest at a discharge of $1.27 \, \text{m}^3 \, \text{s}^{-1}$, and between pool-midpoint and riffle-midpoint at a discharge of $2.47 \, \text{m}^3 \, \text{s}^{-1}$—could be identified, neither of which persists at the succeeding discharge. These probably reflect changing patterns of flow distribution within the channel as discharge rises, while the differences between the two point measures is indicative of flow complexity in the vertical, too. In view of this complexity, it is unlikely that interpretation of hydraulic contrasts between riffles and pool obtained at a single point will be meaningful.

Local Shear Stress

Figure 2.9 illustrates the cross-section distribution of local shear stress derived from equation (2.1) using data from verticals A–E at each pool and riffle section, at the four discharges when full measurements were available. The plots are notable in two respects.

First, many points for which measurements were made could not be plotted because velocity gradients were found to be negative, even where three repetitions of measurement were tried at each point. The number of anomalous points decreased with increasing discharge, confirming the limitation on equations such as (2.1) to flow

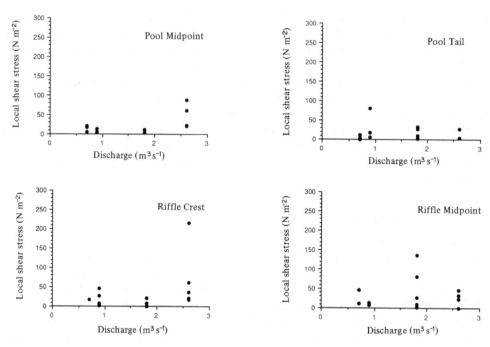

Figure 2.9 Local shear stress measured at verticals A–E at pool and riffle cross-sections, River Quarme

depths where the depth to grain size ratio is greater than *ca.* 3 (Bray, 1980), and possibly also reflecting the changing importance of stress-induced secondary circulations (Bathurst, 1979) arising from the interaction of stagnant water close to pool margins, and a high velocity 'core' close to the channel centre. Second, there is a wide scatter in the points, with order of magnitude differences at the same discharge in several cases. The scatter in the values for local shear stress obviously are not related to discharge as might be expected if depth–grain size effects were dominant. Changes in the cross-section velocity distribution again may be responsible. These differences make overall trends difficult to substantiate for any given section, and given the scatter of values, and the likely errors involved, it would be unwise to comment further on the possibility of trends in the data, except to observe the general tendency for local shear stress to increase with discharge, and again confirm extreme variability even in a narrow cross-section. Obviously, reliable comparisons derived from aggregating profile shear velocity/stress will be difficult to obtain, although in larger rivers with deeper water and more regular cross-sections, comparisons based upon aggregated local shear velocities are possible, especially in straight reaches (Carling, 1991).

Discussion

Both Rubey (1938) and Keller (1971) have argued that near-bed velocity is a better guide to local competence, free of restrictive assumptions regarding the vertical velocity distribution. Profile-average velocities taken from measurements at 60% of the depth and local shear stress have been used to characterize riffle–pool hydraulics (Jackson and Beschta, 1981; Ashworth, 1987; Petit, 1987). Again, the rationale for using these measures is that they are better reflections of the competence through riffles and pools at the points where the measurements are taken. However, the general conclusion from the results presented here is that the spatial representativeness of single measurements is difficult to interpret with respect to the pattern of likely sediment transport through either the riffle or the pool section. Since the flow distribution changes with discharge at all sections, measurements taken at a single, fixed point should be interpreted in different ways at different sections and at different times. Several reports indicate that this flow pattern is not unique to the River Quarme. Cherkauer (1973) for example, used the concept of a narrow, 'active' flow zone through pools, whereas over riffles, the 'effective' width more closely approximated the total width of the channel. Low flow asymmetry is a consequence of converging flow into the pool from the upstream riffle. Asymmetry of velocity distributions through pool sections at high discharge is specified by Keller (1971) in his original account of the velocity reversal hypothesis. As a consequence, a reversal or its absence can be demonstrated simultaneously depending upon the cross-section chosen for comparison between riffles and pools, the kind of measurement made, and the point within the section at which the measurement is made. It is not appropriate, therefore, to infer representative riffle or pool behaviour from such data. Given the degree of variability associated with point-specific measurements, an alternative is to rely on averaged measures to compare cross-sections, in the hope that gross correlations between these and sediment movement can be made. The behaviour of mean section velocity and mean boundary shear stress are examined below.

Section-Averaged Measures

Mean Velocity

There is a strong indication from comparison of the hydraulic geometry relations at the riffle and pool sections (Table 2.1) that as discharge rises, mean velocities *converge* rapidly. Both pool sections possess velocity exponents that are greater than both riffle sections (exponents of 0.41 and 0.32 compared with 0.08 and 0.10). Figure 2.10a shows that at low flow, the velocities follow the anticipated hierarchy: the pool midpoint is the lowest, the riffle crest the highest with pool tail and riffle midpoint sections having almost identical mean velocities. All sections experience a rapid initial increase in mean velocity with discharge up to a value of approximately $1\,\mathrm{m}^3\,\mathrm{s}^{-1}$. Thereafter, however, riffle and pool sections are distinguished by the levelling-off of mean velocity increases at riffle sections while mean velocity continues to increase, but at a reduced rate at pool sections. At bankfull, pool midpoint and pool tail have converged to mean velocities intermediate in value between those at the riffle crest and the riffle midpoint. Significantly, again, either a reversal or a convergence may be appropriate to describe the situation depending upon which riffle cross-section is taken as representative. Inasmuch as the riffle crest is the section that performs the hydraulic function of the riffle, a reversal is not evident.

More observations are required before any definite comment can be made with respect to the causes of non-linearities in the log–log plots (Figure 2.10b), but the data

Table 2.1 Hydraulic geometry exponents at pool and riffle cross-sections, River Quarme (b, width; f, depth; m, velocity)

Section	b	f	m
Pool midpoint	0.14	0.43	0.41
Pool tail	0.12	0.56	0.32
Riffle crest	0.03	0.90	0.08
Riffle midpoint	0.14	0.76	0.10

(a)

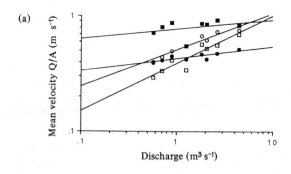

$$V_{pm} = 0.38Q^{0.41} \qquad R^2 = 0.87$$

$$V_{pt} = 0.50Q^{0.32} \qquad R^2 = 0.94$$

$$V_{rc} = 0.76Q^{0.08} \qquad R^2 = 0.11$$

$$V_{rm} = 0.42Q^{0.10} \qquad R^2 = 0.71$$

(b)

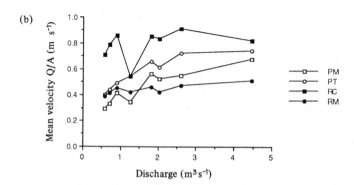

Figure 2.10   Mean section velocities for pool and riffle cross-sections, River Quarme: (a) hydraulic geometry relationships; and (b) arithmetic plot

confirm Carling's (1991) conclusion that the use and extrapolation of conventional hydraulic geometry relationships is inappropriate for describing the flow behaviour through riffle and pool sections, even within the bankfull range. At the transition from bankfull to overbank flow, a further significant disruption to the dynamics of the flow within the channel is to be expected, as redistribution of momentum takes place from the channel to the floodplain (Baryshnikov, 1967; Knight and Hamed, 1984; Keller and Rodi, 1988). A sudden reduction in competence associated with the transition to overbank flow, following a competence reversal, has been used by Hirsch and Abrahams (1981) to explain the coarser and better sorted riffle sediments, and it is unfortunate in the present study that overbank measurements could not be made. Nevertheless, the differing 'm' exponents between pool and riffle shown in Table 2.1 and the behaviour shown in Figure 2.10 require explanation, and three possibilities arise.

If continuity is to be maintained, the existence of a section-averaged reversal in velocity requires some combination of the following circumstances: (1) that riffle and pool possess systematically varying channel geometries, such that the discharge-dependent rate of change in cross-sectional area at the riffle is greater than that at the pool; (2) that contrasting hydraulic geometries result principally from a changing distribution of flow resistance through the sequence; and (3) that scour takes place at higher discharges in the pool and/or fill at the riffle to reduce or reverse contrasts in channel cross-section area.

The geometric argument is the basis of a long-standing explanation for the maintenance of alternating crossings and deeps observed in large, sand-bedded channels (Lane and Borland, 1953). Crossings (riffles) and deeps (pools) are approximated, respectively, by wide rectangular and narrower triangular sections, which exhibit greater and lesser increases in area with increase in stage if flow resistance and water surface slope are constant. Provided that the width of the crossing is considerably greater than that at the pool, for a given increase in stage, the increase in cross-section area will be greater at the rectangular than the triangular cross-section (Lane and Borland, 1953, p. 1075), accounting for a convergence, and possible reversal, in mean section velocity as discharge rises. In gravel-bedded channels, there is a statistical tendency for the bankfull widths of riffles to be approximately 15% greater than pools (Richards, 1976b), and low flow undercutting as the flow is forced to pass around the medial or cross-channel bar form of the riffle (Richards, 1978b) might account for approximately rectangular riffle cross-section shapes. Knighton (1981), however, points out that regularities in the association between bed elevation and cross-section form are frequently present only in downstream reaches of gravel-bedded streams, notwithstanding the presence of well-developed riffle–pool topography. In fact, a variety of pool and riffle shapes within and between river sections is observed (Milne, 1982) and the variation in cross-section asymmetry is associated closely with the rate and kind of planform change (Milne, 1983). In braided environments, primary flow channels are often unconfined, and no increase in average velocity can be expected (Ashworth, 1987). Reversals in mean velocity that entirely rely on geometric differences between riffles and pools cannot, therefore, be considered as general explanations for maintenance of the sequence (Bhowmik and Demissie, 1983), and may be restricted to laterally confined sections.

Teleki (1972) points out that where a changing hierarchy in form resistance is required to initiate a reversal in mean velocity, attention must be directed to the Froude-dependent flow regime, and Froude-dependent bedform changes are the basis for Andrews' (1979) account of mean velocity reversals in the East Fork River, Western Wyoming. Initially, at low flow, pool beds are observed to have high amplitude dune features, which are 'probably flattened and . . . definitely eliminated at the time of the flood crest' (Andrews, 1979, p. 18). The wash-out of pool bedforms coincides with the reversal in mean velocity, whereas over riffles, sand from the pool is thought to accumulate, and larger bedforms grow to replace pre-existing small amplitude forms, thereby reducing velocity. No account is taken of possible supply limitations in Andrew's study, however, and an alternative explanation for contrasts in flow resistance behaviour might be pool scour to a gravel bed. Under these conditions, total resistance through the pool is lower as grain resistance now replaces bedform drag.

Circumstantial support for this explanation is provided by the large-scale pattern of sediment transport in the East Fork River. This is unusual, since coarse sand and fine gravel travel down the river in the form of composite dune fields, which appear at a single point as wave-like pulses in the sediment transport rate (Meade, 1985). Sand supply is augmented from temporary storage in a tributary, Muddy Creek, and changes in bed configuration, and therefore flow resistance, observed from one day to the next during active sediment transport are at least an order of magnitude greater than changes observed from one year to the next during low-water seasons (Meade, Emmett and Myrick, 1981). Data from a pool section on the Rio Grande, which is also a mixed-load river, but where material supply is not augmented, demonstrate that as discharge rises, finer material from the pool is scoured, and flow resistance *increases* to a constant maximum value as the gravel bed is exposed (Nordin and Beverage, 1965). Here, too, because competence is great enough to transport material in the cobble size range *throughout* the sequence, not net scour is recorded in the pool.

In coarse-grained upland rivers, where the population of fine material is much more restricted, sand-sized sediment may undergo redistribution as fine material is winnowed from riffles into the next downstream pool (Milne, 1982), but this is usually only of limited significance in terms of changes in bed elevation (Hack, 1957, pp. 54–55). Therefore, grain resistance from the coarse material should always dominate. Since at high discharges, even under 'reversal' conditions, there is no suggestion that riffle velocities will decrease in *absolute* terms, the supply limitation implies that no continuous sheet of fine material is likely to occur, and riffles kept free of fine sediment at low flow also will be free at high flow. Flow resistance differences caused by changing bedforms seem, therefore, unlikely given the restricted supply of fine sediment in the River Quarme. Instead, a more plausible explanation for the large depth exponent at the riffle midpoint lies in the pattern of scour and fill occurring in the study section with changing discharge, which is discussed below.

Mean Boundary Shear Stress

The behaviour of mean boundary shear stress is plotted in Figure 2.11. Pool shear stress is less than riffle shear stress at lower *and* higher discharges, but there is a 'reversal' at an intermediate value. Reference to Table 2.2. indicates that the

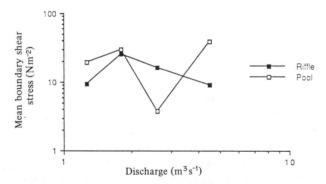

Figure 2.11   Mean boundary shear stress at pool and riffle sections, River Quarme

dominant influence in changing shear stress is the water surface slope (hydraulic gradient) which varies over an order of magnitude with changing discharge when measured between the riffle crest and riffle midpoint. At bankfull discharge, contrary to previous assumptions of an evening-out of the water surface profile (Leopold, Wolman and Miller, 1964, p. 206; Richards, 1982, p. 178), slopes are most varied, and particularly interesting is the recording of a negative water surface slope over the riffle at high discharge. The conditions under which this might occur are detailed by Bathurst (1982a, pp. 330–331), who notes that as flow continues to rise, the riffle becomes drowned, and is essentially a protrusion into the flow. Acceleration takes place on the upstream slope of the riffle counterslope causing drawdown of the water surface, before downstream deceleration occurs into the next pool. As a consequence, riffle slopes are now very gentle, or even negative towards the downstream side of the riffle, which is consistent with the measured water surface slope between the pool tail and riffle midpoint shown in Table 2.2. If the local slope continues to be measured at points fixed at low flow (on the downstream side, where formerly slope was greatest), a reversal in boundary shear stress from equation (2.2) will result, but in the near-boundary zone, however, acceleration of the flow on the upstream face of the riffle is likely to be associated with a steeper velocity gradient than on the downstream face. Variation in near-bed shear stress will again, therefore, be location-dependent within a single riffle unit.

Table 2.2 Water surface slopes between pool and riffle cross-sections in the River Quarme study reach. Slopes were obtained by survey of the water surface profile (for location and name of cross-section see text)

Q	$RC:RM$	$PM:RC$	$PT:RM$
1.27	0.005	0.006	0.005
1.81	0.010	0.008	0.007
3.61	0.005	0.001	0.004
4.47	0.002	0.008	−0.002

Discussion

Pattern of Scour and Fill. The magnitude of scour and fill at a point was obtained by subtracting local changes in depth from the overall change in stage, on the assumption that bed slope changes are much greater than water surface slope changes. The sequence of elevation change determined in this way is shown in Figure 2.12 for each control vertical at each cross-section. A mixed pattern is obtained, but there is evidence of fill at pool sections, especially the pool tail, and scour at riffle sections, especially the riffle midpoint. In this figure, lighter symbols refer to the left-hand side of the channel looking upstream, while darker symbols refer to the opposite side of the channel. At the pool, fill is dominant on the left-hand side, whereas at the riffle, scour is dominant on the left-hand side. Referring back to the cross-section topography in Figure 2.4, it is evident that the riffle–pool unit in this case is composed of an asymmetrical riffle bar form, and that the pattern of scour and fill is consistent with an extension of this unit, possibly involving upstream growth, as suggested by Ashworth (1987) and Petit (1987). Temporary reductions in elevation contrasts through riffle–pool sequences have been observed by Lisle (1982), following the inundation of small streams by fine material generated during a very large magnitude flood, but riffle scour and/or pool fill has never been evaluated as responsible for convergence/reversal in mean velocity at high discharge, possibly because the implication for riffle–pool maintenance is the opposite of that normally considered. The average elevation contrast (Figure 2.13) between riffle and pool cross-sections reduces at bankfull, in contrast to the assumptions of the reversal-based mechanism, although the post-flood bed topography showed little difference to the pre-flood survey. Scour and fill appear, therefore, to be transitory phenomena, suggesting that the maintenance of riffle–pool elevation contrasts occurs at *intermediate/low stages*. In fact, two further complementary pieces of evidence which suggest that flow–sediment interactions at both riffle and pool are more complicated than the reversal hypothesis suggests arise from further examination of mean boundary shear stress.

Flow Complexity and Mean Boundary Shear Stress. Calculations using equation (2.2) at a single riffle and single pool section on the East Fork River have been used by Lisle (1979) to substantiate the reversal hypothesis with respect to mean boundary shear stress, and Keller (1983) argues that the almost constant slope through riffle–pool reaches at high discharges must imply a reversal in τ_0 if depth contrasts between riffles and pools are maintained. However, because of spatially varied flow acceleration, it is unlikely that the force available for sediment transport at the bed will be well-indexed by the shear stress calculated from equation (2.2). Indeed, Milhous' (1987) suggestion that reversals occur at both low and high discharges is dismissed for this reason by Beschta (1987), and recalling the nature of the bed topography and flow distribution, it is inevitable that a more complex secondary circulation will be found in riffle–pool sequences, thus making application of equation (2.2) unacceptable.

Variation in the total energy slope (Table 2.3), which reflects energy losses caused by all possible resistance effects, provides support for a complex secondary flow circulation. Unfortunately, total energy slope is difficult to obtain because it relies on simultaneous measurement of water surface, local bed elevation changes, calculated mean velocity and the velocity distribution coefficient, all of which are subject to

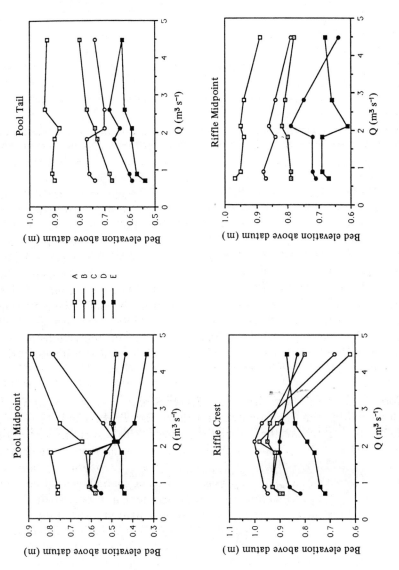

Figure 2.12 Pattern of scour and fill with changing discharge at riffle and pool sections, River Quarme

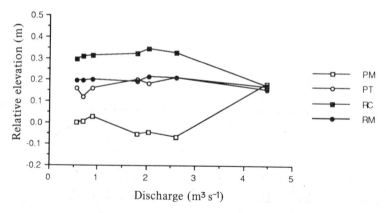

Figure 2.13 Average scour and fill at riffle and pool cross-sections, River Quarme. Elevations are relative to zero datum at the pool midpoint for the lowest discharge shown

error. Errors must be suspected in this data, because total energy head does not always decrease in a downstream direction, but two possible physical explanations for at least some of the discrepancy may lie in the phenomenon of large contra-rotating eddies, which occur at the channel margins in the pool section and involve local *upstream* flow (Hiller and Stavrakis, 1982), and in the possibility of intra-gravel flow of sufficient magnitude to affect estimates of section discharge (Cherkauer, 1973). Once again, these draw attention to the difficulty (and to some of the dangers) involved in conceptualizing the hydraulics of riffle–pool sequences in terms of cross-sections obeying simple continuity constraints. Given the additional problem of defining spatially representative slope measures over spatially varied topography (Bray, 1980), and the possibly dominant influence of sediment supply conditions, it seems clear that the use of mean boundary shear stress is flawed both as a spatially and a physically realistic measure of sediment transport through riffle–pool sequences. Inappropriate application of critical tractive force formulae through riffle–pool sequences together with supply limitations probably accounts for the absence of a correlation between mean boundary shear stress and sediment size reported by Nordin and Beverage

Table 2.3 Total energy heads at pool and riffle cross-sections, River Quarme (for details see text)

Q	PM	PT	RC	RM
0.57	0.35	0.39	0.33	0.38
0.70	0.38	0.42	0.36	0.41
0.89	0.41	0.45	0.39	0.43
1.81	0.74	0.63	0.65	0.60
2.04	0.67	0.70	0.63	0.70
2.61	0.76	0.78	0.65	0.73
4.49	1.52	1.03	0.92	1.02

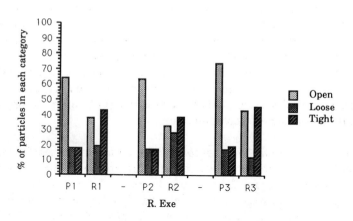

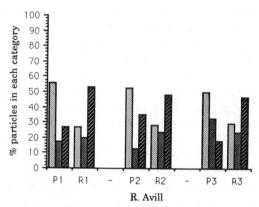

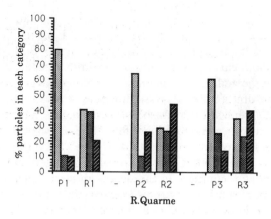

Figure 2.14 Proportions of surface particles in open, loosely structured, and tightly structured positions at nine riffle and pool locations on the River Exe, River Quarme and River Avill

(1965, p. F23) in the Rio Grande, and for the absence of riffle bed material entrainment in the East Fork, even beyond estimated threshold boundary stress values (Lisle, 1979). It is possible to envisage a pattern of sediment transport through riffles and pools that may be consistent with a reversal in mean boundary stress derived from the hydraulic gradient, but for the reasons outlined in the following section, which is entirely coincidental to any analysis of mean shear stress. Measures of stream power, which have been used to suggest the occurrence of reversals in competence between riffles and pools at high discharge (O'Connor, Webb and Baker, 1986) would, of course, be affected similarly.

Importance of Structural Arrangement of Bed Sediments. At riffle sections, the relationship between sediment transport and shear stress also may be complicated for an additional reason. Figure 2.14 shows the results of an analysis of the spatial incidence of sediment arrangement between nine paired riffle–pool locations on the River Quarme and the neighbouring River Avill and River Exe. Structural arrangement was assessed by assigning all particles (with intermediate axes greater than 0.5 cm) under repeated transects through riffle and pool midpoint locations to one of three structural categories: open bed, loosely interacting or tightly interacting. These categories are similar to those used by Brayshaw (1985), and further details of the research design and methods are given in Clifford (1990).

While the usual assumption is that riffle sediments are coarser and better sorted than those in adjacent pools, the results of this analysis demonstrate that another consistent measure to differentiate between sediments in riffles and pools in the degree of development of bedform microtopography. Riffles possess significantly greater proportions of bed sediments occupying structured or interacting positions on the bed, and therefore, bed sediments there should be more stable over most of the flow range than pool sediments. Primarily, structuring imparts greater mechanical stability to sediments, since the topographic position of an individual particle alters the direction and magnitude of the force experienced by it. For a particle within a structure, the area available for lift and drag is reduced at the same time that its effective weight is increased by additional frictional contact. 'Equal mobility' relationships (e.g. Andrews, 1983) break down under these conditions (Bathurst, 1987), so that variation from threshold curves are the opposite of those expected in open bed conditions. Also, because the additional roughness that structures generate is responsible for increased profile drag (Robert, 1988) and increased energy loss by vortex shedding (Clifford, 1990), the stress available for sediment transport is further reduced. The existence of bedform microtopography is thus associated with a delay in entrainment and with sudden, erratic transport (Brayshaw, 1985; Laronne and Carson, 1976), which might help account for the recording of scour only at higher discharges at the riffle. Spatial contrasts in the degree of structuring also may be a factor in controlling sediment routing between successive riffle–pool units, but again, no simple correspondence between hydraulic and sediment behaviour at a single cross-section can be expected: Andrews (1979) and Hey (1986) note that riffles begin to scour before pools, for example.

CONCLUSIONS

Simultaneous measurement of point velocity and shear stress across a wide flow range in a single riffle–pool unit gives some insight into the complexity of hydraulic behaviour that occurs with changing discharge. The point measurements demonstrate that locally, higher near-bed velocity or local shear stress can be found at the pool midpoint than at the riffle crest, but that this is not an appropriate basis for inferring reversals across entire cross-sections. Since the flow distribution is discharge-dependent, measurements taken at a single point may be interpreted in different ways at different sections and at different times. A reversal, or its absence, can be demonstrated simultaneously, depending upon the cross-section chosen for riffle–pool comparison, the kind of measurement made, and the point at which the measurement is taken. In addition, the range of results from the different point measures demonstrates that the various studies that have used different hydraulic parameters to demonstrate or discount the existence of a reversal are not necessarily mutually reinforcing, as at first might be expected.

Closer attention to the behaviour of section-averaged velocity and shear stress also is not supportive of the case for a reversal. Reversals in average velocity may occur in mixed bedded rivers, where sudden decreases of flow resistance arising from removal of dune bedforms at high flow can increase velocity significantly in pool sections. Reversals in average velocity also may occur where channel shape and size is restricted by geological controls. In the alluvial channel at the study reach, either a reversal or a convergence could be shown to occur depending upon which riffle section was taken as representative. In as much as the riffle crest performs the hydraulic function of the riffle, a reversal was not found. Generally, a convergence in mean velocity did occur, but this is explained with respect to filling of the pool and scour of the riffle at higher discharges, which is the opposite of behaviour expected under the assumptions of the reversal hypothesis. The evidence points to the need for an alternative mechanism to explain the maintenance of relative elevation contrasts. The changing pattern of local sediment supply/delivery between upstream and down-stream riffle–pool units may be more crucial in determining the pattern of scour and fill than a simple reversal in transport competence. Reversals in mean boundary shear stress and stream power may or may not occur, but such is the non-uniformity of flow that behaviour of these parameters is likely to be incidental to the observed pattern of sediment transport.

More generally, this study demonstrates the need to formulate explanations of the maintenance of riffle–pool sequences that are sensitive to (1) local variation and (2) the existence of spatially distributed form–process feedbacks. The interaction of channel form and channel flow at any point within a riffle–pool unit depends in part on flow and sediment behaviour in upstream and downstream units, as well as local variables. If anything, explanations relying on cross-section averages complicate, rather than clarify, the characteristics of flow and form interaction. An analogy may be drawn with the 'meander problem' identified by Dietrich (1987), that is, how to formulate a physically realistic model linking sediment transport, bedform/bar charac-teristics and channel morphology. In previous reversal-based accounts of riffle–pool maintenance, an 'answer' corresponding to the low flow sediment pattern and topography can be obtained with a simple model, but a complete, more complicated theory is necessary to explain and predict the behaviour of bed morphology during stage change. In this case, further investigation is required before such a complete model can be identified, but systematic differences in the structural arrangement of bed sediments appear to warrant closer attention in terms of their implication for sediment–flow interactions.

ACKNOWLEDGEMENTS

This research was undertaken while N. J. Clifford was in receipt of a NERC studentship. The manuscript was prepared using facilities at Portsmouth Polytechnic.

REFERENCES

Andrews, E. D. (1979). Scour and fill in a stream channel, East Fork River, Western Wyoming. *U. S. Geological Survey Professional Paper*, **1117**, 49pp.

Andrews, E. D. (1983). Entrainment of gravel from naturally sorted riverbed material. *Geological Society of America Bulletin*, **94**, 1225–1231.

Ashworth, P. J. (1987). *Bedload transport and channel change in gravel-bed rivers*, Unpublished PhD thesis, University of Stirling.

Baryshnikov, N. B. (1967). Sediment transportation in rivers with flood plains. *International Association of Hydrological Sciences Publication*, **75**, 404–413.

Bathurst, J. C. (1979). Distribution of boundary shear stress in rivers. In D. D. Rhodes and G. P. Williams (eds), *Adjustments of the fluvial system*, Kendall-Hunt, pp. 95–116.

Bathurst, J. C. (1982a). Channel bars in gravel-bed rivers. Discussion. In R. D. Hey, J. C. Bathurst and C. R. Thorne (eds), *Gravel-bed Rivers*, Wiley, pp. 330–331.

Bathurst, J. C. (1985). Flow resistance estimation in mountain rivers. *American Society of Civil Engineers*, **111**, 625–643.

Bathurst, J. C. (1987). Measuring and modelling bedload transport in channels with coarse bed materials. In K. S. Richards (ed.), *River channels environment and process*, Institute of British Geographers Special Publication 17, Blackwell Scientific Publications, pp. 272–294.

Beschta, R. L. (1987). Conceptual models of sediment transport in streams. In C. R. Thorne, J. C. Bathurst and R. D. Hey (eds), *Sediment transport in gravel-bed rivers*, Wiley, pp. 387–419.

Bhowmik, N. G. and M. Demissie (1983). Bed material sorting in pools and riffles. *Journal of the Hydraulics Division, American Society of Civil Engineers*, **108**, 1227–1231.

Bray, D. I. (1980). Evaluation of effective boundary roughness for gravel bedded rivers. *Canadian Journal of Civil Engineering*, **7**, 392–397.

Brayshaw, A. C. (1985). Bed microtopography and entraintment thresholds in gravel-bed rivers. *Geological Society of America Bulletin*, **96**, 218–223.

Brookes, A. (1988). *Channelized Rivers: Perspectives for Environmental Management*, Wiley, 336pp.

Carling, P. A. (1991). An appraisal of the velocity-reversal hypothesis for stable pool–riffle sequences in the River Severn, England. *Earth Surface Processes and Landforms*, **16**, 19–31.

Cherkauer, D. S. (1973). Minimization of power expenditure in a riffle–pool alluvial channel. *Water Resources Research*, **9**, 1612–1627.

Clifford, N. J. (1990). *The formation, nature and maintenance of riffle–pool sequences in gravel-bedded rivers*. Unpublished PhD thesis, University of Cambridge.

Chow, Ven Te (1981). *Open-channel Hydraulics*, International Student Edition, McGraw-Hill, 680pp.

Cowx, I. G., G. A. Wheatley and A. S. Mosley (1986). Long-term effects of land drainage works on fish stocks in the upper reaches of a lowland river. *Journal of Environmental Management*, **22**, 147–156.

Dietrich, W. E. (1987). Mechanics of flow and sediment transport in river bends. In K. S. Richards (ed.), *River Channels Environment and Process*, Institute of British Geographers Special Publication 17, Blackwell Scientific Publications, pp. 179–227.

Gilbert, G. K. (1914). Transportation of debris by running water. *U. S. Geological Survey Professional Paper*, **86**, 263pp.

Graesser, N. W. C. (1979). How land improvement can damage Scottish salmon fisheries. *Salmon and Trout Magazine*, **215**, 39–43.

Hack, J. T. (1957). Studies in longitudinal stream profiles in Virginia and Maryland. *U. S. Geological Survey Professional Paper*, **294-B**.

Hey, R. D. (1986). River mechanics. *Journal, Institute of Water Engineers and Scientists*, **40**, 139–158.

Hiller, N. and N. Stavrakis (1982). Reversed currents over a point bar on the Great Fish River. *Transactions of the Geological Society of Africa*, **85**, 215–219.

Hirsch, P. J. and A. D. Abrahams (1981). The properties of bed sediments in pools and riffles. *Journal of Sedimentary Petrology*, **51**, 757–760.

Hulsing, H., W. Smith and E. D. Cobb (1966). Velocity-head coefficients in open channels. *U. S. Geological Survey Water Supply Paper*, **1869-C**, 45pp.

Jackson, W. L. and R. L. Beschta (1981). A model of two-phase bedload transport in an Oregon Coast range stream. *Earth Surface Processes and Landforms*, **7**, 517–527.

Jarrett, R. D. (1984). Hydraulics of high-gradient streams. *Journal of the Hydraulics Division, American Society of Civil Engineers*, **110**, 1519–1539.

Keller, E. A. (1971). Areal sorting of bed-load material: the hypothesis of velocity reversal. *Geological Society of America Bulletin*, **83**, 915–918.

Keller, E. A. (1972). Areal sorting of bed-load material: the hypothesis of velocity reversal. A reply. *Geological Society of America Bulletin*, **83**, 915–918.

Keller, E. A. (1978). Pools, riffles and channelization. *Environmental Geology*, **2**, 119–127.

Keller, E. A. (1983). Bed material sorting in pools and riffles. Discussion. *Journal of the Hydraulics division*, American Society of Civil Engineers, **109**, 1243–1245.

Keller, R. J. and W. Rodi (1988). Prediction of flow characteristics in main channel/flood plain flows. *Journal of Hydrological Research*, **26**, 425–441.

Knight, D. W. and M. E. Hamed (1984). Boundary shear in symmetrical compound channels. *Journal of the Hydraulics Divison, American Society of Civil Engineers*, **110**, 1412–1430.

Knighton, A. D. (1981). Local variations of cross-sectional form in a small gravel-bed stream. *New Zealand Journal of Hydrology*, **20**, 131–146.

Lane, E. W. and W. M. Borland (1953). River-bed scour during floods. *Transactions of the American Society of Civil Engineers*, **2712**, 1069–1089.

Laronne, J. B. and M. A. Carson (1976). Interrelationships between bed morphology and bed material transport for a small gravel-bed channel. *Sedimentology*, **23**, 67–85.

Leopold, L. B., M. G. Wolman and J. P. Miller (1964). *Fluvial Processes in Geomorphology*, Freeman, 522 pp.

Lisle, T. E. (1979). A sorting mechanism for a riffle–pool sequence. *Geological Society of America Bulletin*, **90**, 1142–1157.

Lisle, T. E. (1982). Effects of aggradation and degradation on riffle–pool morphology in natural gravel channels, northwestern California. *Water Resources Research*, **18**, 1643–1651.

Meade, R. H. (1985). Wavelike movement of bedload sediment, East Fork River, Wyoming. *Environment Geology and Water Science*, **7**, 215–225.

Meade, R. H., W. W. Emmett and R. M. Myrick (1981). Movement and storage of bed material during 1979 in East Fork River, Wyoming, USA. *International Association of Hydrological Sciences Publication*, **132**, 225–235.

Midgeley, H. G. Petts and D. Walker (1986). *Streamflow measurement*, British Standards institute, BSI PP7316, 28pp (+ discussion).

Milhous, R. T. (1987). Discussion—Conceptual models of sediment transport in streams. In C. E. Thorne, J. C. Bathurst and R. D. Hey (eds), *Sediment transport in gravel-bed rivers*, Wiley, pp. 387–419.

Miller, B. A. and H. G. Wenzel (1985). Analysis and simulation of low flow hydraulics. *Journal of the Hydraulics Division, American Society of Civil Engineers*, **111**, 1429–1446.

Milne, J. A. (1982). Bed-material size and the riffle–pool sequence. *Sedimentology*, **29**, 267–278.

Milne, J. A. (1983). Variation in cross-sectional asymmetry of coarse bedload river channels. *Earth Surface Processes and Landforms*, **8**, 503–511.

Mosley, M. P. (1985). River channel inventory, habitat and instream flow assessment. *Progress in Physical Geography*, **9**, 494–523.

Nordin, C. F. and J. P. Beverage (1965). Sediment transport in the Rio Grande, New Mexico. *U. S. Geological Survey Professional Paper*, **462-F**, 35pp.

O'Connor, J. E., R. H. Webb and V. R. Baker (1986). Paleohydrology of pool and riffle pattern development: Boulder Creek, Utah. *Geological Society of America Bulletin*, **97**, 410–420.

Petit, F. (1987). The relationship between shear stress and the shaping of the bed of a pebble-loaded river, La Rulles – Ardenne. *Catena*, **14**, 453–468.

Richards, K. S. (1973). Hydraulic geometry and channel roughness – a non-linear system. *American Journal of Science*, **273**, 877–896.

Richards, K. S. (1976a). The morphology of riffle–pool sequences. *Earth Surface Processes*, **1**, 71–88.

Richards, K. S. (1976b). Channel width and the riffle–pool sequence. *Geological Society of America Bulletin*, **87**, 883–890.

Richards, K. S. (1978a). Simulation of flow geometry in a riffle–pool stream. *Earth Surface Processes*, **3**, 345–354.

Richards, K. S. (1978b). Channel geometry and the riffle–pool sequence. *Geografiska Annalar*, 60-A, 23–27.

Richards, K. S. (1982). *Rivers: Form and Process in Alluvial Channels*, Methuen, 358 pp.

Robert, A. (1988). *Statistical modelling of sediment bed profile and bed roughness properties in alluvial channels*. Unpublished PhD thesis, University of Cambridge.

Rubey, W. W. (1938). The force required to move particles on a stream bed. *U. S. Geological Survey Professional Paper*, **189-E**, 121–141.

Rutherford, J. E. and R. J. Macay (1985). The vertical distribution of hydropsychid larvae and pupae (Trichoptera: Hydropsychidae) in stream substrates. *Canadian Journal of Zoology*, **63**, 1306–1315.

Sear, D. A. (in press). The effects of river regulation for hydro-electric power on the sediment transport within riffle–pool sequences. *Proceedings, 3rd International Gravel-bed Rivers Workshop*, Florence, Italy.

Seddon, J. A. (1900). River hydraulics. *Transactions of the American Society of Civil Engineers*, **43**, 179–243.

Stuart, T. A. (1953). Spawning migration, reproduction and young stages of loch trout. *Freshwater and Salmon Fisheries Research*, **5**, 39 pp.

Teissyre, A. K. (1984). The River Bobr in the Blazkowa study reach (central Sudetes): a case study in fluvial processes and fluvial sedimentology. *Geological Sudetica*, **19**, 7–71.

Teleki, P. G. (1972). Areal sorting of bed load material: the hypothesis of velocity reversal. Discussion. *Geological Society of America Bulletin*, **83**, 911–914.

3 Velocities, Roughness and Dispersion in the Lowland River Severn

KEITH BEVEN
Centre for Research on Environmental Systems, Lancaster University
and
PAUL CARLING
Institute of Freshwater Ecology, Ambleside

INTRODUCTION

The flow of water and associated sediments and contaminants in lowland channels is still not well understood. This is a result of the effects of the complex geometry and the seasonal effects of vegetation growth on the bed and banks interacting with changing discharges. Most measurements of velocity distributions, flow roughness and shear stresses have been made at the scale of individual vertical profiles or channel cross-sections of the flow, but for many water resource management problems, the scale of interest is that of extended channel reaches. At this scale, three-dimensional variations in channel shape and vegetation characteristics may have an important effect on flow and transport processes by creating zones of increased shear stress at the bed or banks, or 'dead-zones' separated from the main flow by localized lines of lateral shear.

In a narrow gravel-bed river of low competence, such as the Severn, the primary features of three-dimensional flow can be attributed to grain resistance and bank or channel alignment effects rather than small-scale bedform development (Griffiths, 1989). In turn, macrophyte development may mimic form irregularities being well developed in areas of slow or recirculating flow. Although the influence of channel shape and composite roughness on flow resistance in laboratory channels is known (e.g. Naot, 1984), procedures to disaggregate the various effects of bed material and form resistance in natural streams are poorly developed (Prestegaard, 1983) and pay scant regard to vegetative roughness, despite the fact that in some rivers this latter component may dominate (Dawson, 1978). Hey (1979) presented an approach to the problem of calculating flow resistance where the near-bank roughness is different to that in the central channel section, but noted that the lack of information on the vegetated banks precluded any allowance to be made for that effect.

There is a reasonably extensive literature on the roughness characteristics of grass-lined channels, but much less on the effects of aquatic plants and trees and bushes on

Lowland Floodplain Rivers: Geomorphological Perspectives. Edited by P. A. Carling and G. E. Petts
© 1992 John Wiley & Sons Ltd

the river banks and flood berms. Recent reviews of this subject are provided by Watson (1987) and Hydraulics Research (1988), and some particular examples of the effects of vegetation on roughness are given by Powell (1978), Newson (1984) and Watson (1987). No generally applicable theory exists for estimating the hydraulic effects of vegetation growth, although the analysis of Petryk and Bosmajian (1975) relating change in roughness to the area of vegetation exposed to the flow might be applicable in some limited circumstances. Consequently, empirical studies are required covering a range of river morphologies and macrophyte communities.

However, it is not normally possible to carry out a sufficiently detailed survey of point velocity measurements in the field to fully characterize the spatial variations in mean velocities at a steady discharge within a river reach. In contrast, tracer measurements allow the complex velocity variations within the reach to be integrated to the reach scale, but without revealing any details of the flow structures that lead to the observed tracer velocities through time. In this study, we report the preliminary results of a series of measurements at two sites on the River Severn, comparing velocity distributions and channel roughness coefficients derived from Eulerian cross-sectional current metering measurements and Lagrangian reach-scale tracer measurements at a number of different discharges.

STUDY REACHES

The chosen reaches on the River Severn are at Montford and Leighton close to Shrewsbury in Shropshire, UK (Figure 3.1). At Montford the river is up to 40 m wide, incised in bentonitic clays (Dury, 1984) with steep, well-defined, cohesive, grass and tree-lined banks. The channel is stable, slightly sinuous with a distinct pool and riffle sequence with a mean spacing of around 300 m (Dury, Sinker and Pannett, 1972). The chosen reach is 1 km in length, within which two pool and two riffle sections have been examined in detail.

At Leighton the river is up to 80 m wide, in a classic meandering sequence with alternate steeply cut banks and gravel points bars. The banks consist of poorly indurated terrace gravels (Dury, 1983) and the river has an active alluvial floodplain. Cross-sections at a pool and riffle downstream of a stable straight reach, and a riffle–pool–riffle sequence within a meander were chosen for detailed study (Figure 3.1) within a study reach 1.13 km long.

The valley gradient at Leighton is higher than at Montford and in both cases the riffles are slightly wider than the pools. Width to depth ratios for bankfull flow at Montford are about 12 for the riffles and six for the pools, while at Leighton these ratios are 18 and 12, respectively. Initial analyses of cross-section average velocity data at the study reaches have been published by Carling (1991), while Reynolds, Carling and Beven (1991) have examined the relationship between local flow structures in these reaches and riverine phytoplankton populations.

EXPERIMENTAL METHODS

Velocity and depth profiles were recorded at 2 m intervals across surveyed cross-sections for a range of discharges up to bankfull stage. An array of six Ott current

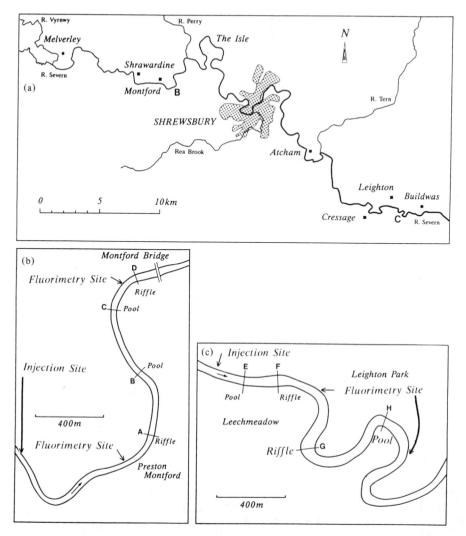

Figure 3.1 (a) General location of the River Severn in Shropshire, central England with test reaches shown. (b) Location of the test sections at Montford. (c) Location of the test sections at Leighton

meters was used, equally spaced over the lower 0.5 m of a 1 m rod lowered by cable from a boat. The array was raised sequentially to obtain a complete velocity profile from within 40–100 mm of the bed to within 100 mm of the water surface. Directional stability was provided by a vane running the length of the array. Close to bankfull, instrument malfunction at depths greater than 3.5 m prevented the collection of near-bed velocities for some profiles, although mid-depth and surface velocity data were obtained for all profiles (Carling, 1991).

Additional velocity and dispersion measurements at the reach scale were obtained using Rhodamine WT tracer and the computer controlled logging fluorimeter apparatus described by Wallis, Blakeley and Young (1987). At both reaches the tracer was

(a)

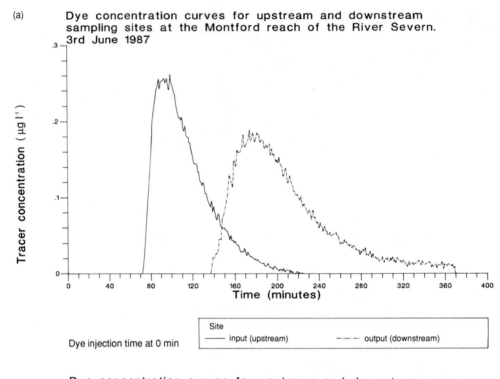

Dye concentration curves for upstream and downstream sampling sites at the Montford reach of the River Severn. 3rd June 1987

Dye injection time at 0 min

Site		
—— input (upstream)	—·—·— output (downstream)	

(b)

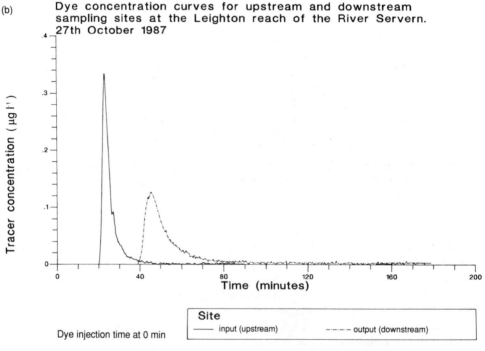

Dye concentration curves for upstream and downstream sampling sites at the Leighton reach of the River Servern. 27th October 1987

Dye injection time at 0 min

Site		
—— input (upstream)	—·—·— output (downstream)	

Figure 3.2 Examples of the tracer data obtained for A. Montford and B. Leighton reaches

injected as an instantaneous input at a point sufficiently far upstream to ensure thorough mixing at the entry to the study reaches. The injection and measurement sites are shown in Figure 3.1. Tracer measurements were carried out over a range of discharges. The input and output data were analysed within an Aggregated Dead Zone model framework (Wallis, Young and Beven, 1989) using the Lancaster University ADZ-ANALYSIS computer program. The analysis yields estimates of the travel time lag, τ, from the difference in the rise times of the two curves; the mean travel time, t, and velocity, U_L, within the reach from the difference in the centroids of the two curves; and a dispersion parameter, the Dispersive Fraction. Figure 3.2 shows a typical example for each of the reaches. Other tracer experiments in lowland rivers have been reported by Sabol and Nordin (1978) and Bauwens, Bellon and van der Beken (1982).

VELOCITY DISTRIBUTIONS AT THE CROSS-SECTION AND REACH SCALES

Some examples of variations in cross-sectional flow velocities for different discharges are shown for selected sites in Figure 3.3. Mean flow velocities, U_E, have been calculated from the point current meter measurements for each of the nine cross-sections in the study reaches and their relationship with changing discharge is shown in Figure 3.4 for the two sites. Regression analysis has been used to calculate the exponent m in the at-a-site power law relationship

$$U_E = k \, Q^m \tag{3.1}$$

where k is a coefficient and Q is discharge.

Reach-scale mean velocities, U_L, may be calculated from the mean travel time, t, of the tracer cloud between the input and output sites. The change in U_L with discharge at the two sites also is plotted in Figure 3.4. A similar regression analysis was carried out, fitting equation (3.1), using the U_L and Q data. The results are given in Table 3.1. Figure 3.4 shows that for both sites, as would be expected, the pool velocities are significantly lower than the reach velocities but that the tracer mean velocities were nearly always lower even than the average pool velocities at the chosen cross-sections.

The tracer measurements also yield the reach-scale time lag, or advective time delay. This also changes with discharge and the results of the current measurements are shown in Figure 3.5.

An alternative discharge–reach mean velocity relationship has been used by Beven (1979) and Wallis, Young and Beven (1989) in analysing mean travel times derived from tracer experiments. This takes the form:

$$t = a + b/Q \tag{3.2}$$

which is equivalent to

$$U_L = \frac{L\,Q}{(a\,Q + b)} \tag{3.3}$$

where L is the length of the reach, a and b are coefficients and Q is discharge. This

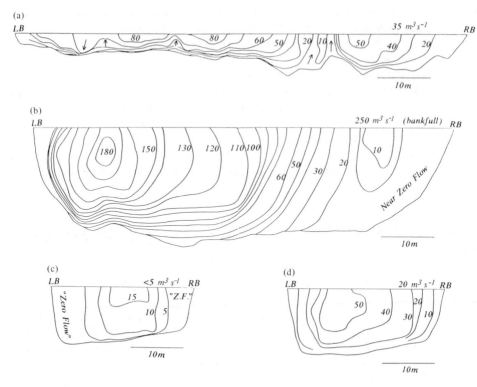

Figure 3.3 Examples of two and three-dimensional flow structure and the development and breakdown of dead-zone structure in the River Severn. (a) Flow structure across a riffle section in the Leighton meanders (section G in Figure 3.1). Note the well-mixed three-dimensional structure evident during low flow with two main flow cores and upwelling and downwelling of currents (arrowed). (b) Close to bankfull, flow cuts across the bend so that the single flow core is close to the left bank and flow is nearly two-dimensional. A region of slow flow with a high residence time develops near the outer bank amid submerged bankside trees. (c) In the narrow Montford pool (section C) during low flow, the isovels are regularly distributed and zero-flow zones are well developed amongst marginal vegetation. (d) As discharge increases in the Montford research, the main flow shifts towards the left bank and dead zone structure is destroyed

relationship has a form such that at high flows the mean velocity asymptotically approaches a constant value (L/a). It also can be shown (e.g. Bates and Pilgrim, 1982) that equation (3.3) is equivalent to a relationship of the form:

$$Q = \alpha \ (V - V_0) = \alpha \ L \ (A - A_0) \tag{3.4}$$

where where V is a reach volume and A a mean cross-sectional area in the reach, $\alpha = 1/a$ and $V_0 = b/(aL)$. V_0 may be interpreted as the effective storage below which discharge is effectively zero, or alternatively the term $(A - A_0)$ as an effective cross-sectional area for flow in the reach. The concept of effective storage or cross-sectional area also has been examined by Kellerhals (1970) and Newson and Harrison (1978) for some smaller reaches. Equation (3.3) also has been fitted to the tracer-derived velocity data and the resulting parameters are reported in Table 3.1. With only a small

Table 3.1 Results of fitting the relationship $U = k\,Q^m$ to the cross-section and reach-scale average velocities

		Number of measurements	Equation (3.1)			Equation (3.3)		
			k	m	r^2	a	b	r^2
			(ms^{-})			(s)	(m^3)	
Cross-sections								
Montford	Upper pool	4	0.047	0.696	0.98			
Montford	Upper riffle	6	0.240	0.358	0.88			
Montford	Lower pool	8	0.047	0.656	0.98			
Montford	Lower riffle	6	0.181	0.425	0.84			
Leighton	Upper pool	8	0.106	0.451	0.97			
Leighton	Upper riffle	7	0.231	0.276	0.75			
Leighton	Lower pool	1	—	—	—			
Leighton	Lower riffle	4	0.156	0.359	0.96			
Reaches								
Montford		7	0.047	0.645	0.93	1900	24 748	0.77
Leighton		6	0.108	0.398	0.89	1408	33 530	0.65

number of points at high flows the fit is not very good and seriously underestimates the mean velocity at high flows. Knighton and Cryer (1990) also provide one example for which equation (3.3) did not give good results where higher flows were not well represented, but, in general, they show similar goodness of fit for this form, the power law form and a quadratic log discharge form of the velocity–discharge relationship. For the data presented here, simple linear and quadratic relationships also gave reasonably good fits to the data (Table 3.2). There is, of course, no reason to favour one functional relationship over another for non-uniform flow conditions in natural channels.

This type of simple empirical relationship between mean velocity and discharge might be expected to break down for overbank flows. The study of Knight, Shiono and Pirt (1990) is relevant in this respect. They present an application to the Montford gauging station site of an analytical approach to predicting overbank flows, based on a solution of the depth-averaged Navier–Stokes equation given specified roughness characteristics.

ROUGHNESS COEFFICIENTS AT THE CROSS-SECTION SCALE

Total resistance to flow in natural gravel-bed rivers, such as the Severn, may be considered to be made up of four components. These are:

(1) grain resistance, that is the resistance afforded by the roughness of the bed material or bedload;
(2) form resistance, from the undulations associated with small-scale bedforms;

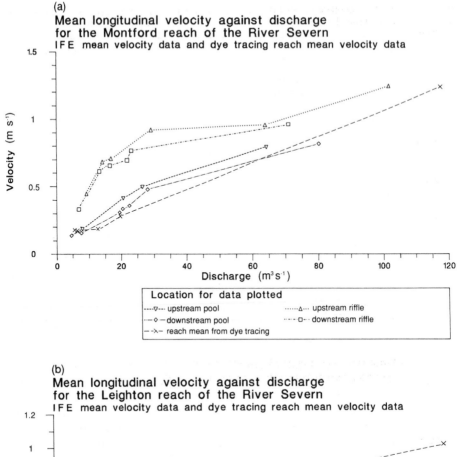

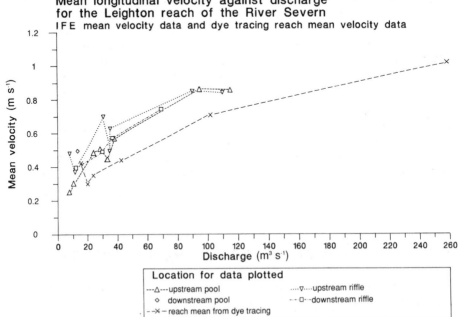

Figure 3.4   Mean velocity–discharge relationships for the cross-section and reach-scale measurements: (a) Montford, (b) Leighton. Institute of Freshwater Ecology data

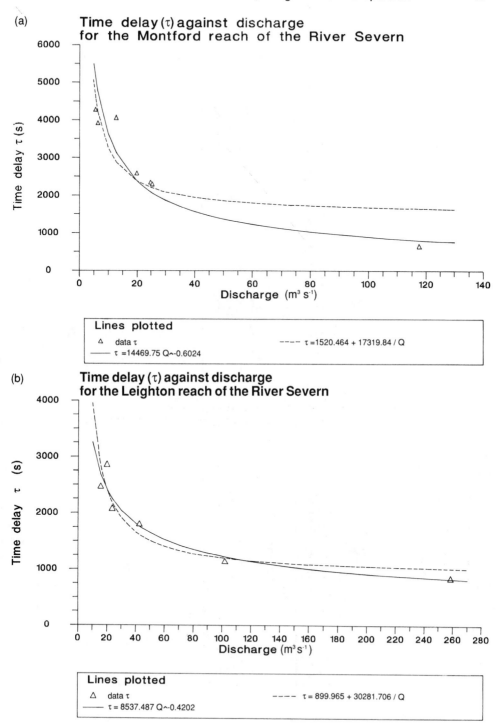

Figure 3.5 Advective time delay τ–discharge relationships for the Montford and Leighton reaches

Table 3.2 Results of fitting different velocity–discharge relationships to the reach-scale mean velocity data

	Number of measurements	r^2 values				
		$U = kQ^m$	$U = nQ^{(p\,+\,q\ln Q)}$	$U = LQ/(aQ + b)^a$	$U = cQ + d$	$U = eQ + fQ^2 + g$
Montford	7	0.926	0.988	0.646	0.997	0.998
Leighton	6	0.888	0.912	0.767	0.942	0.968

[a]Fitted in the form of equation (3.2).

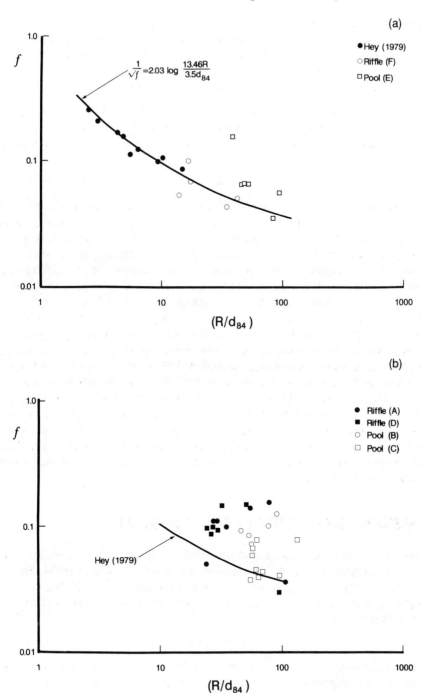

Figure 3.6   Relationship between the friction factor (*f*) and the relative roughness (R/d_{84}) for: (a) the broad pool and riffle at Leighton (sections E and F of Figure 3.1); and (b) for the narrow-tree lined pools and riffles at Montford. Also shown is the predictive equation and curve developed by Hey (1979) for broad riffles on the Severn

(3) internal distortion owing to the three-dimensional nature of the flow induced by longitudinal variation in the planform and cross-section of the channel;
(4) 'spill' resistance associated with rapid local flow acceleration and deceleration. Spill resistance is significant primarily at high Froude numbers (> 0.5) and may be neglected in the present study where the Froude number was typically 0.2.

In complex flows, it always will be difficult to quantify the effects of these different contributions to the roughness coefficient. The effective roughness of a steady uniform flow may be calculated by inverting the Darcy–Weisbach uniform flow equation to give:

$$f = 8gRS/U^2 \tag{3.5}$$

where R is the hydraulic radius ($= A/P$, S is the reach-averaged bedslope or water surface slope, P is the wetted perimeter of the flow, A is area, U is a mean velocity and g is acceleration due to gravity. Normally, equation (3.5) is applied at the scale of a cross-section, with local values of hydraulic radius and cross-section mean velocity, U_E. Plots of the changing effective roughness coefficient for the different cross-sections at Montford and Leighton are shown in Figure 3.6, where it may be seen that the data for the broad pool and riffle on a straight section at Leighton are in accord with the general function proposed for riffles on the Severn by Hey (1979) with decreasing flow resistance with increasing depth. In contrast, however, in the relatively narrow Montford sections where the banks are partially tree-lined, no such accordance is noted. Resistance is much higher than is the case in the broad Leighton sections and tends to increase with stage over the riffles as original bank roughness is inundated. There is some indication that at the highest discharges, the vegetative resistance is in part overcome with friction factors more closely aligned with the predictive curve. The calculated roughness coefficients are of the same order as those calculated for the Montford reach by Knight, Shiono and Pirt (1990) on the basis of a theoretical model derived from flume experiments of the change of roughness in the transition as bankfull discharge is exceeded.

ROUGHNESS COEFFICIENTS AT THE REACH SCALE

Equation (3.5) also may be applied at the reach scale, using the tracer-derived mean velocity, U_L, to give an effective or apparent roughness coefficient at the reach scale. There is a difficulty at this scale, however, in estimating the average hydraulic radius. This difficulty may be overcome by assuming that the channel is rectangular in cross-section so that by continuity

$$Q = U\,W\,D \tag{3.6}$$

Mean channel width is measured relatively easily, so that for a given Q with $U = U_L$, an effective mean depth and corresponding hydraulic radius ($= WD/(W + 2D)$) can be calculated. Furthermore, for channels in which the change of velocity with depth can be described by equation (3.1), combining equations (3.1), (3.5) and (3.6) gives

$$f = \frac{8\,g\,S\,W\,Q^{(1-3m)}}{k^3\,W^2 + 2\,k^2\,Q^{(1-m)}} \tag{3.7}$$

Table 3.3 Flow parameters of equation (3.7) for reaches shown in Figure 3.7

Reach number	Reference	Reach	Slope	Width (m)	k (m/s^{-1})	m
1	Kellerhals (1970)	BR1-2	0.074	0.89	1.99	0.66
2		BR2-3	0.349	0.99	1.46	0.72
3		PL1-2	0.083	2.75	0.56	0.66
4		PL2-3	0.036	3.16	0.54	0.69
5		PL3-4	0.034	7.02	0.27	0.53
6		BL1-3	0.047	12.80	0.31	0.52
7		BL3-5	0.039	11.10	0.30	0.47
8		BL5-4	0.095	12.90	0.34	0.56
9		PH1-2	0.031	11.50	0.32	0.46
10		PH2-3	0.049	12.60	0.33	0.54
11		PH3-4	0.064	12.60	0.35	0.52
12		PH4-6	0.099	12.30	0.33	0.60
13	Beven, Gilman and Newson (1979)	Crimple	0.039	1.78	0.98	0.57
14		Severn 1	0.017	3.93	0.71	0.50
15		Severn 2	0.025	3.28	1.13	0.66
16		Severn 3	0.007	3.31	1.49	0.73
17		Severn 4	0.038	2.91	1.10	0.58
18		Severn 5	—	—	0.82	0.87
19		Severn 6	0.079	1.66	1.79	0.78
20		Severn 7	0.079	1.47	1.81	0.78
21	Calkins and Dunne (1970)	Sleepers River	0.032	3.72	0.48	0.40
22	Day (1975)	Thomas	0.027	3.76	0.77	0.21
23		Bealy	0.009	20.13	0.40	0.42
24		Porter	0.018	7.71	0.79	0.24
25		Bruce	0.020	7.02	0.64	0.43
26		Craigieburn	0.023	6.97	0.61	0.46
27	This paper	Montford	0.00032	28.0	0.047	0.64
28		Leighton	0.00032	56.0	0.108	0.40

Equation (3.7) can be modified further for the case where mean channel width is known to vary with discharge, but here it has been applied to a number of reach-scale tracer experiment data sets for which equations of the form of (3.1) together with a constant mean width and reach slope were available. The data sets are summarized in Table 3.3 and the variation of f with Reynolds number is shown in Figure 3.7. These data are mostly from relatively small, steep, upland streams, and the lower Severn reaches fall at the lower end of the range of roughness coefficients calculated in this way from tracer velocity data. It will be noted from this 'Moody' diagram that the effective roughness coefficients determined from tracer experiments take on some extremely high values. This implies that the mean velocities measured at the reach scale are much lower than normally would be expected for uniform smooth channels of the same average bed slope. This results from the effects of channel shape and, in particular, the presence of vegetation and low-velocity pool sections within the reach.

For the Montford and Leighton data sets, effective roughness coefficients calculated using U_L and the average hydraulic radius for the measured cross-sections within each reach, are compared with the mean values obtained for the measured pool and riffle

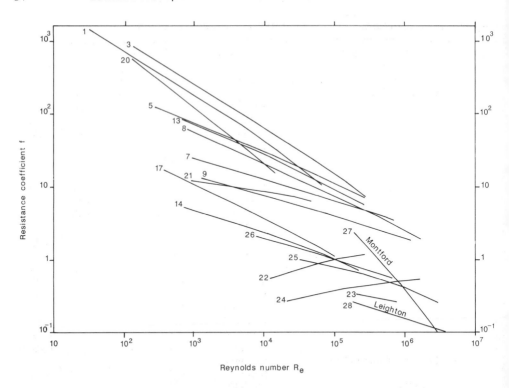

Figure 3.7 Approximate roughness coefficient–Reynolds Number relationships for the reaches of Table 3.3, based on reach-scale mean velocity data as calculated from equation (3.7)

sections in Table 3.4 for dates when both current metering and tracer data were available. Energy slopes calculated from the velocity profile data for both pools and riffles converge on a value of 3.2×10^{-4} at bankfull. This value is equal to the regional slope of the reach (Dury, 1983) and has been used as a best estimate of the reach-scale slope in the calculation of the effective roughness coefficients for individual discharges. It will be seen that for both reaches the values generally are larger than the cross-sectional (riffle) values. They also are higher than the values calculated by Knight, Shiono and Pirt (1990) for a cross-section at Montford (falling from a maximum of 0.08 at a flow of $20 \, \mathrm{m}^3 \, \mathrm{s}^{-1}$ to a minimum of 0.02 at $197 \, \mathrm{m}^3 \, \mathrm{s}^{-1}$ approximately bankfull flow).

DISPERSION AT THE REACH SCALE

The observed tracer curves typically show skewed distributions, and indicate that the data should not be interpreted in terms of the traditional advection–dispersion equation, as it clearly has not yet reached its full mixing length. In fact measurements from many rivers (e.g. Day, 1975; Sabol and Nordin, 1978; Wallis, Young and Beven, 1989) indicate that such skewed distributions persist for considerable distances down-

Table 3.4 Effective roughness values for the River Severn reaches for dates when both current metered and tracer measurements of mean velocities are available

	Discharge	Effective roughness coefficients		
	(m³ s⁻¹)	^{f}riffle	^{f}pool	^{f}reach
Montford	6.4	0.0496	0.0709	0.8410
	12.7	0.1446	0.0664	0.3112
Leighton	9.4	0.0685	—	0.3416
	20.1	0.0484	—	0.0842
	101.9	0.0405	—	0.0572

stream, in the case of the Rhine, for example, for hundreds of kilometres. Such behaviour justifies the use of alternative, more realistic, models for analysing the data. The River Severn tracer experiments have been analysed using the Aggregated Dead Zone (ADZ) model (Beer and Young, 1983; Wallis, Young and Beven, 1989; Wallis, Guymer and Bilgi, 1990), which is based upon a general linear time series analysis approach that allows the model structure, as well as the parameters, to be determined from discrete time experimental data. The general form of the model may be written as

$$O_k = \frac{b_0 + b_1 Z^{-1} + \ldots b_m Z^{-m}}{1 + a_1 Z^{-1} + \ldots a_n Z^{-n}} I_{k-\tau} \tag{3.8}$$

where O_k is the output tracer concentration at time step k, I is the input concentration, τ is the discrete advective time delay, and Z is the backwards difference operator, i.e. $Z^{-1} O_k = O_{k-1}$. The a and b parameters in equation (3.8) are calibrated by a simplified recursive least-squares technique (Young, 1984) using the ADZ-ANALYSIS program. The fit of different model structures to the data is evaluated using two criteria: the coefficient of determination (R^2) which shows how well the model fits the data, with a maximum value of one for a perfect fit; and the Young Information Criterion (YIC), which is a function of the standard error of the residuals and the standard errors of the fitted parameter values. The YIC increases rapidly in magnitude as the models tested become overparameterized (see Young, 1989).

In fact, in most applications it has been found that only a first-order ADZ model is required to obtain an adequate fit to input–output concentration data, where the input concentrations are already well mixed with the flow, although higher order models generally are required for reaches involving the initial mixing of a localized source. The first-order model has the form

$$O_K = \frac{b_0}{1 + a_1 Z^{-1}} I_{k-\tau} \tag{3.9}$$

Models of the form of equation (3.8) can be obtained by combining first-order models in series and parallel (see Wallis, Young and Beven, 1989 for examples of some parallel models of dispersion in small gravel-bedded streams).

Wallis, Young and Beven (1989) show that the parameters of the first-order model also can be interpreted in terms of different residence times in the reach. The mean residence time for tracer in the reach, t, is the sum of the advective time delay, τ, and the mean residence time, T, of an effective mixing volume V_e, where for a time step of Δt

$$a_1 = -\exp(-\Delta t/T)$$

and, for a conservative tracer,

$$b_0 = (1 + a)/T$$

A discharge, Q, can be calculated from the dilution of a conservative tracer. The effective mixing volume then can be determined from

$$V_e = QT \tag{3.9a}$$

and an estimate of the total volume in the reach from

$$V = Qt \tag{3.9b}$$

so that the ratio of the mixing volume to the total volume is given by

$$V_e/V = T/t = T/(T + \tau) \tag{3.10}$$

This ratio is called the Dispersive Fraction, and is an alternative dispersion parameter to the dispersion coefficient used in the advective–dispersion equation. It has a number of advantages over the dispersion coefficient, in that it has a limited range (0 to 1), it is determined readily from tracer data (even without fitting an ADZ model; by use of equation 3.10), and Wallis, Young and Beven (1989) have shown that, at least for some rivers, it may be almost constant with discharge. The dispersion coefficient on the other hand is known to vary rapidly with discharge (e.g. Whitehead, Williams and Hornberger, 1986). All that is then needed to construct a first-order ADZ (conservative) dispersion model for a reach is the Dispersive Fraction and one of the residence time parameters t or τ. Solute losses also are handled easily within the model if it can be assumed that losses are, to a first approximation, directly proportional to concentration. This requires the specification of an additional loss parameter that may be related to the steady-state gain in the reach, i.e. the ratio of the mass of tracer recorded at the output site to that entering the measurement reach at the input site.

Figure 3.8 shows examples of fitted first-order models for the Montford and Leighton reaches, while Figure 3.9 shows the change in Dispersive Fraction at the two sites. A summary of the results is given in Table 3.5. The values of Dispersive Fraction for these reaches show a wider spread than the rivers reported in Wallis, Young and Beven, (1989) and there seems to be some tendency for a decrease in Dispersive Fraction with an increase towards bankfull discharge, although the correlation coefficients are low. There also is one particularly low value at Leighton at the lowest flow recorded, despite the fact that the 'dead zone' storage might be expected then to be proportionally greater. It is perhaps worth noting, however, that the volume calcula-

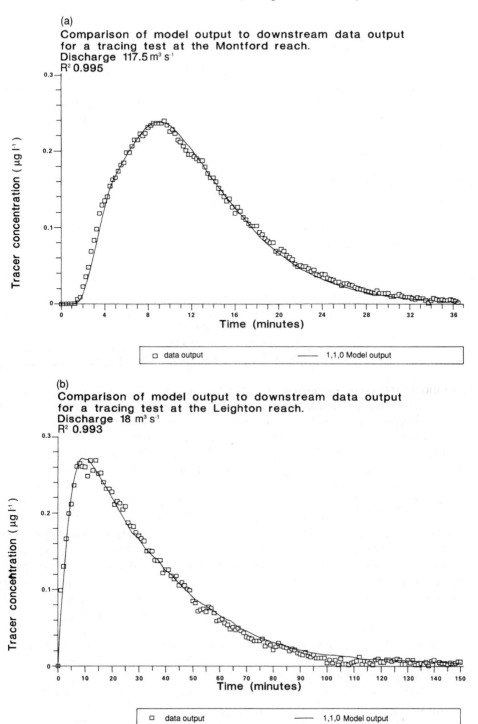

Figure 3.8 Examples of observed downstream tracer concentrations and Aggregated Dead Zone (ADZ) model predictions for (a) Montford and (b) Leighton reaches

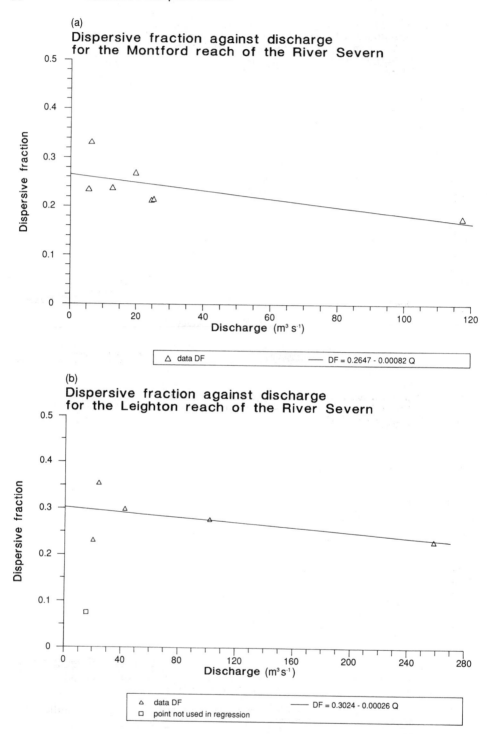

Figure 3.9 Relationship between the ADZ Dispersive Fraction and discharge for (a) Montford and (b) Leighton reaches

Table 3.5 Aggregated Dead Zone model parameters for the Montford and Leighton reach-scale measurements

Discharge (m³ s⁻¹)	Time delay, τ (s)	Mean travel time, t (s)	ADZ residence time, T (s)	Dispersive fraction T/t
Montford reach				
5.6	4287	5602	1315	0.235
6.4	3832	5738	1906	0.332
12.7	4066	5358	1292	0.241
19.8	2595	3550	955	0.269
24.7	2354	2996	642	0.214
25.3	2315	2948	633	0.215
117.5	665	808	143	0.177
Leighton reach				
9.4	2472	2739	267	0.097
18.0	1896	2495	599	0.240
20.1	2868	3728	860	0.231
23.9	2076	3219	1142	0.355
42.5	1841	2585	744	0.288
101.9	1150	1590	440	0.277
258.4	852	1110	228	0.205

tions of equations (3.9) take no account of any volume of water in the reach that effectively takes no part at all in the dispersion process.

CONCLUDING DISCUSSION

This paper has presented the results of a study of two 1 km reaches of the River Severn at Montford and Leighton near Shrewsbury. Eulerian velocity measurements at specific pool and riffle cross-sections and reach-scale Lagrangian velocity measurements made using tracers are compared for a number of discharges. The reach-scale velocities are, in general, lower than the cross-sectional mean velocities, suggesting that there are large-scale flow structures within the reach resulting in retardation of the tracer owing to exchange across internal shear zones into volumes of slow moving water. This difference in velocities also is reflected in the calculated effective roughness coefficients, which were higher for the reach-scale measurements, and in the case of the Leighton reach, considerably higher. The high values of effective resistance coefficients at the reach scale were placed in the context of values calculated from other data sets based on tracer velocity measurements in smaller streams, some of which demonstrate extreme values of the resistance coefficient.

It can be argued (see Clifford and Richards, this volume) that hydraulic data derived from sectionally averaged data do not characterize the section adequately when spatial variation in velocity is excessive. This, of course, is true for the extreme examples given in Figure 3.3 and for the data presented for the meander sequence,

Table 3.6 Regression relationships for the Aggregated Dead Zone model parameters for the Montford and Leighton reach-scale measurements

	Montford			Leighton		
		N	r^2		N	r^2
Advective time delay (s)	$\tau = 14454.4\ Q^{-0.602}$	7	0.92	$\tau = 8537.5\ Q^{-0.42}$	6	0.96
	$\tau = 1520.5 + 17319.84/Q$	7	0.74	$\tau = 899.96 + 30281.7/Q$	6	0.86
Dispersive Fraction	$DF = 0.265 - 0.00082\ Q$	7	0.42	$DF = 0.303 - -0.00026\ Q$	5	0.23

but detailed plots of the cross-channel data for the straight sections at Leighton and for the dates when comparative tracer and current meter data are available showed a roughly square distribution of parameters. The difference in the roughness coefficients, comparing methods in these reaches, indicates that the velocity profile data during low flows largely integrate the effects of the local grain roughness and underrepresents the effect of other roughness effects, such as vegetative roughness along the banks. This is a useful finding that should allow reasonable assessments to be made of grain roughness in similar rivers. It is in any case difficult to obtain meaningful profiles in the strongly upwelling flow over the sloping bank topography and considerable perseverance is required. At Montford, during high flows, however, the distortion of the f versus R/d_{84} data trend demonstrates that vegetated bank effects influence sectionally averaged velocity profile data. The methods used in combination, therefore, provide a tool to separate local grain roughness from local vegetative roughness and reach-scale roughness.

Analysis of the tracer data in terms of the Aggregated Dead Zone model for dispersion in rivers showed that very good fits could be obtained for all the measurements by a first-order ADZ model (Table 3.6). Values of the Dispersive Fraction, the basic dispersion parameter of the ADZ model, show some tendency in these reaches to decrease with an increase discharge up to bankfull stage. One measurement, at Leighton, gave an anomalously low value of Dispersive Fraction at the lowest discharge measured.

ACKNOWLEDGEMENTS

The Director and staff of the Preston Montford Field Centre are thanked for their hospitality during the fieldwork. We are indebted to Mark Glaister, Ben James, Chris Hutchings, Kev Buckley, and Hannah Green for help in the field and in the analysis of the data. The project has been funded primarily by the DOE.

REFERENCES

Bates, B. C. and D. H. Pilgrim (1982). Investigation of storage-discharge relations for river reaches and runoff routing models. Institute of Civil Engineers of Australia, *Hydrology and Water Resources Symposium*, Melbourne, pp. 120–126.

Bauwens, W., J. Bellon and A. Van der Beken (1982). Tracer measurements in lowland rivers. In J. A. Cole (ed.), *Advances in Hydrometry*, International Association of Hydrological Sciences Publication No. 134.

Beer, T. and P. C. Young (1983). Longitudinal dispersion in natural streams. Journal of Environmental Engineers American Society for Civil Engineers, **109**, 319–331.

Beven, K. J. (1979). On the generalised kinematic routing method. *Water Resources Research*, **15**, 1238–1242.

Beven, K. J., K. Gilman and M. Newson (1979). Flow and flow routing in upland channel networks. *Hydrological Sciences Bulletin*, **24**, 303–325.

Calkins, D. and T. Dunne (1970). A salt tracing method for measuring channel velocities in small mountain streams. *Journal of Hydrology*, **11**, 379–392.

Carling, P. (1991). An appraisal of the velocity reversal hypothesis for stable pool–riffle sequences in the River Severn, England. *Earth Surface Processes and Landforms*, **16**, 19–31.

Dawson, F. H. (1978). The seasonal effects of aquatic plant growth on the flow of water in a stream. In *Proc. European Weed Research Society 5th International Symposium on Aquatic Weeds*, Vol. 5, 5–8th September 1978, pp. 71–78.

Day, T. J. (1975). Longitudinal dispersion in open channels. *Water Resources Research*, **11**, 909–918.

Dury, G. H. (1983). Osage-type underfitness on the River Severn near Shrewsbury, Shropshire, England. In K. J. Gregory (ed.), *Background to Palaeohydrology*, Wiley, pp. 399–412.

Dury, G. H. (1984). Abrupt variation in channel width along part of the River Severn, near Shrewsbury, Shropshire, England. *Earth Surface Processes and Landforms*, **9**, 485–492.

Dury, G. H., C. A. Sinker and D. J. Pannett (1972). Climatic change and arrested meander development on the River Severn. *Area*, **4**, 81–85.

Griffiths, G. A. (1989). Form resistance in gravel channels with mobile beds. *Journal of the Hydraulics Division, American Society of Civil Engineers*, **115**, 340–355.

Hey, R. D. (1979). Flow resistance in gravel-bed rivers. *Journal of the Hydraulics Division, American Society of Civil Engineers*, **105**, 365–379.

Hydraulics Research (1988). *Assessing the Hydraulic Performance of Environmentally Acceptable Channels*, Report EX 1799, Hydraulics Research Ltd, Wallingford, 201 pp.

Kellerhals, R. (1970). Runoff routing through steep natural channels. *Journal of the Hydraulics Division, American Society of Civil Engineers*, **96**, 2201–2218.

Knight, D. W., K. Shiono and J. Pirt (1990). Prediction of depth, mean velocity and discharge in natural rivers with overbank flow. In R. A. Falconer, P. Goodwin and R. G. S. Matthew (eds), *Hydraulic and Environmental Modelling of Coastal, Estuarine and River Waters*, Gower Technical, pp. 419–428.

Knighton, A. D. and R. Cryer (1990). Velocity–discharge relationships in three lowland rivers. *Earth Surface Processes and Landforms*, **15**, 501–512.

Naot, D. (1984). Response of channel flow to roughness heterogeneity. *Journal of the Hydraulics Division, American Society of Civil Engineering*, **110**, 1568–1587.

Newson, M. D. (1984). River process and form. In G. Lewis and G. Williams (eds), *Rivers and Wildlife Handbook: A Guide to Practices which Further the Conservation of Wildlife on Rivers*, Royal Society for the Protection of Birds and Royal Society for Nature Conservation, 295 pp.

Newson, M. D. and J. G. Harrison (1978). *Channel Studies in the Plynlimon Experimental Catchments*, Institute of Hydrology Report 47, Wallingford.

Petryk, S. and G. Bosmajian (1975). Analysis of flow through vegetation. *Journal of the Hydraulics Division, American Society of Civil Engineers*, **101**, 871–884.

Powell, K. E. C. (1978). Weed growth—a factor of channel roughness. In R. W. Herschy (ed.), *Hydrometry: Principles and Practice*, Wiley-Interscience, pp. 327–352.

Prestegaard, K. L. (1983). Bar resistance in gravel bed streams at bankfull stage. *Water Resources Research*, **19**, 472–476.

Reynolds, C. S., P. A. Carling and K. J. Beven (1991). Flow in river channels: new insights into hydraulic retention. *Archiv fuer Hydrobiologie*, **121** 2, 171–179.

Sabol, G. V. and C. F. Nordin (1978). Dispersion in rivers as related to storage zones. *Journal of the Hydraulics Division, American Society for Civil Engineers*, **104**, 695–708.

Wallis, S. G., C. Blakeley and P. C. Young (1987). A microcomputer-based fluorometric data logging and analysis system. *Journal of the Institute of Water Engineers and Scientists*, **41**, 122–134.

Wallis, S. G., I. Guymer and A. Bilgi (1990). A practical engineering approach to modelling longitudinal dispersion. In R. A. Falconer, P. Goodwin and R. G. S. Matthew (eds), *Hydraulic and Environmental Modelling of Coastal, Estuarine and River Waters*, Gower Technical, pp. 291–300.

Wallis, S. G., P. C. Young and K. J. Beven (1989). Experimental investigation of the aggregated dead zone model for longitudinal solute transport in stream channels. *Proceedings of The Institute of Civil Engineers, Part 2*, **87**, 1–22.

Watson, D. (1987). Hydraulic effects of aquatic weeds in UK Rivers. *Regulated Rivers: Research and Management*, **1**, 211–227.

Whitehead, P. G., R. Williams and G. M. Hornberger (1986). On the identification of pollutant or tracer sources using dispersion theory. *Journal of Hydrology*, **84**, 273–286.

Young, P. C. (1984). *Recursive Estimation and Time Series Analysis: an Introduction*, Springer-Verlag, 299 pp.

Young, P. C. (1989). Recursive identification, forecasting and adaptive control. In C. T. Leondes (ed.), *Control and Dynamic Systems*, Vol. XXX, Part 3, Academic Press, pp. 119–165.

4 Bend Scour and Bank Erosion on the Meandering Red River, Louisiana

COLIN R. THORNE

Department of Geography, University of Nottingham

INTRODUCTION

This study is concerned with a reach of the Red River extending from Index, Arkansas to Shreveport, Louisiana (Figure 4.1). This reach of river has undergone intense morphological change during the last 120 years. Prior to that time, the river below Shreveport was blocked by a massive accumulation of timber called the Great Red River Raft. At the time of its removal the raft occupied the channel from near the Arkansas–Louisiana State Line to a point some 250 km downstream. It reduced flow velocities and raised water levels in the river so much that a multi-channel or anastomosing pattern resulted from the frequent overbank events. The raft was removed under the direction of Captain Shreve, with the work essentially being completed by 1873, although new jams formed and were removed for another 25 years. A detailed review of the raft and its removal may be found in McCall (1988).

The removal of the raft allowed the river to form a single channel and concentration of the stream's energy led to approximately 4.5 m of bed degradation in the vicinity of Shreveport (Veatch, 1906). Degradation became progressively less severe going upstream, the bed lowering being about 1.5 m at Index. Response of the channel width to lowering of the bed was also spectacular. Prior to raft removal the channel width was of the order of 110 m. This increased to over 330 m. Channel sinuosity decreased following raft removal, from about 2.3 in 1886, to about 1.5 in 1968. Since 1968 a programme of channel realignments undertaken by the Corps of Engineers has further reduced sinuosity to its present value of 1.36. A study of the geomorphology of the Red River from Denison Dam to Shreveport has recently been performed by Water Engineering and Technology Inc. (1987).

The historical evidence indicates that the response of the channel in the study reach to raft removal is essentially complete and that the channel is now in a state of dynamic equilibrium. That is, the average depth, width and slope of the channel are time invariant, although channel migration continues. This is possible because erosion of the outer banks in bendways is just balanced by inner bank deposition, so that the channel sweeps back and forth across its meander belt without a significant change in overall cross-sectional size. Also, in a dynamically stable river, reductions in sinuosity through natural cut-offs of meanders are just balanced by the growth of other

Lowland Floodplain Rivers: Geomorphological Perspectives. Edited by P. A. Carling and G. E. Petts
© 1992 John Wiley & Sons Ltd

meanders, so that the overall sinuosity of the reach is steady. This is not entirely the case on the Red, however, because several cut-offs have been initiated artificially, leading to reduction of sinuosity, which is on-going. The river may be expected to respond to this reduction in sinuosity, and this is discussed later on. Even so, it may be concluded that, at least in comparison with the previous 80 years, the Red River has been dynamically stable during the last decade. However, the present configuration of the channel, and particularly the nature of the stream banks, are direct results of the recent history of channel instability.

The aim of this paper is to examine the link between the contemporary erosion processes and failure mechanisms that are responsible for bank retreat, and medium

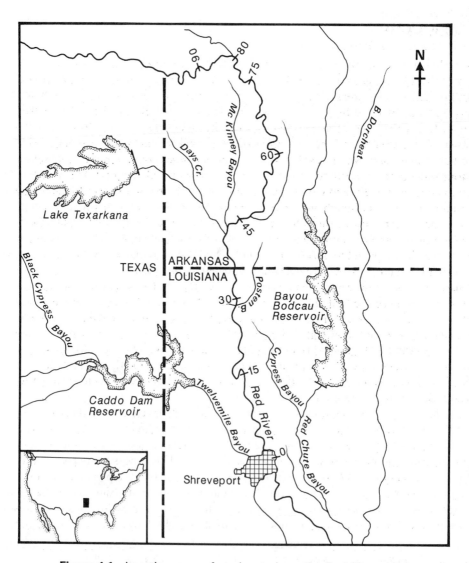

Figure 4.1 Location map of study reach on the Red River, USA

to long-term meander evolution. To attempt this, it is first necessary to understand how and why the geometry, geotechnical properties and characteristic modes of erosion and failure of the banks vary along the study reach. The links between bed scouring, bank retreat and the resulting bend migration then may be considered in detail. Bed scour and bank retreat in bends are closely related and attempts to explain bend geometry and planform evolution must account for the influence of bank properties on both scouring and lateral erosion. The results of this study have important implications for further bank stabilization schemes and these are considered in conclusion.

BANK STUDIES

The Mechanics of Bank Retreat

Serious bank retreat usually occurs by a combination of flow erosion of intact bank material and mass failure of the bank owing to gravity, followed by basal clean-out of the failed material from the bank toe. The relative importance of these two components of bank retreat depends on the processes responsible for erosion and the geometry, scale and geotechnical properties of the bank. Whilst the long term rate of

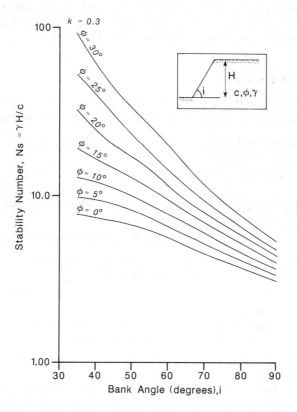

Figure 4.2 Generalized stability charts for river banks

retreat of the bank depends on the rate of basal clean-out and scouring at the toe (and therefore is controlled primarily by near-bank sediment transport capacity), the configuration of the bank and the critical mechanism of failure are functions of the geotechnical properties of the bank materials. A thorough review of bank erosion processes and mechanisms of failure may be found in Thorne (1982).

Osman and Thorne (1988) developed bank stability analyses that can be used to predict the factor of safety of a bank with respect to either slab-type or rotational slip failure. Their analyses show that banks formed in weakly cohesive sediments tend to be low and steep, failing mostly as slabs. Conversely, strongly cohesive banks tend to be higher, to slope more gently, and fail by rotational slip. This is the case because factors of safety with respect to slab-type failures are more critical for banks steeper than about 60° and those for rotational slips more critical for banks flatter than 60°.

The Osman–Thorne analyses for rotational slip and slab-type failures was used to produce the dimensionless stability chart of the stability number (N_S) as a function of the bank material friction angle (ϕ) and the bank angle (i), shown in Figure 4.2.

In producing the chart, it was necessary to assume a value for the characteristic tension crack depth for slab-type failures of the banks in the study reach. As no measured data were available, a ratio of crack depth to bank height of 0.3 was used to generate the charts. Field inspection of failed banks indicated that this should be a reasonable approximation to actual values at failure. A more precise identification of the tension crack depth is not essential, because the results of a sensitivity analysis showed that the factor of safety with respect to slab-type failure was relatively insensitive to variation in the tension crack depth (Thorne, 1988).

Red River Bank Characteristics

The banks of the Red River in the study reach are formed in a variety of materials that have been classified broadly on the basis of their origin and age as follows:

(1) meander belt alluvium (MBA)—material deposited in the present floodplain, under the present flow regime;
(2) clay plug deposits (CP)—fine material deposited in cut-off meanders and oxbow lakes;
(3) back-swamp deposits (BSD)—sediments laid down in the nineteenth century floodplain by overbank flows;
(4) Pleistocene deposits (Pl)—lithified materials in high terraces and the valley sides.

These different materials have quite different geotechnical properties and, therefore, might be expected to have different characteristic bank geometries and modes of failure. This thesis was investigated by a combination of field reconnaissance, documentary study and theoretical analysis.

Field observations made during the summer of 1988 supported the hypothesis that slab-failures predominated on the less cohesive meander belt alluvium, while rotational slips occurred most often in the more strongly cohesive clay plug and back-swamp banks. Typical examples are shown in Figures 4.3 (a and b).

The documentary study was based on the 1980–1981 hydrographic survey. Cross-sections were selected that included stable and eroding banks from each of the bank material types. The cross-sections were computer plotted and the plots then were used

(a)

(b)

Figure 4.3 Bank failures on the Red River near Plain Dealing, Louisiana. (a) Slab-type failure in meander belt alluvium. (b) Rotational slip in clay plug deposits

to derive data on the bank heights and angles. The complete data set is too large to be included here, but indicated that banks formed in Pleistocene materials (Pl) are mostly in the range of 7.5–10.5 m high, those in back-swamp deposits (BSD) are in the range 6–9 m, clay plugs (CP) 4.5–7.5 m and meander belt alluvium (MBA), 3–9 m high. The terrace banks (Pleistocene and back swamps) should be higher than those formed by contemporary fluvial processes (clay plugs and meander belt alluvium), given the origin of the bank types and the degradational history of the river over the last 100 years. Hence, these results are as expected.

These data represent a snap-shot of the banks at a particular moment. It is unknown whether the geometry of the bank at that time was very stable, or close to failure. Also, it is impossible to know to what extent the surveyed bank was composed of low-angle slump material that had not yet been removed by the flow and was masking the angle of the intact bank. Consequently, the data cannot be taken to represent the critical bank geometries with respect to mass failure. Finally, few banks steeper than 45° appear in the surveys for the practical reason that such banks are very hazardous to work on. Nonetheless, the bank geometry data do allow a coarse test of the bank stability analysis.

Geotechnical Properties

To allow application of the Osman–Thorne bank stability analyses, data were required on the properties of the bank materials. Data on the engineering properties of Red River sediments were supplied by the Foundations and Materials Branch of the Vicksburg District, US Army Corps of Engineers, on the basis of soils tests run on samples from site investigation borings all along the study reach. The data, which refer to 'worst case' conditions of high moisture content and low strength, are listed in Table 4.1.

The form of the data supplied did not yield the characteristics of the clay plug and meander belt alluvium materials directly. Examination of the relevant size distribution curves from sampling of banks in the study reach suggested that the clay plugs generally are made up of about 50% clay and silt and 50% fine sand. Meander belt alluvium banks seem to be about 15% clay and silt, and 85% sand. On this approximate basis figures for the bank properties in Table 4.2 were generated by weighting the sedimentary data in Table 4.1.

Table 4.1 Sediment properties of the Red River

Bank Material	Cohesion (kPa)	Friction angle (degrees)	Bulk unit weight (kN m^{-3})
Clay	23.94	0	18.07
Silt	14.36	20	18.07
Sand	0.00	30	18.85
Back swamp	19.15	0	18.07
Tertiary	47.88	10	18.85

Table 4.2 Bank material properties of the Red River

Bank Material	Cohesion (kPa)	Friction angle (degrees)	Bulk unit weight (kN m^{-3})
Meander belt			
Alluvium	2.87	27	18.85
Clay plug	9.58	20	18.54
Back swamp	19.15	0	18.07
Tertiary	47.88	10	18.85

Stability Analysis

The geotechnical data were used to calculate critical stability numbers and bank heights for Red River banks formed in different materials. The resulting curves are plotted in Figure 4.4 (a–d). The observed bank heights and angles also were plotted onto the dimensionless stability charts for comparison with the lines of limiting stability derived from the theoretical analyses and the geotechnical data.

Discussion of Results

The stability chart is divided into three zones, following the results of studies on other waterways in this region (Thorne, 1988). Banks that plot above the line of limiting stability for worst case conditions are unstable and should be expected to fail when wet. Banks that plot below the phi = 0 line should be stable under all circumstances. Banks that plot between these two lines pose a problem. They should not fail under the worst case conditions of high moisture content, but could fail if the operational friction angle for the bank material was reduced toward zero by positive pore pressures. Positive pore pressures may be generated by a combination of saturated conditions in the bank plus rapid drawdown in the adjacent channel. This circumstance is unlikely to occur frequently, but is nonetheless sufficiently likely to make the stability of banks plotting in this zone unreliable.

In theory, the points for the observed banks should plot in the 'unreliable' zone of the stability chart with the line of limiting stability for worst case conditions forming an upper envelope to the data cloud. This follows because the survey data come from banks that were actively retreating, but were stable at the actual time of observation.

The most complete data set is for the MBA banks (Figure 4.4a), because the geotechnical properties are well defined and there are a sufficient number of bank angle and N_S points to establish a trend. The data plot as expected, in the unreliable zone, with the line of limiting stability as an upper bound. However, confirmation that Figure 4.4a is a true representation of the stability of banks formed in MBA materials in the study reach requires data from banks having angles steeper than 45°. In view of the limitations on bank data derived from hydrographic surveys, field measurements probably are the only way to obtain these data.

The results for clay plug banks (Figure 4.4b) also support the form of the stability chart. Most banks plot in the unreliable zone, as expected, and the trend of the

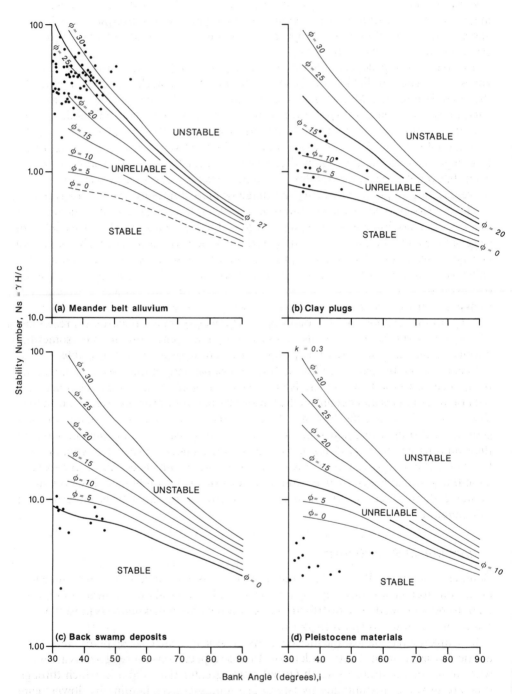

Figure 4.4   Stability charts for Red River banks

limiting stability line does follow that of the data. However, the line looks a little high, suggesting that the friction angle of 20° approximated from the geotechnical data might be an overestimate. The stability chart distribution indicates that a friction angle of 15–17° might be more appropriate. More data are required to confirm this, especially for steep banks ($i > 50°$).

The back-swamp deposit stability chart (Figure 4.4c) shows stable and unstable zones only. The unreliable zone is absent because the geotechnical data supplied by the Foundations and Materials Branch specify a zero friction angle for worst case conditions. Field data are sparse for banks of this type, but those available suggest that the friction angle of the BSD materials may have been underestimated. An operational value of 5° might be appropriate. Alternatively, the cohesion might be a little high. Decreasing the assumed cohesion by a relatively small proportion would bring the points just below the line of limiting stability for a friction angle of zero. More data are required across the full range of bank angles to resolve this issue.

The Pleistocene plot is the least satisfactory of the four (Figure 4.4d). The data are so sparse that no firm conclusions can be drawn, but it appears that the cohesion value suggested by the Foundations and Materials Branch is much too high for use in analysing river bank stability. The explanation for this statement is that although the value is representative of the strength of these materials in the ground, it is an overestimate of their strength in a river bank, where they are subject to weathering, weakening processes and tensile cracking owing to stress relief at the bank face.

An operational cohesion of around 20–25 kPa would appear to better represent the Pleistocene river bank materials. This would rank them close to, but somewhat stronger than, the back-swamp deposits, which does not seem unreasonable.

In conclusion, the bank stability analyses of Osman and Thorne (1988) do appear to be applicable to the banks of the Red River in the study reach, although more field data on bank angles and heights in the various bank units are required to confirm this. The analyses highlight the fact that the geometry of the banks is a function of their geotechnical properties, and suggests that retreating banks are maintained close to their limiting geometry with respect to mass failure under gravity. This supports the theory that bank retreat in the study reach takes place by fluvial erosion of the bank to a limiting state, followed by mass failure under worst case conditions. Basal clean-out of slump debris then occurs, prior to further erosion of intact bank materials and subsequent failures.

Distribution of Bank Retreat

Bank erosion on the Red River in the study reach is associated mostly with meander evolution. Reconnaissance trips by boat made at low flow allowed reaches of eroding bank to be identified. The distribution of erosion within a bed depends mostly on the radius of curvature to width (R_c/w) ratio.

In 'conventional' bends, the locus of bank erosion is located at the outer bank at, or a little downstream of, the meander exit. In this respect, conventional bends are those with radius of curvature to width (R_c/w) ratios greater than about 2, which through time are growing in amplitude by lateral erosion, and are migrating by downstream progression. However, this is by no means the only form of meander present in the study reach. Several meanders have very low R_c/w ratios of the order of unity. These bends are found where a meander comes up against resistant outer bank materials, so that its normal growth and/or progression are disrupted. Examples of such situations occur where a meander reaches the outer limit of the current alluvial meander belt and impinges on older and more resistant terrace deposits, or where a clay plug within the alluvial deposits is encountered. In these very tight bends, the distribution of bank erosion is quite different from that in the conventional meanders. Bank erosion is concentrated at the inner bank, the outer bank being an area of flow separation and sediment deposition during formative flows. This leads to bench or berm building at the outer bank, a tendency for the bend to reduce rather than increase its amplitude with time, and more rapid downstream progression of the up-valley limb of the meander than the down-valley limb, leading to neck cut-off in due course. In geomorphological terms, the meander works its way around the obstruction presented by the resistant material at the outer bank. As a result, the distribution of bank erosion and the pattern of bend evolution are related closely to the distribution of clay plugs within the meander belt, and the occurrence of terraces. These terrace deposits consist of two major categories of material: back-swamp deposits in a terrace associated with the nineteenth century floodplain, and much older deposits of Pleistocene and Tertiary age in terraces and in the valley sides above the valley floor (Water Engineering and Technology Inc. 1987).

RIVER BEND STUDIES

Background

As long ago as 1945, Friedkin observed experimentally that stabilization of the eroding outer bank in a free, alluvial meander led to excess scouring of the bed adjacent to that bank. Recently, Thorne and Osman (1988) put forward a theoretical explanation of this phenomenon on the basis of considerations of bank stability and sediment flux at the toe of the outer bank.

In a migrating bend, erosion of the outer bank and scour at the bank toe keep the bank close to its critical height for mass failure under gravity. This limits the scour pool depth adjacent to the bank, because if the flow attempts to further scour the bed and increase the bank height, bank failures are generated and the rate of bend migration increases instead. In this way the balance of sediment input to the toe area by bank erosion and mass failures is adjusted to satisfy the capacity of the flow in this area to entrain and remove sediment.

When the bank is stabilized with a revetment, bank erosion ceases and gravity induced failures no longer occur. The impact on bed scour adjacent to the toe is twofold. First, stabilization of the bank allows the scour pool depth to increase because toe scour no longer induces gravity failures. Second, the input of sediment from bank erosion and failures is cut off, so that the bed becomes the only available source of sediment to satisfy the transport capacity of the near-bank flow. Consequently, toe scour is promoted and scour pools in bends with erosion resistant outer bank materials are expected to be deeper than those in geometrically and hydraulically similar free meanders.

River Bend Geometry Data

Data from the 1980–1981 hydrographic survey were used to characterize the three-dimensional geometry of meander bends and crossings between Index and Shreveport. The complete set of data that was compiled is available in a report to the US Army Research Office, London (Thorne, 1989).

Not every bend was suitable for analysis. Some were rejected because of their complex geometry, location near a channel confluence, or owing to uncertainties either concerning the nature of the outer bank materials or date of stabilization by revetment construction. Each bend that was used is defined by the location of its apex in river miles on the 1980–1981 hydrographic survey.

Definition of the geometry of a channel (width, depth, radius of curvature and so on) demands that some consistent value be used for the discharge. Usually, the dominant discharge is used for this purpose. On this basis the 2-year flow was selected as the representative flow for this study, as analysis of long-term records from the gauge at Shreveport by Biedenharn, Little and Thorne (1987) has established this as approximating the dominant flow for this reach of the Red River. All measurements of channel geometry and scour depth are then referenced to the 2-year flow line, as established from runs of the HEC-2 water surface profile programme undertaken by the Corps of Engineers in the course of their own hydraulic investigations. Bed elevations relative to the Vicksburg District gauge datum were converted to depths below the 2-year water surface profile using a computer program.

The nature of the outer bank at each bend was established from geological maps and field surveys. The distributions of meander belt, clay plug, back swamp and Pleistocene materials in the floodplain and terraces of the Red River valley in this reach, were marked onto aerial photographic mosaics. US Army Corps' records were used to establish the extent and date of installation of revetments in the study reach. Using the marked aerial photographs, each bend was classified by outer bank type as being a free meander (M), clay plug (C), back swamp (B), Pleistocene (P), or revetted (R) bend.

To investigate whether the nature of the outer bank in a bend does influence the scour depth, the base data were processed to produce the dimensionless parameters listed in Table 4.3.

The purpose of dividing bend scour depths (BD_{max}) by the average of the mean channel depth at the crossings immediately upstream and downstream of the bend in

Table 4.3 Parameters used to analyse bend scour

Definition of parameter	Equation	Symbol
1. The radius of curvature to width ratio	Bend radius/width	R_c/w
2. Maximum scour depth at the bend	2 year flow elevation – maximum scour elevation	BD_{max}
3. Mean depth at the upstream and downstream crossings	2 year flow elevation – average scour elevation	XD_{bar}
4. Ratio of maximum bend scour depth to mean crossing depth	Parameter 2/Parameter 3	BD_{max}/XD_{bar}

question (XD_{bar}) was to remove any scale effects. This produced a relative measure of the excess depth at a bend over that at a straight reach, owing to bend scour.

Bend Scour Depth – Outer Bank Type Relations

To investigate relations between the scour depth and the outer bank type at bends, the data (Table 4.3) were grouped by bank type.

Non-linear regression analyses were then performed on the data. R_c/w was the independent variable and BD_{max}/XD_{bar} was the dependent variable. A number of transformations were tried and that giving the best coefficients of determination was selected. This involved the use of the parameter $[(R_c/w) - 2]$ to predict scour pool depth at a bend. Some justification of the use of this parameter is relevant here.

A great body of theoretical, experimental and observational evidence indicates that as the ratio of radius of curvature to width in a bend is reduced to a value of about 2, gross changes occur to the flow patterns in the bend. These result in marked differences in the bend's morphology and its rate and direction of migration. Notable among the researchers to investigate these phenomena are Bagnold (1960) and Nanson and Hickin (1983). A plot of bank caving rate versus R_c/w for the study reach produced in a parallel study (Biedenharn *et al.*, 1989) confirms the significance of a ratio of 2 here also (Figure 4.5). This plot shows that the upper bound of caving rate *increases* steeply as R_c/w reduces from 5 to 2. However, further reduction to a value below 2 produces a rapid *reduction* in the upper bound for the caving rate. This probably is associated with the generation of large separation zones at the outer bank,

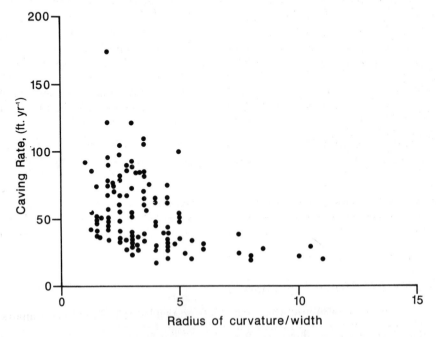

Figure 4.5 Plot of caving rate of outer bank versus radius of curvature to width ratio for the study reach of the Red River (adapted from a figure in Biedenharn *et al.*, 1989)

Table 4.4 Regression results for maximum bend scour depth

Outer bank type	Correlation coefficient (r)	Coefficient of determination (r^2)	a	b	Number of points
Free meander	0.90	0.81	1.98	−0.17	20
Clay plug	0.65	0.42	2.12	−0.10	8
Back swamp	—	—	—	—	3
Pleistocene	0.85	0.72	2.07	−0.20	8
Revetted	0.83	0.69	2.15	−0.27	31
All bends	0.80	0.64	2.07	−0.19	66

[a]Regression equation $(BD_{max}/XD_{bar}) = a + b \log_e (R_c/w - 2)$

and a switch to erosion of the inner bank, as observed by Reid (1984) on the Connecticut River, and by Markham (1990) on the Fall River and the River Roding.

On the basis of the available evidence it was decided that it was justifiable to take $R_c/w = 2$ as a lower boundary to the regression analysis, and this was supported by the good fit between the data and the regression lines that resulted. The results are listed in Table 4.4.

Regression analysis could not be carried out for the back-swamp bends because there were only three bends in this category and this is too few for meaningful results. The numbers of bends in the clay plug and pleistocene categories also are undesirably low, but are acceptable. The two groups of primary interest in this study, free meanders and revetted meanders, had 20 and 31 observations, respectively: ample to perform analyses of this type.

All of the r values listed in Table 4.4 are statistically significant at the 95% level of significance. The clay plug relation has a much lower correlation coefficient than the other bend types, indicating more scatter in the relationship between bend scour depth and R_c/w. This probably stems from the fact that the maximum scour depth in these bends is controlled only indirectly by the bend geometry. It is the prominence and position of the clay plug in the bend, rather than the bend geometry, that primarily controls maximum scour depth. As this is not accounted for in the analysis, this weakens the correlation coefficient.

Values of the coefficient of determination (r^2) indicate the proportion of the total variance explained by the regression equation. Unexplained variance is attributable to errors and to the effects of other independent variables. It was suspected that some variance could be explained by the different arc angles of the bends. However, no improvement in explanation could be found when this variable was added to the analysis.

Generalized Bend Scour Depth Relations

The regression lines in Figure 4.6 summarize the results of the regression analyses. They show a systematic and explainable distribution. For tight meander bends the ratio of bend maximum scour depth to mean crossing depth increases from alluvial meanders to clay plugs to Pleistocene to revetted bends. This is exactly as would be

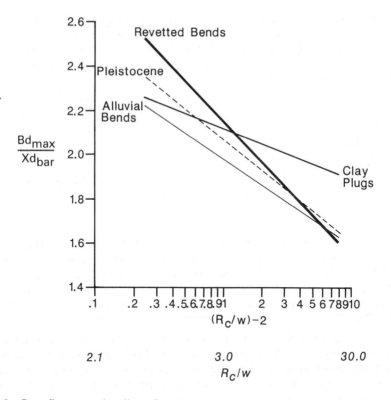

Figure 4.6 Best-fit regression lines for maximum scour depth to mean crossing depth ratio versus radius of curvature to width ratio for all bend types in the study reach of the Red River

expected from the experimental results of Friedkin (1945) and the theory of Thorne and Osman (1988). The slopes of the best-fit lines for alluvial, Pleistocene and revetted bends also increase systematically, so that the lines converge as R_c/w increases. This shows that the influence of outer bank properties decreases as bends become longer and less strongly curved. In fact the line for revetted bends actually crosses the other two at R_c/w of about 5. The reason for this is that in relatively straight revetted reaches the cross-section tends to be more trapezoidal, so that depth is more uniform than in an equivalent alluvial reach. Consequently, the ratio of maximum to mean depth in the revetted channel is actually lower than that in the alluvial reach, although average depth may itself be greater.

The best-fit line for clay plug bends cuts across the other lines at a much lower slope, showing that in these bends the ratio of maximum scour depth to mean crossing depth decreases much less quickly with increasing R_c/w. This is a result of the nature of clay plugs. These are resistant deposits of limited extent in the outer bank. They characteristically produce 'hard points' that limit the rate of bank retreat, but also disturb the flow and generate local scour of the bed adjacent to, or just downstream of, the plug. In such bends, the deepest pool is associated with local scour at the clay plug and this may be at any point in the bend, depending on the location of the hard point.

By contrast, the deepest pool in free meanders invariably was close to or just downstream of the bend exit, where fully developed bend flow attacks the outer bank and the bed at the bank toe. Hence, maximum scour depth at a clay plug bend depends strongly on the depth of local scour generated by the hard point or points in the bend, rather than the bend geometry itself. It is quite possible for deep scour to be generated by a particularly effective hard point in a long radius bend, or possibly even in a straight reach in some circumstances, and this is the cause of the low gradient of the clay plug line in Figure 4.6.

From the plot and relation for free meanders it appears that the scour pool depth approaches a value of about 1.5 times the mean depth as the R_c/w becomes large (for example $R_c/w = 25$), that is, in almost straight reaches. Comparison of observed maximum depths with mean depths for straight reaches (crossings) yields a ratio of 1.46 for 140 data points. This indicates that the scour pool/mean depth ratio for a bend does approach that for a straight reach as bend effects disappear, as would be expected. For revetted bends the ratio of scour pool depth to mean depth at a high value of $R_c/w = 25$, is 1.3, suggesting that in nearly straight reaches there is a tendency for revetted reaches to have more uniform cross-sections than free alluvial reaches. This again is as expected.

It should be remembered that all these depths are based on bed elevations observed in a low flow hydrographic survey. Usually, the bed is scoured deeper during high flows, with scouring being strongest in the pools of bends and weakest at crossings. Consequently, since no allowance has been made for bed scour during floods, these relationships probably are conservative. It would be interesting to predict the distribution of scour during floods using a suitable mobile-bed model for water and sediment routing and then add the effects of flood scouring on to the figures here. However, this would require considerable research effort.

The results of this part of the study do seem to make sense and support the hypothesis that as the erosion resistance of the outer bank increases, so does the depth of the scour pool, the amount of extra scouring being a systematic function of R_c/w. However, the coefficients of determination indicate that other factors also may be important and could contribute significantly to scour pool depth variation.

IMPLICATIONS FOR RIVER TRAINING

Background

The US Army Corps of Engineers is currently undertaking a multi-million dollar project to make the Red River navigable up to Shreveport. This involves construction of five locks and dams to provide a 9 ft (2.75 m) deep navigation channel. Because they lower the water surface gradient, the dams tend to induce deposition of sediment carried by the flow, and this sometimes requires dredging to maintain the navigation channel. As the sediment load of the Red River at Shreveport is high, it is expected that the dredging requirement may be considerable when all the locks and dams are in place in the mid-1990s. For comparison, some serious sediment-related problems have been encountered on the Arkansas River, which for 15 years has had a navigation scheme somewhat similar to that being constructed on the Red. The sediment load concentrations in the Red River are an order of magnitude greater than those on the Arkansas.

It presently is not planned to extend navigation upstream of Shreveport into the study reach, but it has been estimated that the sediment yield from bank erosion in the study reach may be of the order of 65 million metrics tons per year (Biedenharn, pers. comm., 1988). Although a large percentage of this sediment goes into point and middle bar storage (Water Engineering and Technology Inc., 1987), it is logical to conclude that bank erosion is a major source of material in the sediment load at Shreveport. This is consistent with the observed increase in the annual sediment discharge from 20 to 33 million metric tons per year between Fulton and Shreveport (Raphelt, pers. comm., 1986). However, it is notoriously difficult to estimate sediment deposition or erosion from the difference between two rating curves (Water Engineering and Technology Inc., 1987) and so Raphelt's conclusion should be accepted cautiously.

It also is noteworthy that the grain size of sediment derived from bank erosion is considerably smaller than that derived from bed scour. Material in transport in a river that is finer than that found in the bed is termed 'wash load'. In a study of wash load in the Red River, Yu and Wolman (1986) found that 90% of this type of sediment came from bank erosion. Recent experience at Lock and Dam 1 on the Red River suggests that it is the finer fraction in the sediment load that poses the most severe problems for efficient operation and maintenance of the navigation project. Consequently, a sound understanding of the processes and mechanisms controlling the bank erosion that inputs this sediment is important when selecting the best approach to controlling and reducing the dredging requirement in the navigation reach.

Theoretically, one way to reduce sediment-related problems would be to reduce the sediment load at the head of navigation at Shreveport, by stabilizing the banks in the Index–Shreveport reach. Intuitively, this follows because bank stabilization cuts off the source of the sediment. However, bank stabilization might not lead directly to sediment load reduction, if some side effects cause an increase in sediment supply from sources other than the banks. There are three major impacts of bank stabilization that may promote scour that could offset the reduction in sediment availability from the banks:

(1) increased bed scour in revetted bends;
(2) reduced sediment storage capacity in crossing bars;
(3) enhanced sediment transport capacity owing to channel realignment.

Each of these impacts is considered briefly in the following sections.

Increased Bend Scour in Revetted Bends

For a typical bend with an $R_c/w = 3$, and a mean crossing depth (XD_{bar}) of 5.2 m, the maximum scour pool depth would be 10.3 m in a free meander but 11.2 m in a revetted meander. This represents an increase of about 9%. For a long radius bend, with $R_c/w = 6$, the relevant depths would be 9 m and 9.2 m, respectively, a difference of only 2%. For a tight meander, however, with $R_c/w = 2.1$, the scour depths rise to 12.2 m and 14.3 m, respectively, an increase of 17% for the revetted bend. As a first approximation, the maximum scour depth in revetted bends might be expected to be about 10–20% greater than that for the equivalent free meander, the difference decreasing as R_c/w increases. Unless this increased toe scour has been accounted for correctly in the design of the revetment, failure may occur by launching.

If one of the main purposes of revetting the banks and stabilizing the meanders is to reduce the sediment load, then the question arises as to whether bank stabilization will simply substitute bed sediments for bank sediments with little short-term reduction in sediment production.

Reduced Sediment Storage Capacity in Crossing Bars

Often, a series of bends is stabilized by revetting of the outer banks in each bend. Current practice is to continue the revetment well downstream of the bend exit, because experience shows that the point of maximum attack on the outer bank migrates downstream almost to the meander inflection point at times of high in-bank flow. Usually then, the revetment for one bend overlaps with that for the next bend downstream. Under these circumstances, Friedkin's 1945 flume study demonstrated that the capacity of the channel to store sediment in the mid-channel bar located in the crossing between bends is reduced. Consequently, crossing bed elevations are lowered with the benefits of lowered water levels during floods and increased navigation depths at low flows, but at the cost of the more rapid transmission of sediments downstream through the system. This means that the capacity of the channel to *store* sediment being input from upstream is reduced, so that the sediment *output* downstream may increase.

If the primary aim of revetting bends in the study reach is to reduce the sediment load at Shreveport, then if significant, the effect of reducing crossing storage of sediment would be counterproductive.

Enhanced Sediment Transport Capcity Owing to Channel Realignment

Sometimes bends are revetted along their existing alignment, but more usually the channel is realigned to produce a lower sinuosity and a shorter crossing between bends. This may involve a relatively small reduction in sinuosity, through channel training, or a major reduction in sinuosity, through artificial neck cut-off of an entire meander loop. Realignment complicates the relation between bank stabilization and reduced sediment load for two reasons.

First, it increases the channel gradient. Sediment transport capacity is known to be a power function of channel slope. Hence increasing the slope of a channel usually produces a large increase in sediment load, the bed and banks of the channel being eroded to supply the sediment. In the case of the Red River, bank stabilization precludes bank erosion and so bed scour alone would result. This could more than offset the tendency for a reduction in sediment load owing to bank protection, so that the net effect of bank stabilization and channel realignment might be an increase in sediment load at Shreveport.

Second, it reduces flow resistance owing to channel non-uniformity. Meander bends are known to cause resistance to flow (Bagnold, 1960), and this can be a considerable contribution to overall resistance in tortuous bends. Hence, removing the bends reduces overall flow resistance and increases the flow velocity (and, therefore the sediment transport capacity) even more than would be expected from the increase in channel gradient. When a tortuous channel is straightened, velocity increases of the order of 50% have been observed, 20% being attributed to increased gradient and 30% to reduced flow resistance (Biedenharn, pers. comm., 1989).

Discussion

It appears that bank stabilization in bends of the Red River should be expected to add between 5 and 20% to the scour pool depths in those bends. This is not a negligible amount and it could add significantly to the sediment load output from a bend while the additional scouring is actually occurring. Once bed levels have restabilized the sediment output should decrease somewhat, but it seems unlikely that it will ever return to its pre-revetment level.

Transmission of the sediment downstream involves its movement through further crossings and bends. Present experience is that increased storage at crossings will not occur; neither is there clear evidence of increased scour (Thorne, 1989). Storage in subsequent bends also is strictly limited. This is the case because point-bar growth in revetted bends ceases when retreat of the outer bank opposite prevents further advance of the inner bank, through constricting the flow. From these considerations, it must be concluded that the bed material load derived from extra pool scouring will make its way downstream through the reach relatively quickly, arriving in due course at the head of the navigation reach.

The calibre of the sediment yielded from bed scour rather than bank erosion may be quite different. Generally, bed sediment in this reach of the Red River is much coarser than bank sediment. Consequently, although its movement will be quick for bed material load, it will still move more slowly than the present 'wash load' from bank sediments and will, therefore, take longer to reach the navigation reach. Also, its coarser size will cause it to be deposited in different locations in the system. Specifically, bed material load should be expected to form deltaic type deposits at the head of navigation pools, while the wash load falls out in lock chambers and behind dams. If it is lock and dam sedimentation that poses the more serious problem, then this should be a benefit.

If the stabilization of the banks also involves a significant increase in channel slope and a decrease in flow resistance through the reduction of sinuosity by channel realignment, then sediment transmission will be accelerated. In addition, there may be further bed scour associated with increased transport capacity owing to the increased slope and decreased resistance, even after that caused by bank stabilization has ceased. The question of channel response to realignment is an important one, but it is beyond the scope of this study. However, the results reported here do suggest lines of research that might form the basis for a predictive model of channel response to realignment for large lowland rivers.

CONCLUSIONS AND RECOMMENDATIONS

(1) On the Red River bank retreat in the study reach takes place by basal erosion, which triggers mass failure, followed by basal clean-out of failed material.

(2) Bank geometry and processes are affected by bank material properties. On this basis banks may be classified according to their origin as: meander belt alluvium, clay plug, back-swamp deposit and Pleistocene.

(3) The dominant failure mechanisms are rotational slips and slab-failures. Generally, slabs occur on meander belt alluvium and clay plug banks and rotational slips on back swamp and Pleistocene banks, but with many exceptions to this rule.

(4) The stability analyses of Osman and Thorne can be used to predict the cricital geometry (height and angle) for the different bank types using reasonable estimates of the engineering properties. However, it appears that the operational strength of the Pleistocene materials is about half that suggested from conventional engineering analyses.

(5) A relationship can be demonstrated between outer bank properties and scour depth at a bend. Scour depth increases as bank resistance increases from alluvium to clay plug to Pleistocene to revetted banks. The data were insufficient to analyse back-swamp banks in this regard.

(6) Generally, the influence of outer bank properties on scour depth increases as the radius of curvature to width ratio decreases (Figure 4.6). There is a marked discontinuity at $R_c/w = 2$, consistent with the observation made elsewhere that the flow pattern at a bend changes radically in such tight bends. $R_c/w = 2$ therefore forms a real, physical boundary to the applicability of these results.

(7) Maximum scour depths in revetted bends are greater than those in the equivalent free meanders. As a rule, maximum depths are 10–20% greater.

(8) Stabilization of the remaining active meanders of the Red River in the study reach can be expected to promote bend scour and to increase scour depths in pools at bends. The sediment so derived will be transmitted downstream to arrive in the navigation reach. Its coarser size will promote pool-head deposition rather than sedimentation in locks and at dams.

(9) Significant increase of slope and decrease of flow resistance caused by realignment of the channel could generate additional scour over and above that associated with bank stabilization alone. This should be taken into account when planning engineering measures in the reach if the aim is to reduce the sediment supply to the navigation reach.

SUMMARY

Studies of bank and bend processes were undertaken on the Red River between Index, Arkansas and Shreveport, Louisiana. The aims were to determine: the nature of the bank materials and the dominant mechanisms of bank failure; the role of outer bank type in affecting the scour depth of meander bends; and the implications for bank stabilization and river engineering schemes.

The results show that the banks are formed in materials of four different origins: meander belt alluvium and clay plug materials associated with the present floodplain; back-swamp deposits in a terrace left from the nineteenth century floodplain; and Pleistocene/Tertiary materials in the valley walls. Slab failures and rotational slips are the dominant failure modes.

The Osman–Thorne analyses of bank stability were found to be useful in predicting the critical geometry for failure of banks formed in the different materials.

A relationship between outer bank type and bend scour depth was found. Generally, scour depth increases with outer bank resistance to erosion and failure, especially for tight bends of low radius of curvature to width ratio. Maximum scour pool depths for revetted bends with non-erodible outer banks are 5–20% greater than those in an equivalent free, alluvial meander.

The US Army Corps of Engineers is currently constructing a multi-million dollar scheme to make the Red River below Shreveport navigable. Concerns have been expressed regarding sedimentation problems in the navigation reach caused by the heavy sediment input from upstream. It is believed that bank erosion in the study reach between Index, Arkansas and Shreveport, Louisiana is a major source of this sediment. The results of this study suggest that some additional bend scour should be expected as a consequence of stabilizing the remaining free meanders in the study reach, promoting delta type deposition of coarse sediment at the head of pools in the navigation reach downstream, but reducing sedimentation of fines in the lock chambers and behind the dams.

ACKNOWLEDGEMENTS

The work reported here was undertaken while the author was a visiting scientist at the Hydraulics Laboratory of the Waterways Experiment Station (WES), Vicksburg, Mississippi on leave from Queen Mary College, University of London. The help and encouragement of the staff at WES, and particularly Dr Bobby Brown and Terry Waller, is gratefully acknowledged. Much time was spent at the Vicksburg District Offices and there the staff of the Hydraulics Branch all lent me support and advice. I am particularly grateful to Phil Combs, David Biedenharn, Tim Hubbard, Charlie Little, Charlie Montague, John Watkins, Glenda Hill and Clara Pinkstone. My postgraduate students, Lisa Hubbard and Andrew Markham, both took time off from their own research projects to help me with this one, and their help was invaluable.

REFERENCES

Bagnold, R. A. (1960). Some aspects of the shape of river meanders. *U.S. Geological Survey Professional Paper*, **282-E**, 135–144.
Biedenharn, D. S., C. D. Little and C. R. Thorne (1987). Magnitude and frequency analysis in large rivers. In *Hydraulic Engineering*, Proceedings of the Williamsburg Symposium, American Society of Civil Engineers, August 1987, pp. 782–787.
Biedenharn, D. S., P. G. Combs, G. J. Hill, C. F. Pinkard and C. B. Pinkstone (1989). Relationship between channel migration and radius of curvature on the Red River. In S. S. Wang (ed.), *Sediment Transport Modeling*, Proceedings of the International Symposium, American Society of Civil Engineers, Hydraulics Division, New Orleans, August 1989, pp. 536–541.
Friedkin, J. F. (1945). *A Laboratory Study of the Meandering of Alluvial Rivers*, US Army Corps of Engineers, MS 39180, 40 pp.

Markham, M. J. (1990). *Flow and sediment processes in gravel-bed river bends.* Unpublished PhD thesis, Queen Mary College, University of London, 415 pp.

McCall, E. (1988). The attack on the Great River Raft. *American Heritage of Invention and Technology*, **Winter**, 10–16.

Nanson, G. C. and E. J. Hickin (1983). Channel migration and incision on the Beatton River. *Journal of the Hydraulics Division, American Society of Civil Engineers*, **109**, 327–337.

Osman, A. M. and C. R. Thorne (1988). Riverbank stability analysis: Part I Theory. *Journal of the Hydraulics Division, American Society of Civil Engineers*, **114**, 125–150.

Reid, J. B. (1984). Artificially induced concave bank deposition as a means of floodplain erosion control. In *River Meandering*, Proceedings of the Conference on Rivers '83, American Society of Civil Engineers, New Orleans, pp. 295–305.

Thorne, C. R. (1982). Processes and mechanisms of bank erosion. In R. D. Hey, J. C. Bathurst and C. R. Thorne (eds), *Gravel-Bed Rivers*, Wiley, pp. 227–271.

Thorne, C. R. (1988). *Analysis of Bank Stability in the DEC Watersheds, Mississippi*, Final report to the US Army European Research Office, London, England, under contract No. UA45-87-C-0021, 40 pp.

Thorne, C. R. (1989). *Bank Processes on the Red River between Index, Arkansas and Shreveport, Louisiana*, Final report to the US Army European Research Office, London, England, under contract No. DAJ45-88-C-0018, 45 pp.

Thorne, C. R. and M. A. Osman (1988). Riverbank stability analysis: Part II Applications. *Journal of the Hydraulics Division, American Society of Civil Engineers*, **114**, pp. 151–172.

Veatch, A. C. (1906). Geology of the underground water resources of northern Louisiana and southern Arkansas. *US Geological Survey Professional Paper*, **46**, 394 pp.

Water Engineering and Technology Inc. (1987). *Geomorphic and hydraulic analysis of the Red River from Shreveport, Louisiana to Denison Dam, Texas*, Report to Vicksburg District, Corps of Engineers, under contract number DACW38-86-D-0062, 207 pp.

Yu, B. and M. G. Wolman (1986). Bank erosion and related washload transport on the Lower Red River, Louisiana. *Third International Symposium on River Sedimentation*, Jackson, Mississippi, March 1986, pp. 1277–1285.

5 Process Dominance in Bank Erosion Systems

D. M. LAWLER
School of Geography, The University of Birmingham

INTRODUCTION

'The major controls on bank erosion remain unclear at present' (Hasegawa, 1989, p. 219), and this is reflected in the divergence of conclusions concerning the dominant mechanisms of bank retreat, as embodied in the following quotations:

> Banks retreated primarily by mass failures of overheightened and oversteepened banks.
>
> > (Little, Thorne and Murphey, 1982, p. 1321)

> The shearing of bank material by hydraulic action at high discharges is a most effective process, especially on non-cohesive banks and against bank projections.
>
> > (Knighton, 1984, p. 61)

> The erosion of a (river) bank is not the result of erosion by high-velocity water, whether in a concave bank on a curve or in a straight section. Rather, for effective erosion to occur, the material must be loosened—which in this stream, is done by formation of ice crystals in winter.
>
> > (Leopold, 1973, p. 1850)

Many authors argue that it is *combinations* of processes that are important (e.g. Hooke, 1979; Thorne, 1982; Lawler, 1987). Perhaps, in view of the wide range of alluvial materials, riverine forms and hydroclimatic environments encountered, a variety of conclusions is to be expected. However, it is argued below that the following three factors also have to be considered when trying to disentangle competing hypotheses of process-dominance in river bank erosion systems:

(1) limitations of present field monitoring techniques;
(2) temporal change in bank erodibility;
(3) downstream change in bank erosion processes.

Lowland Floodplain Rivers: Geomorphological Perspectives. Edited by P. A. Carling and G. E. Petts
© 1992 John Wiley & Sons Ltd

LIMITATIONS OF BANK EROSION MONITORING SYSTEMS

Nature of the Problem

Considerable progress has been achieved in bank erosion process work through building and testing models of bank retreat rates in relation to *spatially distributed* hydraulic variables in meander bends (e.g. Pizzuto and Meckelnburg's (1989) test of the Parker, Diplas and Akiyama (1983) form of the St Venant equation on the Brandywine Creek, Pennsylvania). Another route to explanation, though, is through the *temporal* approach, in which short-term (e.g. individual event, monthly, or seasonal) changes in erosion rate at individual sites are interpreted in terms of temporal fluctuations in the stresses applied by the suspected influential variables (e.g. Hooke, 1979; Kesel and Baumann, 1981; Gardiner, 1983; Lawler, 1986).

One major problem with the latter methodology at present, however, is that no *automatic*, quasi-continuous, erosion monitoring technique has been available to quantify the erosional and depositional impact of each event that affects the site. Existing methods of field monitoring (see Thorne (1981) and Lawler (1991a) for reviews) are based on some form of *manual* resurvey method (e.g. repeated bank profiling (Hudson, 1982; Pizzuto and Meckelnburg, 1989) or erosion pins (Lewin, 1981; Lawler, 1991a)) and simply reveal the net retreat or advance of a bank face since the previous measurement. It is often impossible to ascribe, on a routine basis, the geomorphological work done by a specific flow event, as there may have been many events in a particular measurement interval. By definition, then, these techniques yield data of much lower temporal resolution than commonly is available for the suspected controlling variables (e.g. discharge). This makes process inference and model building extremely difficult, and the calibration and testing of event-based bank erosion models virtually impossible.

The Photo-Electronic Erosion Pin (PEEP) System

To tackle this problem, the author recently has developed the Photo-Electronic Erosion Pin (PEEP) monitoring system, which provides, for the first time, automatic and quasi-continuous information on bank erosion and deposition (Lawler, 1989, 1991b,c). The PEEP sensor is an optoelectronic device consisting of a row of solar cells connected in series and encased in a transparent acrylic tube. A reference cell allows outputs to be normalized for changing ambient illumination (Lawler, 1991c). The sensor is inserted into the eroding bank—like the traditional erosion pin—with the cables taken out through the bank interior to avoid possible bank-surface fouling. As erosion proceeds and the bank face retreats so more solar cells are exposed to light. This increases the voltage output to a data logger, which, housed in an environmental enclosure nearby (Figure 5.1), is programmed to scan the sensors at a desired frequency (e.g. 30 min intervals). Peaks, ramps and troughs in the logged signal thus reveal the magnitude, frequency and timing of erosional and depositional activity much more precisely than has been possible hitherto, and reduces the need for time-consuming repetitive field visits. Preliminary work, the design and installation of PEEP systems, and examples of usage are discussed, respectively, in Lawler (1989; 1991c; 1991b). A prototype PEEP sensor, together with a thermistor for continuous information on bank-surface thermal regime, can be seen in Figure 5.2.

Figure 5.1 The PEEP bank erosion monitoring site on the River Arrow at Studley, Warwickshire. Visible portion of bank is around 2 m in height, and flow is from right to left. Note erosional notch just above present river level

Figure 5.2 The PEEP prototype sensor (no. 1) emerging from bank face (lower left) at River Arrow monitoring site, with thermistor for bank surface temperature measurement (to right). Lens cap for scale is 55 m in diameter

Installation of Prototype Sensors

Following field trials in spring 1988, two sensors (PEEP2 and PEEP3) were installed in January 1989 at two different heights on the right bank of the River Arrow, Warwickshire, UK (Figure 5.1). Sensor PEEP2 was inserted around 20 cm above 'normal' winter river level (at stage 80 cm) to detect the erosional and depositional impact of relatively frequent events, with PEEP3 approximately 40 cm higher, and just upstream, at stage 120 cm. The river here has a gravel bed and is cut into fine-grained, cohesive, material that overlies stiff clay basal units (Figure 5.1). Drainage basin area at the monitoring site is 98 km^2, bankfull channel width is around 12 m, and bankfull channel depth is around 2.2 m (Figure 5.1). Reconnaissance indicated that a wide range of flows can cause bank erosion here: note the clear notch (of about 25 mm depth) less than 30 cm above low water level (Figure 5.1).

Using PEEP Systems to Detect Erosion

Figure 5.3 illustrates the typical diurnal data series generated by a PEEP sensor during an erosion event. The reference cell signal indicates reasonably uniform peak light incidence during the 7-day sequence (although light reception is extinguished almost completely when the sensor is submerged by highly turbid water during the flood event (Figure 5.3)). The cell series output, however, which shows how much sensor

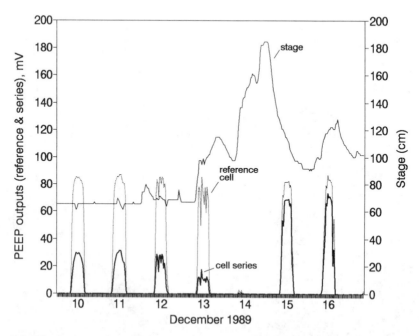

Figure 5.3 Typical erosion event detected by the PEEP system at the Arrow site. Note fairly constant diurnal peaks in reference cell outputs, but marked increase in cell series outputs because of greater tube exposure (equivalent to bank retreat of 40 mm, as 1 mV ~ 1 mm erosion) following the flow event of 14–15 December 1989

tube is revealed at any given moment, after initially declining on the 13 December (probably owing to the collection of rain-washed material on the tube), rises dramatically after the passage of the main flow event on the 15 December. The difference between pre- and post-flood outputs represents an erosion event of around 40 mm (Figure 5.3). One great advantage of this monitoring system, then, is its ability to define the erosional or depositional impact of every event that affects the site, and this can substantially enhance magnitude–frequency analyses, which normally and necessarily are built around temporally 'lumped' data. In future, therefore, it should be possible to perform a time-based test of the 'excess shear stress' erosion-rate predictions (as opposed to the spatial approach of Parker, Diplas and Akiyama (1983) and Pizzuto and Meckelnburg (1989)) in which variable flow stresses at single sites, and not just the bankfull condition, can be handled.

A further advantage of the system is its capability to define, within the constraints of the scan-interval selected and periods of data loss at night or during submergence in turbid water, the precise moment of material removal. This facility should be very useful in disentangling competing hypotheses of process-dominance in a given situation. Note how in Figure 5.4, for example, material removal is known to have started by 0630 GMT on the 10 August 1989 and that all 22 mm of the bank retreat recorded was complete by 0700 GMT—only 1 h after the flow peak. On this occasion, therefore, we can discount time-lagged 'falling-stage' failure processes associated with drawdown conditions (e.g. Bradford and Piest, 1977), and look to other mechanisms (e.g. direct fluid entrainment).

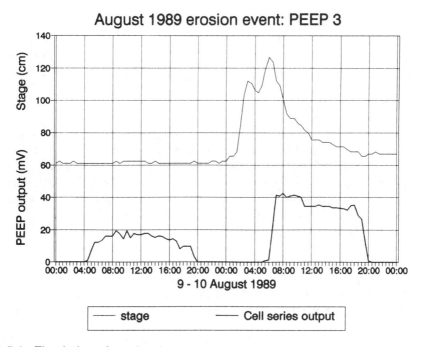

Figure 5.4 The timing of erosional activity revealed by the automatic PEEP monitoring system at the Arrow site. The erosion event is complete by 0700 GMT on 10 August 1989. Reference cell outputs removed for clarity

TEMPORAL CHANGE IN BANK ERODIBILITY

Force and Resistance in Bank Erosion Systems

Recent vigorous research activity on the measurement and modelling of bend flow hydraulics and the distribution of fluid stresses on river banks (e.g. Ikeda, Parker and Sawai, 1981; Bathurst, 1982; Hey and Thorne, 1984; Lapointe and Carson, 1986; Odgaard, 1986a,b; Ikeda and Parker, 1989; Pizzuto and Meckelnburg, 1989) has not been matched by a similar level of interest in the *resistance* side of the bank erosion equation. This relative neglect of bank erodibility, and especially the patterns and controls of spatio-temporal change in erodibility, urgently needs to be addressed. This is particularly important in the case of cohesive materials, whose erodibility is a function of a complex combination of physico-chemical bonding processes that act to control the resistance of a particle or aggregate to detachment by fluid shear applied at the sediment surface. Useful discussions of cohesive material erodibility appear in Thorne (1978, pp. 67–81), Grissinger (1982), Kamphuis and Hall (1983), Nickel (1983), Osman and Thorne (1988) and Thorne and Osman (1988). This work demonstrates the importance of, for example, the temperature, pH and salinity of the pore and eroding fluids, as well as the more commonly appreciated variables of material water content, grain-size profile and soil mechanics properties.

Heterogeneity in alluvial materials will of course introduce considerable vertical and reach- and basin-scale *spatial* complexity in bank erodibility, which can influence longer term patterns of meander development (Parker, Diplas and Akiyama, 1983; Lewin, 1987; Pizzuto and Meckelnburg, 1989). Less attention, however, has been paid to *temporal* variations in erodibility at individual sites, associated with, for example, seasonal vegetation growth, bank moisture changes (Knighton, 1973; Hooke, 1979), desiccation processes (Bello *et al.*, 1978) and freeze–thaw processes (Leopold, 1973; Gardiner, 1983; Lawler, 1986,1987) (Table 5.1). The research gap is particularly noticeable for the water quality parameters mentioned above. Scatter in the temporal relationships between flow variables and at-a-site bank erosion response is common (e.g. Knighton, 1983, Figure 3; Hooke, 1977, Figure 6.16; Lawler, 1984,1986). As Knighton (1973, p. 421) put it: 'discharges of different frequency and magnitude could produce the same amount of erosion'. Some random scatter will, of course, be introduced by, for example, the problem of progressive incremental undercutting of cantilevers, which eventually may fail in association with quite small events, i.e. 'carry over effects' (Lawler, 1986). However, it is clear also that other variables are at play that influence the resistance that a given bank material can offer at a given time to a given level of applied stress.

Detecting Erosion Variability

Because the erosional impact of each flow event can be defined automatically with the PEEP system, the technique is well-placed to help unravel some of this complexity. Note how in Figure 5.5a, for example, a *single* flow event on 10 August 1989 effects 22 mm of bank retreat around PEEP sensor 3, as shown by the sharp jump in cell series outputs. Four months earlier, however, a whole *series* of similarly high stage rises failed to cause any erosion at all (Figure 5.6).

Table 5.1 Bank erosion studies demonstrating the importance of freeze–thaw processes

Reference	River	Location	Country	Drainage basin area (km^2)
Hill (1973)	Clady and Crawfordsburn	Northern Ireland	UK	3.4
Curr (1984)	Corston Brook	Avon, England	UK	4.1
Lawler (1986, 1987)	Ilston	Gower, South Wales	UK	6.8–13.2
Stott et al. (1986)	Kirkton Glen	Balquhidder, Scotland	UK	<7.7
Wolman (1959) and Leopold (1973)	Watts Branch	Maryland	USA	9.6
Blacknell (1981)	Afon Crewi	Mid-Wales	UK	35.5
McGreal and Gardiner (1977) and Gardiner (1983)	Lagan	Northern Ireland	UK	85

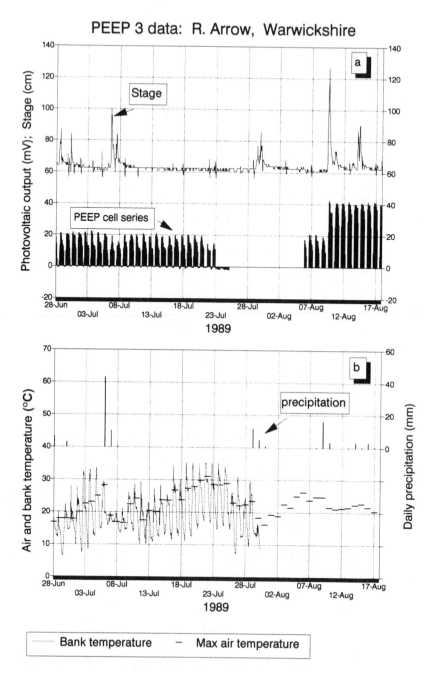

Figure 5.5 Antecedent conditions for 10 August 1989 event: (a) stage and PEEP 3 cell series outputs (some data lost from 26 July to 6 August); (b) bank surface temperature data at the Arrow site, with daily precipitation and maximum air-screen air temperature for Edgbaston meteorological station, 20 km to the north. Note that the bank temperature sensing system has a 'ceiling' response of 35° C

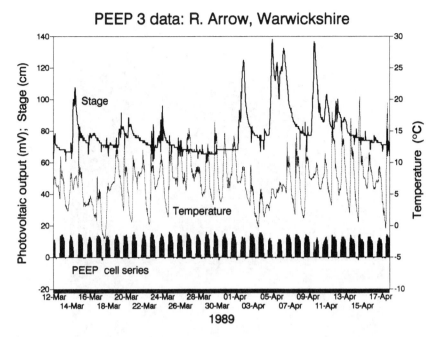

Figure 5.6   An absence of erosional activity in March–April 1989, shown by constant diurnal peaks in PEEP signals, despite a sequence of high-flow events. Bank temperature trace is shown also

Although firm conclusions must await further data from networks of such sensors, it is tempting to explain this difference with reference to erodibility concepts. Summer 1989 was exceptionally hot and dry in southern Britain (Jones and Hulme, 1990). Bank surface temperatures regularly rose above 30° C, and occasionally over 35° C, especially in late July, 2 weeks before the erosional event (Figure 5.5b). The cohesive materials of the bank surface became intensely desiccated, with considerable polygonal cracking evident and much aggregate material spalled from the upper bank units to collect in mini-talus slopes towards the bank toe. Examples of such effects from summer 1990 are shown in Figures 5.7 and 5.8. Two accumulation events around a PEEP sensor in 1989 are shown in Figure 5.9 by pronounced drops in the cell series outputs as light to the photodiodes is first reduced (event A, at around 0930 GMT on 22 July), then shut off altogether (event B, at 1330 GMT on 23 July). This appears to be in response to intense early morning heating of these east facing upper bank faces, often at a rate of 7° C h^{-1}, which leads to diurnal temperature ranges in excess of 20° C (Figure 5.9). It is believed that these trends are reasonably representative of warm summer conditions in British lowland rivers, but similar continuous data on bank thermal regimes from other rivers are not available for comparison at present. The large cracks that widen as summer progresses are accompanied by exfoliation processes, which cause large slabs of material to become partially detached from the main bank (Figure 5.7). This activity probably allows water in subsequent flow events to flow behind the exfoliating surface, and enhances entrainment of material in the form of whole peds or particle aggregates, rather than as individual grains.

Figure 5.7 Highly cracked and desiccated bank face at the River Arrow site in summer 1990. Approximately 1 m height of bank is visible

In the period prior to the April 1989 flow events, however, bank surface temperatures seldom exceeded 15° C or dropped below 0.0° C, and precipitation receipts were regular, but not large (Figure 5.6). It is tentatively suggested, therefore, that to account for the zero erosional response to the multiple flow events depicted in Figure 5.6, the banks were offering maximum entrainment resistance at this time. This low erodibility may be related to moderate bank moisture contents and subdued thermal cycling in the antecedent period, which prevented either desiccation (Bello *et al.*, 1978) or freeze–thaw (Hill, 1973; Gardiner, 1983; Lawler, 1986,1987; Stott *et al.*, 1986) processes to operate effectively. Also, an absence of erosion, even with the later events of the sequence (Figure 5.6), suggests that fluvial pre-wetting is ineffective in this system (cf. Knighton, 1973).

If preconditioning processes are accepted as significant in the dynamics of cohesive material bank erosion, then two interesting questions arise, namely: which prepara-

Figure 5.8 Accumulations of particle-aggregates in the vegetated bank foot zone in summer 1990. This debris is produced by desiccational activity on the upper slopes but, if not fluvially entrained shortly after deposition, subsequent rainfall can induce amalgamation into the lower bank units. Approximately 30 cm of tape is visible

tory processes or erodibility factors are the most important for a given event or site, and how do micrometeorological, hydrological, fluvial and sedimentological conditions interact to control temporal change in their relative efficacy? Again, automatic field erosion monitoring systems should help us to build, calibrate and test appropriate models more rigorously than before.

For instance, if it were desirable to test the hypothesis that pre-wetted banks were essential for significant erosion then one could examine the impact of complex multi-peaked flow events (difficult to do this routinely in the absence of automatic methods). For us to accept the hypothesis, the later events in the sequence, acting on banks whose moisture stores had been replenished by immediately preceding stage rises (through bank storage—see Sharp (1977)), should be associated with the greatest erosion. Conversely, the early events would be characterized by the highest erosion if other preparational processes (e.g. desiccation) were important. Finally, if preconditioning mechanisms were not important at all (and the problem reduces simply to one of excess shear stress) then comparable flow events should effect broadly similar amounts of bank retreat, nowithstanding random scatter and measurement deficiencies. It should be stressed that the idea of 'comparable' flow events is more complex than might at first appear: although in terms of, say, peak stage or flow duration, two events may be judged to be 'similar', (1) the distribution of shear on the bank surface may be quite different for each, because of intervening channel

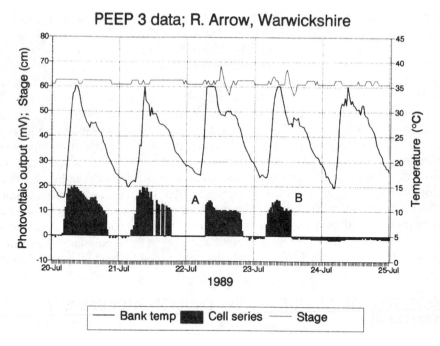

Figure 5.9 Two desiccation–accumulation events (A and B) picked up by the PEEP system as cell series output reductions, in relation to strong bank surface heating in July 1989. Note that the temperature sensing system has a 'ceiling' response of 35° C

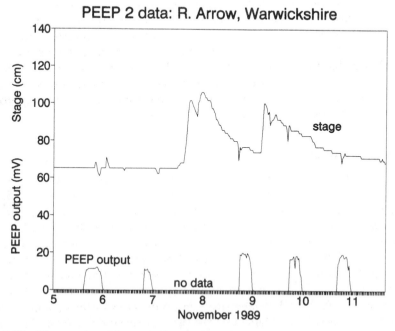

Figure 5.10 The erosional impact of a double-peaked hydrograph event in November 1989: note that all the 10 mm of bank retreat is achieved by the first flow event

geometry or micro-roughness changes, and (2) the pH, temperature and salinity ratios of the pore and eroding fluids may vary between events.

It is too early to attempt a definitive discriminatory explanation from the River Arrow data set as yet, but one example serves to outline some interesting possibilities. Figure 5.10 shows how, for a double-peaked hydrograph in November 1989, all of the resultant erosion is associated with the *first* event—evidence that, along with Figure 5.6, would help to refute the pre-wetting hypothesis. It would also support the idea of bank erosion being a supply-limited rather than transport-limited process here. Interestingly, the pattern of Figure 5.10 correlates well with the oft-cited suspended sediment exhaustion effect through closely spaced storms (e.g. Walling and Webb, 1981, their Figure 5.20), and raises the exciting possibility of coupling networks of automatically recording turbidity meters and PEEP systems in the same catchment to investigate the dynamic relationships between suspended sediment fluxes and changes in channel-side sediment supply. Furthermore, a similar arrangement deployed at the bend scale could help to test the Lapointe and Carson (1986) model of flow trajectories, bank material supply and turbidity currents through individual meander systems.

DOWNSTREAM CHANGE IN BANK EROSION PROCESSES: A HYPOTHESIS

Whether or not the absolute rate of lateral channel change increases in a downstream direction with width and drainage area (Hooke, 1980; Hasegawa, 1989) or peaks in the middle courses (Lewin, 1987), exactly how the pattern is achieved needs to be explored. Is a given set of bank processes operating in the headwater reaches simply scaled up in intensity, frequency and/or duration in a downstream direction to account for enhanced erosion rates, or are new, more dynamic, processes introduced at various points downstream as (scale) thresholds are crossed? Although the data are simply not available to test these alternatives, the latter idea is explored below, in a deliberately speculative way, through an examination of possible downstream change in the relative efficacy of three bank-process groups: subaerial preparation processes, fluid entrainment, and mass failure.

Subaerial Preparation Processes

There are reasonable theoretical grounds to suggest that surface preparation processes may decline in relative importance in a downstream direction, and attain most significance in headwater/smaller channels. Subaerial processes in general, and freeze–thaw and desiccational activity in particular, may be thought of as largely extraneous to the immediate 'fluvial' system. In other words, being related to imposed microclimatic and moisture-balance controls that are dependent only partially on channel hydraulics and hydrology, these processes may not be influenced greatly by internal two-way adjustments between flow properties and channel form. As such, the effects may not be 'scaled up' in intensity as one moves downstream, as mean velocity tends to be (Leopold and Maddock, 1953), in association with declines in bed roughness and increases in hydraulic radius. If absolute erosion *rate* increases downstream, but the contribution from subaerial activity remains essentially constant then, *ceteris paribus*, their relative significance will weaken in a down-valley direction. For freeze–thaw activity, at least, this trend may be reinforced by the effects of altitudinal temperature lapse (although freeze–thaw frequencies may actually decrease with increasing elevation in certain cold environments or if inversions are common (Lawler, 1988)).

Direct Fluid Entrainment

Despite recent detailed research on flow patterns at the bend scale, knowledge at the catchment scale of downstream change in the hydraulic properties of rivers, apart from mean section velocity (e.g. Leopold, 1953; Carlston, 1969) is extremely limited. Very few of the 'real' controlling variables, such as near-bank velocities, differentials between channel centreline and boundary layer velocities (cf. Parker, Diplas and Akiyama, 1983; Pizzuto and Meckelnburg, 1989), mean-section or near-bank shear stresses, or stream power, have been investigated in this way (but see Graf, 1982; Lewin, 1983,1987; Ferguson and Ashworth, 1991).

However, at least in some systems (and notwithstanding considerable noise in the signal), stream power seems likely to increase downstream to a maximum and then decline again. For example, let us consider changes in gross bankfull stream power, Ω (W m^{-1}), given as

$$\Omega = \varrho g Q S \tag{5.1}$$

where ϱ is fluid density (1000 kg m^{-3}), g is gravitational acceleration (9.81 m s^{-2}), Q is bankfull discharge (m^3 s^{-1}) and S is channel (energy) slope (m m^{-1}). If it is assumed that ϱ and g are constants, then combining the individual functions for downstream change in Q and S will yield an equation for the downstream change in Ω. In its simplest form, and ignoring any 'stepping' in the trends, both discharge and slope are made linear functions of channel length, L (km), thus:

$$Q = a + bL \tag{5.2}$$

and

$$S = c - dL \tag{5.3}$$

(i.e. a gently concave profile). Then, multiplying out the right-hand terms of equations (5.2) and (5.3), the QS product is given by:

$$QS = ac + (bc - ad)L - bdL^2 \tag{5.4}$$

It follows then, that the downstream distribution of stream power in this case would be a quadratic function that peaks at some intermediate point in the catchment, the exact locus varying with the relative rates of change indicated by the coefficients in equations (5.2) and (5.3). Differentiating equation (5.4) with respect to L yields the downstream rate of change of QS

$$\frac{\delta(QS)}{\delta L} = bc - ad - 2bdL \tag{5.5}$$

Solving equation (5.5) for $\delta(QS)/\delta L = 0$ determines the turning point of the curve and hence the critical channel length, L_0, at which stream power achieves a maximum:

$$L_0 = (bc - ad)/2bd \qquad (5.6)$$

Perhaps a more realistic scenario, however, involves combining discharge as a power function of L

$$Q = kL^m \qquad (5.7)$$

with slope as a negative exponential function of L (Rana, Simons and Mahmood, 1973):

$$S = S_0 e^{-rL} \qquad (5.8)$$

where S is channel slope at any point downstream, S_0 is initial slope at some upstream reference section, and r is the coefficient of slope reduction. Combining equations (5.7) and (5.8) as before gives

$$QS = (kL^m)(S_0 e^{-rL}) \qquad (5.9)$$

which, when differentiated, yields the downstream rate of change, thus

$$\frac{\delta(QS)}{\delta L} = (mkL^{m-1})(S_0 e^{-rL}) + (kL^m)(-rS_0 e^{-rL}) \qquad (5.10)$$

which simplifies to:

$$\frac{\delta(QS)}{\delta L} = kL^m S_0 e^{-rL} \left[\frac{m}{L} - r \right] \qquad (5.11)$$

As only the square bracketed expression in equation (5.11) can assume a zero value, these are the only terms we have to deal with in determining the turning point of the curve, i.e. the critical channel length, L_0, at which stream power achieves a maximum. So,

$$0 = \frac{m}{L_0} - r \qquad (5.12)$$

and

$$L_0 = \frac{m}{r} \qquad (5.13)$$

which is simply the ratio of the two rates of change in the constituent relationships in equations (5.7) and (5.8).

As an example of using the latter approach, the relevant curves have been plotted up in Figure 5.11 using $k = 0.03$ and $m = 1.8$ in equation (5.7) and $S_0 = 0.06$ and $r = 0.15$ in equation (5.8). Note the peak in gross power at a channel length of 12 km, as predicted by equation (5.13). A hypothetical width series (in metres), of the form $w = xL^y$ (where $x = 0.7$ and $y = 0.8$ here), has been added to allow the derivation of unit stream power (Ω/w), in W m^{-2}, which also reaches a maximum just a few kilometres downstream (Figure 5.11).

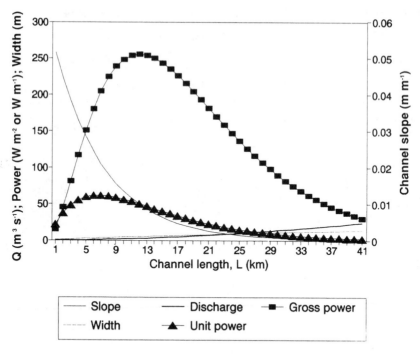

Figure 5.11 Hypothetical downstream stream power distribution, based on combining a negative exponential slope function with power functions for discharge and channel width. See text for details of equation coefficients. Note peaking of both gross power and unit power in intermediate positions

Further numerical experiments using a range of combinations of discharge and slope functions have indicated that a peaking of stream power in mid-basin appears to be the more general case. Of course, if Q and S are both power functions of L, Ω will not peak in mid-basin. Empirical evidence supporting the existence of a stream power maximum in the middle reaches has been published by Lewin (1982). He suggested that, in Wales and the Borderland, 'generally speaking both stream power and rates of floodplain reworking reach a peak in the middle courses of rivers: stream power is less upstream where discharges are very small, and may be less downstream or in lowland rivers where gradients become minimal' (Lewin, 1982, pp. 24–26). Lewin (1983) added further data from the Severn catchment which demonstrated that a peak in unit stream power (i.e. gross power/channel width) can be reached in the middle parts of some basins, a zone that tended also to boast maximum river mobility (Lewin, 1987). Graf (1982) also found unit stream power and tractive force to peak in the middle reaches of a semi-arid catchment. Even the gross stream power data of Graf (1983, his Figure 8), although fitted with a log–linear equation the limitations of which are recognized, show a tendency to peak in mid-basin, and probably are best fitted by a low-order polynomial. It is likely, however, that only a small and variable portion of this energy becomes available for bank erosion (Lewin, 1983).

Mass Failure Processes

Retreat of river banks by mass failure processes (i.e. the delivery of sizeable blocks of material to the stream rather than in the form of individual particles or aggregates) has been modelled previously using slope stability theory (Schofield and Wroth, 1968; Thompson, 1970; Thorne and Tovey, 1981; Little, Thorne and Murphey, 1982), much of it originally developed for hillslopes, dams, embankments and canal banks. Many of the stability formulae contain some kind of scale element whereby, for a slope in a given material of given mechanical properties lying at a particular angle, a maximum stable height may be identified. If this height is exceeded, then the mass of soil above a potential failure surface becomes too great to be supported by the shearing resistance of the material, and instability results. In a fluvial context, and assuming a gradual downstream increase in channel depth, there should be a point at which bank height exceeds the critical value for the particular boundary material, and mass failure begins. Upstream of this point other processes must be responsible for bank erosion and width adjustment, while downstream, bank retreat increasingly becomes a geotechnical problem of slope instability.

One example of a simple stability equation, the Culmann formula (cited in Selby, 1982, p. 138), is based on total, rather than effective, stress principles (ignoring pore water pressure phenomena) and assumes a planar shear surface along which slab or wedge failure takes place:

$$H_c = \frac{4_c}{\gamma} \frac{\sin\alpha \, \cos\phi}{[1 - \cos(\alpha - \phi)]} \qquad (5.14)$$

where H_c is the critical height of slope (m), c the cohesion of material (k Pa), γ the *in situ* unit weight of material (kN m^{-3}), α the slope angle (degrees), and ϕ the friction angle (degrees).

In the absence of high pore-water pressures (not unreasonable assumptions—see Thorne (1978)) it may be possible to predict the height of river bank needed before slab failure will take place. Lohnes and Handy (1968), for example, found good agreement between observed and predicted maximum heights of cuts in friable loess. Thorne (1978, pp. 57–60) notes that some of the assumptions of a Culmann analysis, however, may not be justified, particularly the exclusion of the possibility of tension cracking, and presents a related equation for the critical height of a *vertical* bank, H'_c, in which a tension crack may extend to around one-half of total bank height (Thorne, 1982):

$$H'_c = (2c/\gamma)\tan(45 + \phi/2) \qquad (5.15)$$

Using equation (5.15) a family of curves has been constructed in Figure 5.12 that predict, for a range of saturated (i.e. worst case condition) bulk unit weights, cohesions and friction angles, the critical height of river bank needed before wedge failure takes place.

Furthermore, if a survey of channel bank materials and bank angles within a given catchment is available then, theoretically, it should be possible to use the hydraulic geometry equations linking bankfull depth (bank height surrogate) to discharge (or

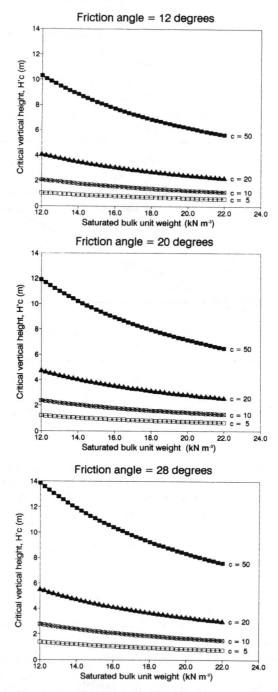

Figure 5.12 Family of Culmann-type bank stability curves, predicting maximum stable vertical banks for given cohesion values (c), friction angles and saturated bulk unit weights

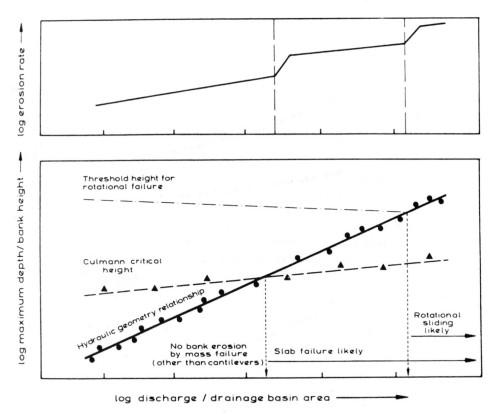

Figure 5.13 Hypothetical use of downstream hydraulic geometry relation with critical bank height information to determine mass failure process zone

some surrogate, e.g. drainage basin area) in order to identify the reach in the catchment where bank heights begin to exceed the critical value for (Culmann-type) failure and where bank retreat might be expected to take place as much by mass failure as by fluvial entrainment (Figure 5.13). The Culmann threshold line may have a slight positive gradient (Figure 5.13) because of increasing cohesion in the bank materials related to progressively decreasing particle size. Further downstream still, bank height exceeds the threshold for rotational failure to occur (Figure 5.13). This threshold line may have a slight negative gradient, reflecting decreasing residual angle of shearing resistance of the bank material owing to a progressively increasing clay content (see Rouse and Farhan, 1976).

Figure 5.14 shows how, by combining geotechnical theory with hydraulic geometry relationships, a rough prediction can be made of the location of the various mass-failure scale threshold(s). For example, for a hypothetical material where $\phi = 20°$, $c = 10$ k Pa, and $\gamma = 19$ kN m^{-3}, vertical banks should be stable up to 1.5 m in height ($H'_c = 1.5$) (Figure 5.14a). To calculate the drainage area 'required' before channel depths exceed critical values, H'_c is inserted into the relevant hydraulic geometry equation. Assuming that, for a hypothetical case (actually drawn from the River

Onny, Shropshire), this is (Figure 5.14b):

$$D_{max} = 0.271\ A^{0.386}\ (r^2 = 86.9\%,\ n = 12,\ p < 0.001) \qquad (5.16)$$

where D_{max} is maximum bankfull depth (m) and A is drainage basin area (km^2) then, solving equation (5.16) for $H'_c = D_{max} = 1.5$, we obtain a predicted threshold drainage area, A_t, of 84 km^2. This would represent a minimum basin area required before mass failure became significant. The graphical solution can be seen in Figure 5.14.

The use of a *traditional* downstream hydraulic geometry relation, however. being built on cross-sections likely to have been taken on straight reaches where bankfull geometry is more readily defined but where bank heights tend to be lower than those adjacent to scour pools in bends, would probably overestimate A_t. A downstream survey of reach-maximum bank heights would be better. The approach ignores, of course, possible downstream changes in bank material geotechnical properties, bank angles, riparian vegetation, and channel-margin hydrology.

Clearly, this approach can be extended to other types of stability analysis, including non-linear slip surfaces, rotational slumping, and those incorporating effective stresses. Schofield and Wroth (1968, p. 249), for instance, obtain an approximate solution for the maximum height of a river bank assuming a circular slip surface in cohesive material. On the other hand, the cantilever instabilities that Thorne (1978), Thorne and Lewin (1979) and Thorne and Tovey (1981) observed to develop after undercutting of the basal gravel layers in the composite banks of the River Severn may well be less sensitive to scale effects. The geometry of any overhang created depends more on the *relative* thickness of the cohesive units that lie above the basal gravels, and the different erosion rates of the two units.

If the general nature of this relationship has some validity, then it represents an interesting example of the mutual influence of channel form on channel process. It also may be the case that, given the sizeable nature of many failed blocks on river

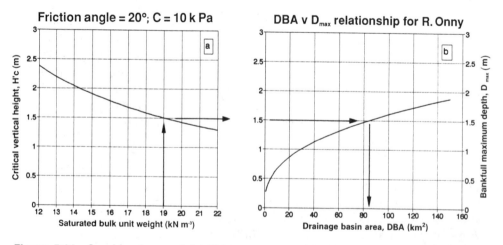

Figure 5.14 Combined use of (a) Culmann-type stable bank-height prediction for a hypothetical material with (b) hydraulic geometry relationship ($D_{max} = 0.271A^{0.386}$) to obtain approximate scale and drainage area thresholds for wedge failure on river banks

banks (Hooke, 1979), the stretch of river at the scale threshold for mass failure is characterized by a sharp increase in the rate of lateral migration, as Figure 5.13 suggests.

Changes in hydrograph shape downstream also may influence the mode of bank retreat. For example, an attenuating flood peak in a down-valley direction, with increased base-time and a more gently sloping recessional limb (Gregory and Walling, 1973), may increase the time of bank wetting for a given flow event, which, in turn, may increase the probability of excess pore-water pressures being generated in lower reaches. Conversely, however, 'drawdown' bank failures (Frydman and Beasley, 1976) often are associated with steep hydrograph recessions, which cause *rapid* removal of external water pressures on the face of the bank. It is not known as yet how these two opposing tendencies may interact.

Process-intensity Domains

Figure 5.15 draws together the hypothesized spatial zoning of the three process groups above. Thus, in upstream sections of low stream power and low banks, subaerial preparation processes are relatively most effective; stream power peaks in the middle courses where fluid entrainment mechanisms prevail; in lower reaches bank heights achieve supercritical values and mass failure dominates. The system is represented, therefore, as a series of overlapping dominance domains (Kirkby, 1980), which also reflect the importance of process combinations. Clearly, in any formal testing of this scheme, changes in other aspects of channel geometry, sedimentology, hydrology and vegetation characteristics would have to be accounted for.

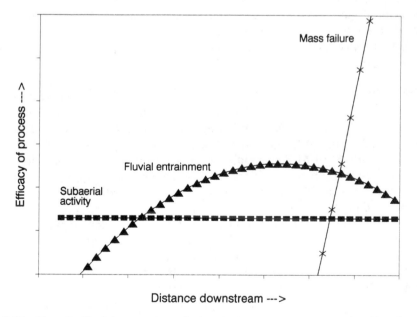

Figure 5.15 Hypothetical downstream change in dominant bank erosion processes. The system is represented as a sequence of overlapping process-intensity domains that also suggest the importance of process combinations

Evidence from the literature provides some support for the scheme in Figure 5.15. For instance, from a near-exhaustive survey (Lawler, 1991a), it is notable that all writers who have commented on the dominance or prominence of frost processes on river banks have been working on 'small' rivers in 'small' catchments ($< 85\,km^2$ in drainage area) in humid temperate environments (Table 5.1, p. 123). Furthermore, the significance of material preparation by other subaerial agents (e.g. rain splash or desiccation) has been noted, in general, within smaller catchments (e.g. Twidale, 1964; McTainsh, 1971; Bello et al., 1978).

The fluvial-entrainment dominance domain is especially difficult to define because of a lack of suitable hydraulic data (e.g. bank shear stress distributions) at the catchment scale and the fact that fluid stresses so often work in conjunction with subaerial preparation processes and mass failure processes (e.g. basal clean-out (Osman and Thorne, 1988)). However, those process studies that have demonstrated the dominance of direct fluid entrainment of bank materials have tended to be conducted in 'middle order' basins. The following examples serve to illustrate the point: Twidale (1964, p. 196) working on the River Torrens in Australia (drainage area = $78\,km^2$) suggests that 'late winter floods are the principal cause of erosion . . . on the river banks'; Knighton (1973, p. 424) found that, on the Rivers Bollin and Dean, Cheshire, UK (channel widths = 3–13 m; depths < 1.5 m), 'the amount of bank erosion at a cross-section was largely a function of the magnitude and variability of discharge and the degree of asymmetry in the distribution of velocity'; Hooke (1979, p. 60) reports that, from a range of sample sites in Devon, UK, 'erosion is mostly by corrasion at high flow' on the River Exe (reaches D and C) and the River Axe (reaches 3 and 4) which, at $620\,km^2$ and $288\,km^2$, are the 'largest' sites in her sample; and Pizzuto and Meckelnburg (1989, p. 1011) observed that, for the medium-sized Brandywine Creek, Pennsylvania, USA (bankfull width = 42.0 m; bankfull mean depth = 2.6 m), 'near-bank velocity alone explains at least 90% of the total variance' in spatial differences in bank erosion rates at the bend scale. Also, we can note the evidence assembled by Lewin (1982,1983,1987) linking peak stream power to maximum change rates in the middle courses of rivers in Wales and the Welsh Borders.

A possible reduction in the efficacy of fluvial entrainment in the lowest reaches of rivers associated with low stream powers might be reinforced by increased clay contents in the bank material, which would increase resistance to fluid shear (see Kamphuis and Hall, 1983) but at the same time decrease resistance to mass failure (Rouse and Farhan, 1976). The middle courses of rivers, where complications of mass failure and preparational activity are perhaps minimized, also may give rise to the strongest spatial and temporal relationships between indices of flow magnitude/ frequency and bank retreat rate.

Finally, the studies that have identified mass failure as the dominant process of bank retreat usually have been undertaken on large systems, such as the Mississippi (Turnbull, Krinitsky and Weaver, 1966; Laury, 1971; Brunsden and Kesel, 1973; Padfield, 1978; Kesel and Baumann, 1981), the Brahmaputra River (Coleman, 1969), the Beatton River (Hickin and Nanson, 1975), and the Yazoo Basin (Little, Thorne and Murphey, 1982). Grissinger (1982, p. 273) noted that, for erosion of the high, steep, banks of the norther Mississippi, 'gravity forces are relatively more significant than hydraulic forces'. Scott's (1978) study of Arctic rivers revealed that bank mass failure was non-existent at the smallest sites (Happy Creek: $88\,km^2$ drainage area), while moderate activity was recorded in the drainage basins of intermediate size (132–450 km^2) with maximum activity occurring in the largest basin (4680–4830 km^2). He concluded that 'there are definite differences in stream behaviour related to size' (Scott, 1978, p. 16). Mosley (1975) recognized that the main retreat mechanism of the banks of the rather smaller River Bollin, UK (mean annual flood = 30 m^3 s^{-1}), was large-scale rotational slumping.

That the scale of the system under consideration is only one factor in a situation that is clearly multivariate, however, is illustrated by contradictory evidence. For example, Hooke (1979) found that, although corrasion was dominant at some of her (moderately sized) sites, slumping or toppling of blocks of material occurred also. She found, in fact, a tendency for slumping to be more common on *lower* banks, although this was thought to be because low banks were composed of material susceptible to this particular process and were subject to more frequent wetting.

Combinations of factors and processes are therefore important for significant bank retreat (Hooke, 1979; Thorne, 1982; Lawler, 1987), as shown by the examples below. First, mass failure itself can be driven by fluvial scour at the bank toe (Thorne, 1982). This can oversteepen the bank or increase its height beyond the critical condition (e.g. Kesel and Baumann, 1981). Secondly, cantilever failures, common on some types of composite banks where the lower units are much more erodible than the overlying ones, are preceded by the preferential fluvial erosion of lower sedimentary units to create overhangs (Thorne and Tovey, 1981). Thirdly, slumped blocks may be subject to fluid erosion: indeed Thorne (1982, p. 233) considers this to be crucial to continued bank retreat through the mechanism of 'basal endpoint control'. Fourthly, fluvial entrainment of the products of frost weathering is necessary for the process to become a really active agent of bank retreat (McGreal and Gardiner, 1977; Lawler, 1987), although some direct sediment transport to the stream is possible (Lawler, 1984, 1987). Fifthly, (fluvial) pre-wetting may be necessary to generate mass instability through excess pore-water pressures or by increasing the weight of material above the potential failure surface. Finally, desiccational cracking of the bank surface may be essential for significant fluid entrainment of cohesive bank material in some instances, by allowing flowing water access behind partly detached slabs, flakes, crumbs or micropeds.

CONCLUSIONS

The following points emerge from the foregoing discussion:

(1) Considerable uncertainty exists with the identification of the dominant processes in any given bank erosion system. 'Hydraulic only' approaches provide only part of the solution to understanding bank erosion processes—at least with cohesive materials and/or in large systems. Even at individual sites, a given flow can effect widely differing erosional responses. Excess shear stress methodologies work best for cohesionless boundary materials. As Pizzuto and Meckelnburg (1989, p. 1012) observe: 'for . . . complex failure processes, simple correlations between erosion rate and near-bank velocity may not exist'.

(2) The automatic bank erosion monitoring system (PEEP) allows quasi-continuous data on bank retreat and accretion to be obtained routinely, and facilitates definition of the erosional impact of individual (flow) events. Methods like this promise to refine process inference, improve magnitude–frequency analyses, and enable a more rigorous relation of bank processes to channel dynamics. This should strengthen considerably our abilities to build, calibrate and test process-based bank erosion and channel-change models.

(3) Resistance of river bank material to fluvial entrainment changes through time with seasonal vegetation growth, freeze–thaw activity, bank moisture status and desiccational activity. The erodibility side of the force-resistance equation requires further research. Particular attention needs to be focused on channel-margin hydrology and bank thermal regime to enhance understanding of desiccation, saturation and freeze–thaw preconditioning. Field tests of recent ideas on the controls of resistance changes in cohesive materials also are needed.

(4) Possible downstream variation in the relative importance of processes of bank erosion, related to changes in channel scale and hydraulic properties (particularly the general demonstration that stream power tends to peak in mid-basin), may help to explain a lack of consensus amongst bank erosion workers on the dominant processes of retreat. Further work is required in this area, a point that echoes the more general statement of Lewin (1987, p. 174) that 'there is a clear research need for considering spatial changes in river systems much more systematically and precisely'.

ACKNOWLEDGEMENTS

I thank Robert Brown for much enthusiastic advice, Mike Dolan and Ali Hasan for assistance with PEEP sensor installation, Bill Green for permission to work at the River Arrow site, and Kevin Burkhill, Simon Restorick and Geoff Dowling for production of some of the figures. I am also extremely grateful to Heather Lawler for help with the fieldwork, data processing and manuscript preparation, and to Paul Carling and John Lewin for some very useful comments on the MS.

REFERENCES

Bathurst, J. C. (1982). Theoretical aspects of flow resistance. In R. D. Hey, J. C. Bathurst and C. R. Thorne (eds), *Gravel-bed rivers*, Wiley, pp. 83–105.

Bello, A., D. Day, J. Douglas, J. Field, K. Lam and Z. B. H. A. Soh (1978). Field experiments to analyse runoff, sediment and solute production in the New England region of Australia. *Zeitschrift für Geomorphologie, Suppleband*, **29**, 180–190.

Blacknell, C. (1981). River erosion in an upland catchment. *Area*, **13**, 39–44.

Bradford, J. M. and R. F. Piest (1977). Gullywall stability in loess-derived alluvium. *Soil Science Society of America Journal*, **41**, 115–122.

Brunsden, D. and R. Kesel (1973). Slope development on a Mississippi River bluff in historic time. *Journal of Geology*, **81**, 576–597.

Carlston, C. W. (1969). Downstream variations in the hydraulic geometry of streams: special emphasis on mean velocity. *American Journal of Science*, **267**, 499–509.

Coleman, J. M. (1969). Brahmaputra River: channel processes and sedimentation. *Sedimentary Geology*, **3**, 129–239.

Curr, R. H. (1984). *The sediment dynamics of Corston Brook*, Unpublished PhD thesis, University of Exeter.

Ferguson, R. and P. Ashworth (1991). Slope-induced changes in channel character along a gravel-bed stream: the Allt Dubhaig, Scotland. *Earth Surface Processes and Landforms*, **16**, 65–82.

Frydman, S. and D. H. Beasley (1976). Centrifugal modelling of riverbank failure. *Journal Geotechnical Engineering Division, Proceedings of the American Society of Civil Engineers*, **102**, 395–409.

Gardiner, T. (1983). Some factors promoting channel bank erosion, River Lagan, County Down. *Journal of Earth Science Royal Dublin Society*, **5**, 231–239.

Graf, W. L. (1982). Spatial variations of fluvial processes in semi-arid lands. In C. E. Thorn (ed.), *Space and Time in Geomorphology*, Allen and Unwin, pp. 193–217.

Graf, W. L. (1983). Downstream changes in stream power in the Henry Mountains, Utah. *Annals of the Association of American Geographers*, **73**, 373–387.

Gregory, K. J. and D. E. Walling (1973). *Drainage Basin Form and Process*, Arnold, 456 pp.

Grissinger, E. H. (1982). Bank erosion of cohesive materials. In R. D. Hey, J. C. Bathurst and C. R. Thorne (eds), *Gravel-bed Rivers*, Wiley, pp. 273–287.

Hasegawa, K. (1989). Studies on qualitative and quantitative prediction of meander channel shift. In S. Ikeda and G. Parker (eds), *River Meandering*, American Geophysical Union, Water Resources Monograph No. 12, pp. 215–235.

Hey, R. D. and C. R. Thorne (1984). Flow processes and river channel morphology. In T. P. Burt and D. E. Walling (eds), *Catchment Experiments in Fluvial Geomorphology*, Proceedings of the International Geographical Union Commission on Field Experiments in Geomorphology, Exeter and Huddersfield, UK, 16–24 August 1981, Geo Books, pp. 489–514.

Hickin, E. J. and G. C. Nanson (1975). The character of channel migration on the Beatton River, Northeast British Columbia, Canada. *Geological Society of America Bulletin*, **86**, 487–494.

Hill, A. R. (1973). Erosion of river banks composed of glacial till near Belfast, Northern Ireland. *Zeitscrift für Geomorphologie*, **17**, 428–442.

Hooke, J. M. (1977). *An analysis of changes in river channel patterns: the example of streams in Devon*. Unpublished PhD thesis, University of Exeter, 452 pp.

Hooke, J. M. (1979). An analysis of the processes of river bank erosion. *Journal of Hydrology*, **42**, 39–62.

Hooke, J. M. (1980). Magnitude and distribution of rates of river bank erosion. *Earth Surface Processes*, **5**, 143–157.

Hudson, H. R. (1982). A field technique to directly measure river bank erosion. *Canadian Journal of Earth Sciences*, **19**, 381–383.

Ikeda, S. and G. Parker (eds) (1989). *River Meandering*, American Geophysical Union, Water Resources Monograph No. 12, 485 pp.

Ikeda, S., G. Parker and K. Sawai (1981). Bend theory of river meanders. Part 1. Linear development. *Journal of Fluid Mechanics*, **112**, 363–377.

Jones, P. D. and M. Hulme (1990). Temperatures and sunshine duration over the United Kingdom during the period May to October 1989 compared with previous years. *Weather*, **45**, 430–437.

Kamphuis, J. W. and K. R. Hall (1983). Cohesive material erosion by unidirectional current. *Journal of the Hydraulics Division, Proceedings of the American Society of Civil Engineers*, **109**, 49–61.

Kesel, R. H. and R. H. Baumann (1981). Bluff erosion of a Mississippi river meander at Port Hudson, Louisiana. *Physical Geographer*, **2**, 62–82.

Kirkby, M. J. (1980). The stream head as a significant geomorphic threshold. In D. R. Coates and J. D. Vitek (eds), *Thresholds in Geomorphology*, Allen and Unwin, pp. 53–73.

Knighton, A. D. (1973). Riverbank erosion in relation to streamflow conditions, River Bollin–Dean, Cheshire. *East Midlands Geographer*, **5**, 416–426.

Knighton, A. D. (1984). *Fluvial Forms and Processes*, Edward Arnold, 218 pp.

Lapointe, M. F. and M. A. Carson (1986). Migration patterns of an asymmetric meandering river: the Rouge River, Quebec, *Water Resources Research*, **22**, 731–743.

Laury, R. L. (1971). Stream bank failure and rotational slumping: preservation and significance in the geologic record. *Geological Society of America Bulletin*, **82**, 1251–1266.

Lawler, D. M. (1984). *Processes of river bank erosion: the River Ilston, South Wales*, Unpublished PhD thesis, University of Wales, 518 pp.

Lawler, D. M. (1986). River bank erosion and the influence of frost: a statistical examination. *Transactions of the Institute of British Geographers*, **11**, 227–242.

Lawler, D. M. (1987). Bank erosion and frost action: an example from South Wales. In V. Gardiner (ed.), *International Geomorphology 1986 I*, Wiley, pp. 575–590.

Lawler, D. M. (1988). Environmental limits of needle ice: a global survey. *Arctic and Alpine Research*, **20**, 137–159.

Lawler, D. M. (1989). Some new developments in erosion monitoring: 1. The potential of optoelectronic techniques. *School of Geography, University of Birmingham Working Paper*, **47**, 44 pp.

Lawler, D. M. (1991a). The measurement of river bank erosion and lateral channel change. *British Geomorphological Research Group Technical Bulletin*, c. 61 pp. (in press).

Lawler, D. M. (1991b). A new technique for the automatic monitoring of erosion and deposition rates. *Water Resources Research*, **27**, 2125–2128.

Lawler, D. M. (1991c). The design and installation of a new automatic erosion monitoring system, *Earth Surface Processes and Landforms* (in press).

Leopold, L. B. (1953). Downstream change of velocity in rivers. *American Journal of Science*, **251**, 606–624.

Leopold, L. B. (1973). River channel change with time: an example. *Geological Society of America Bulletin*, **84**, 1845–1860.

Leopold, L. B. and T. Maddock (1953). The hydraulic geometry of stream channels and some physiographic implications. *U.S. Geological Survey Professional Paper*, **252**, 57 pp.

Lewin, J. (1981). River Channels. In A. Goudie (ed.), *Geomorphological Techniques*, Allen and Unwin, pp. 196–212.

Lewin, J. (1982). British floodplains. In B. H. Adlam, C. R. Fenn and L. Morris (eds), *Papers in Earth Studies*, Geo Books, pp. 21–37.

Lewin, J. (1983). Changes of channel patterns and floodplains. In K. J. Gregory (ed.), *Background to Palaeohydrology*, Wiley, pp. 303–319.

Lewin, J. (1987). Historical river channel changes. In K. J. Gregory, J. Lewin and J. B. Thornes (eds), *Palaeohydrology in Practice*, Wiley, pp. 161–175.

Little, W. C., C. R. Thorne and J. B. Murphey (1982). Mass bank failure of selected Yazoo Basin streams. *Transactions of the American Society of Agricultural Engineers*, **25**, 1321–1328.

Lohnes, R. and R. L. Handy (1968). Slope angles in friable loess. *Journal of Geology*, **76**, 247–258.

McGreal, W. S. and T. Gardiner (1977). Short-term measurements of erosion from a marine and a fluvial environment in County Down, Northern Ireland. *Area*, **9**, 285–289.

McTainsh, G. H. (1971). *Stream bank erosion, Banks Peninsula, New Zealand*. Upublished MA thesis, University of Canterbury, New Zealand, 70 pp.

Mosley, M. P. (1975). Meander cutoffs on the River Bollin, Cheshire 1872–1973. *Revue Geomorp. Dynamique*, **24**, 21–31.

Nickel, S. H. (1983). Erosion resistance of cohesive soils. *Journal of Hydraulics Division, Proceedings of the American Society of Civil Engineers*, **109**(1), 142–144.

Odgaard, A. J. (1986a). Meander flow model. I: development. *Journal of the Hydraulics Division, Proceedings of the American Society of Civil Engineers*, **112**(8), 117–1136.

Odgaard, A. J. (1986b). Meander flow model. II: applications. *Journal of the Hydraulics Division, Proceedings of the American Society of Civil Engineers*, **112**, 1137–1150.

Osman, A. M. and C. R. Thorne (1988). Riverbank stability analysis. I: theory. *Journal of the Hydraulics Division, Proceedings of the American Society of Civil Engineers*, **114**, 134–150.

Padfield, C. J. (1978). *The stability of river banks and flood embankments*. Unpublished PhD thesis, Cambridge University.

Parker, G., P. Diplas and J. Akiyama (1983). Meander bends of high amplitude. *Journal of the Hydraulics Division, Proceedings of the American Society of Civil Engineers*, **109**(10), 1323–1337.

Pizzuto, J. E. and T. S. Meckelnburg (1989). Evaluation of a linear bank erosion equation. *Water Resources Research*, **25**, 1005–1013.

Rana, S. A., D. B. Simons and K. Mahmood (1973). Analysis of sediment sorting in alluvial channels. *Journal of the Hydraulics Division, Proceedings of the American Society of Civil Engineers*, **99**, 1967–1980.

Rouse, W. C. and Y. I. Farhan (1976). Threshold slopes in South Wales. *Quarterly Journal of Engineering Geology*, **9**, 327–338.

Schofield, A. N. and P. Wroth (1968). *Critical State Soil Mechanics*, McGraw-Hill.

Scott, K. M. (1978). Effects of permafrost on stream channel behaviour in Arctic Alaska. *U.S. Geological Survey Professional Paper*, **1068**, 19 pp.

Selby, M. J. (1982). *Hillslope Materials and Processes*, Oxford University Press, 264 pp.

Sharp, J. M. (1977). Limitations of bank storage model assumptions. *Journal of Hydrology*, **35**, 31–47.

Stott, T. A., R. I. Ferguson, R. C. Johnson and M. D. Newson (1986). Sediment budgets in forested and unforested basins in upland Scotland. In R. F. Hadley (ed.), *Drainage Basin Sediment Delivery, International Association of Hydrological Sciences Publication*, **159**, 57–68.

Thomson, S. (1970). Riverbank stability at the University of Alberta. *Canadian Geotechnical Journal*, **7**, 157–168.

Thorne, C. R. (1978). *Processes of bank erosion in river channels*. Unpublished PhD thesis, University of East Anglia, 447 pp.

Thorne, C. R. (1981). Field measurements of rates of bank erosion and bank material strength. In *Erosion and Sediment Transport Measurement, Proceedings of the International Association of Hydrological Sciences Publication*, Florence Symposium, June 1981, **133**, 503–512.

Thorne, C. R. (1982). Processes and mechanisms of river bank erosion. In R. D. Hey, J. C. Bathurst and C. R. Thorne (eds), *Gravel-bed Rivers*, Wiley, pp. 227–259.

Thorne, C. R. and J. Lewin (1979). Bank processes, bed material movement and planform development in a meandering river. In D. D. Rhodes and G. P. Williams (eds), *Adjustments of the Fluvial System*. Kendall/Hunt, pp. 117–137.

Thorne, C. R. and A. M. Osman (1988). Riverbank stability analysis. II: applications. *Journal of the Hydraulics Division, Proceedings of the American Society of Civil Engineers*, **114**, 151–172.

Thorne, C. R. and N. K. Tovey (1981). Stability of composite river banks. *Earth Surface Processes and Landforms*, **6**, 469–484.

Turnbull, W. J., E. L. Krinitsky and F. J. Weaver (1966). Bank erosion in soils of the Lower Mississippi Valley. *Journal of the Soil Mechanics and Foundations Division, Proceedings of the American Society of Civil Engineers*, **92**, 121–136.

Twidale, C. R. (1964). Erosion of an alluvial bank at Birdwood, South Australia. *Zeitscrift für Geomorphologie*, **8**, 189–211.

Walling, D. E. and B. W. Webb (1981). Water quality. In J. Lewin (ed.), *British Rivers*, Allen and Unwin, pp. 126–171.

Wolman, M. G. (1959). Factors influencing erosion of a cohesive river bank. *American Journal of Science*, **257**, 204–216.

6 Estimation of the Fractal Dimension of Terrain from Lake Size Distributions

STEPHEN K. HAMILTON, JOHN M. MELACK
Department of Biological Sciences, University of California

MICHAEL F. GOODCHILD
Department of Geography, University of California

and

WILLIAM M. LEWIS, JR.
Center for Limnology, Department of Environmental, Population and Organismic Biology, University of Colorado

INTRODUCTION

Examination of the scaling of geomorphological features yields information on the processes that shaped the terrain (Church and Mark, 1980; Milne, 1988). In flood-plains, which are often a lake-rich terrain, the lakes reflect the pattern of depressions and the overall relief. The sizes and shapes of floodplain lakes, which can be measured readily by remote sensing, reveal small differences in relief that do not appear on conventional topographic maps. The origin of some kinds of floodplain lakes is apparent from their size and shape. Oxbow lakes and scroll lakes, which are created by migrating river channels (Hutchinson, 1957), are two such examples. Many floodplain lakes are irregular in shape, however, and the geomorphological processes that formed the lakes are less apparent. Comparison of scaling patterns between these different kinds of floodplain lakes could provide insight into the processes that shape the floodplain (Salo, 1990).

The frequency distribution of lake sizes can be used in analyzing the scaling of lake morphometry. Such an analysis is based upon a comparison of the distribution of empirical observations with that which would be expected in the absence of a predominant scale-specific geomorphological process (Goodchild, 1981). Fractal theory provides a stochastic model of scale-invariant relief that is useful as a point of reference or 'null hypothesis' for this comparison. The relation between area and perimeter in a population of lakes also provides information on scaling and can be analyzed for consistency with the stochastic model (Mandelbrot, 1982).

In this study, we analyze the statistical properties of size distributions for floodplain lakes of the Amazon and Orinoco rivers. For the Orinoco floodplain lakes, we also

Lowland Floodplain Rivers: Geomorphological Perspectives. Edited by P.A. Carling and G.E. Petts
© 1992 John Wiley & Sons Ltd

examine the relation between area and perimeter. Our objectives are to evaluate the consistency of these statistical properties with the fractal model, and to explore the usefulness of lake size distributions as indicators of geomorphological patterns and processes on floodplains and other kinds of lake-rich terrain.

CONCEPTUAL BACKGROUND

Mandelbrot (1977, 1982) introduced the term 'fractal' for a spatial or temporal phenomenon whose variation is scale-invariant. The variation in a fractal phenomenon can be described mathematically by a non-integer dimension (the fractal dimension, D). Many geophysical phenomena, including coastlines, topography, island areas, river discharges, and climatic variation, show fractal characteristics over a wide range of observational scales (Mandelbrot, 1982; Burrough, 1985; Kaye, 1989; Takayasu, 1990). Computer simulations of fractal surfaces based on a stochastic process called fractional Brownian motion (FBM) produce images that resemble closely certain types of real terrain (Mandelbrot, 1982; Goodchild, 1988). This resemblance is surprising given the scale-specific nature of most geomorphological processes, and suggests that the types of terrain that are simulated realistically by FBM models are shaped primarily by geomorphological processes that are scale-invariant, and that the terrain can be described by the same fractal dimension that characterizes FBM surfaces.

Mandelbrot used conceptual models to demonstrate that islands created by flooding a fractal landscape have a size distribution that reflects the fractal geometry of the landscape. The statistical properties of the island area–number relationship had been studied earlier by Korčak (1940), who found that an inverse power law fitted empirical data on the size distribution of oceanic islands. If the islands are ranked by area from largest to smallest, then the number of islands (A) above a given size (a) can be denoted as $Nr(A > a)$ and is given by:

$$Nr(A > a) = F'a^{-B}$$

where F' and B are positive constants. Mandelbrot called this the Korčak Empirical Law, and he argued that the exponent B should be equal to $D/2$, where D is the fractal dimension of the relief (i.e. of the topographic contours). Fractal dimension D lies between 1 (Euclidean shapes with smooth boundaries) and 2 (infinitely convoluted boundaries). The graphical diagnosis for goodness-of-fit to the Korčak Law is a log–log plot of rank against area, which will yield a straight line of slope $-B$. This plot is called a Pareto fit; it takes the same form as an asymptotically hyperbolic probability distribution called the Pareto distribution, which describes diverse socioeconomic and biological phenomena and is known particularly for its application in economics for modeling the upper tail of the distribution of income (Badger, 1980; Arnold, 1985; West and Shlesinger, 1990).

Lake areas also reportedly follow the Korčak Law (Mandelbrot, 1982), as noted originally by Korčak (1940), although we have found few empirical demonstrations of this fit in the literature (notable exceptions include Goodchild (1981) and Kent and Wong (1982) who studied lakes on glacial landscapes). Mandelbrot speculated that landscapes with fractal relief will have lake area distributions that follow the Korčak

Law, and that the fractal dimension D therefore can be estimated from B, the exponent of the lake size distribution. Although a multitude of factors in addition to relief may affect the occurrence and size of lakes on a fractal landscape, Mandelbrot hypothesized that these factors may act independently of lake size, with little resultant effect on the original hyperbolic size distribution described by the Korčak Law.

An independent method to determine the fractal dimension of lake-rich terrain is based on the relation between area and perimeter in a population of lakes of varying size that is observed at a single scale of observation (Mandelbrot, 1982; Burrough, 1985). For a population of lakes with smooth boundaries ($D = 1$), the perimeter will vary as area to the one-half power. The perimeter of lakes on fractal surfaces should vary as area to the power $D/2$, because the larger lakes appear to have more convoluted boundaries at a fixed observational scale. The slope of a log–log plot of perimeter against area is constrained to lie between 0.5 (Euclidean shapes with $D = 1.0$) and 1.0 (infinitely convoluted boundaries with $D = 2.0$). Theoretically, the estimate of D obtained in this way will agree with that obtained from the Korčak Law. However, only a few studies have estimated D by both methods. Goodchild (1981) reported disagreement between the two estimates for lakes on Random Island, off the coast of Newfoundland, while Kent and Wong (1982) found reasonable agreement for Canadian Shield lakes in Ontario. The two methods yielded different estimates of D for lakes simulated on FBM surfaces; the Pareto fit produced an infeasible estimate of D (Goodchild, 1988).

The fractal model has been proposed as a point of reference or 'null hypothesis' for the analysis of lake-rich landscapes: comparison of the statistical properties of empirical lake size distributions with those predicted by the FBM model might reveal the presence or absence of scale-specific geomorphological processes that shape the relief and thereby determine patterns in lake morphometry (Goodchild, 1988). If the empirical lake size distribution possesses distinct statistical properties, then we can reject the FBM model and consider particular scale-specific processes that might be responsible for the genesis of the lakes. It should be noted that we use the term 'null hypothesis' only in an informal sense; we are not referring to a statistical test based on probability theory.

STUDY SITES

Amazon Floodplain

The Amazon River, which is the world's largest river, has a mean discharge of ca. $210\,000$ m^3 s^{-1} (Richey et al., 1989) and inundates a floodplain of $170\,000$ km^2, of which 65% is located within Brazil (Melack and Fisher, 1990). We will limit this description to the Brazilian Amazon (Figure 6.1), which is the study site analyzed in this paper. The main stem of the Amazon in Brazil is called the Solimões River above the confluence with the Negro River at Manaus. Along the 2600 km of channel from the Xingú River confluence up-river to the Peruvian border, the floodplain is typically 10–50 km wide and is composed of a mosaic of open waterbodies, dense fields of herbaceous plants that float on the water surface during floods, and forest. At Manaus on the central Amazon, the river stage varies seasonally by ca. 10 m. Many of the

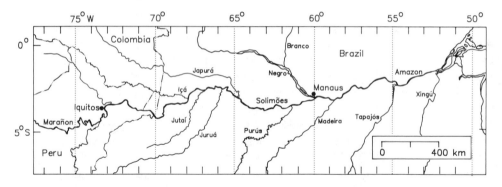

Figure 6.1   Map of the Amazon River

largest lakes are connected with the river throughout the year, and their water levels are controlled directly by the river.

Most of the suspended sediment carried by the Amazon main stem originates in the Andes along the western margin of the basin, and enters the Brazilian Amazon through the main stem and through the Madeira River (Meade *et al.*, 1985). Tributaries that drain lowland basins, including the Negro and Tapajós rivers, carry relatively little material in suspension, and their waters have a black or clear appearance.

Lake outlines for two distinct reaches of the Amazon floodplain are depicted in Figure 6.2. In the upper panel, which is near the confluence with the Japurá River, the floodplain is dominated by scroll-bar topography and contains numerous small, narrow lakes (channel lakes). Channel lakes on the main stem Amazon floodplain occur principally in association with scroll bars; few channel lakes have dimensions that approach those of the main river channel. Mertes (1985) noted that the curvature of meander scrolls on the Amazon floodplain tends to resemble that of small channels (paranás) that carry water from the main channel across the levees and into the floodplain, as opposed to that of the meanders of the main channel. Mertes concluded that the paranás probably create most of the meander scrolls.

The lower panel of Figure 6.2 depicts the floodplain near Manaus, which has both scroll-bar topography and back-swamp deposits with round or irregular lakes. Also, back-flooded tributary valleys (blocked-valley lakes) form dendritic lakes along the floodplain boundary.

The origin of blocked-valley lakes has attracted attention from various authors, and several hypotheses have been advanced. One view holds that eustatic changes in sea level during the Flandrian transgression caused changes in river base level, backing up the Amazon main stem and flooding lateral tributary valleys. The backwater effect resulted in enhanced aggradation of floodplains inundated by turbid river water. Where less turbid influents entered the floodplain of a turbid river, the flooded valleys were not filled at comparable rates and persist today as blocked-valley lakes (Tricart, 1977; Klammer, 1984). Other investigators have provided geomorphological evidence for neotectonic sinking of lower river courses along fault lines, resulting in box-shaped valleys that are flooded permanently (Departamento Nacional da Produção Mineral, 1978). Damming of tributary valleys by fluvial deposition from the main stem also has

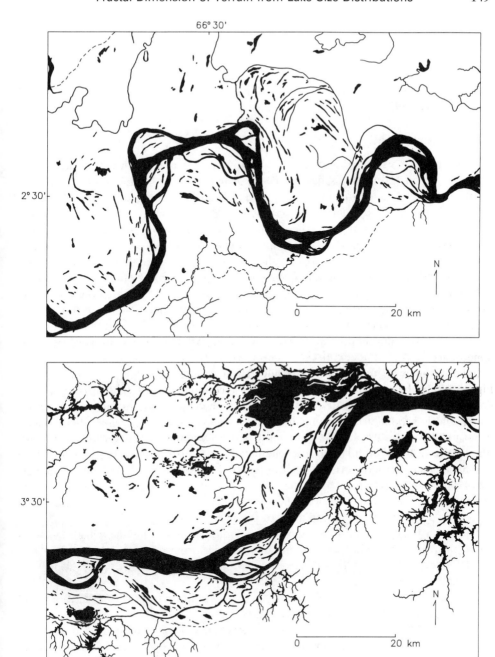

Figure 6.2 Examples of the Amazon floodplain. Solid areas indicate open-water surfaces, and dashed lines show the floodplain boundaries. The upper panel is a reach near the confluence of the Japurá River, and the lower panel is a reach near Manaus. (Adapted from planimetric maps published by Departamento Nacional da Produção Mineral, 1978)

been suggested (Iriondo, 1982). These processes may operate separately or in combination to produce the blocked-valley lakes along the Amazon (Iriondo, 1982). Regardless of the processes that result in formation of blocked-valley lakes, the size distribution of the lakes is expected to be determined by the size distribution of the tributary stream valleys that were produced by fluvial erosion before the valleys became back-flooded.

Orinoco Floodplain

The Orinoco River of Venezuela and Colombia (Figure 6.3) is the world's third largest river in terms of discharge (mean, 36 000 m^3 s^{-1}: Meade *et al.*, 1983). The floodplain that fringes the main stem covers 7000 km^2 along the lower 770 km of channel between the delta and the Meta River (Hamilton and Lewis, 1990a). A more extensive savanna floodplain (the *Llanos*) occurs west of the main stem and is contiguous with the main stem floodplain between the Meta and Apure rivers. Scroll-bar topography is less common on the Orinoco floodplain than on the Amazon floodplain. The vegetation of the Orinoco floodplain resembles that of the Amazon. The level of the Orinoco River fluctuates seasonally by 10–15 m along the lower main stem. However, few lakes on the Orinoco floodplain fluctuate in level as much as the river because the lakes are usually isolated from and perched above the river during low water (Hamilton and Lewis, 1990b). Like the Amazon, the sediment load of the Orinoco is derived almost entirely from the Andes to the west, and enters the Orinoco main stem largely through the Meta and Apure Rivers.

Examples of the Orinoco floodplain are shown in Figure 6.4. The upper panel is a reach between the Meta and Apure rivers, where lakes are particularly abundant along the left bank. The lower panel shows the floodplain in a reach between the Caura and Caroní rivers, where the floodplain is interrupted periodically by outcrops of high ground. Floodplain bordering the main stem is narrower than along the Amazon, and the lakes are generally smaller. Blocked-valley lakes are rare on the Orinoco floodplain.

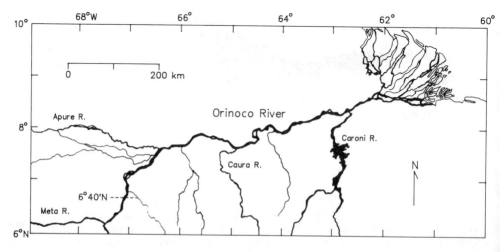

Figure 6.3 Map of the Orinoco River

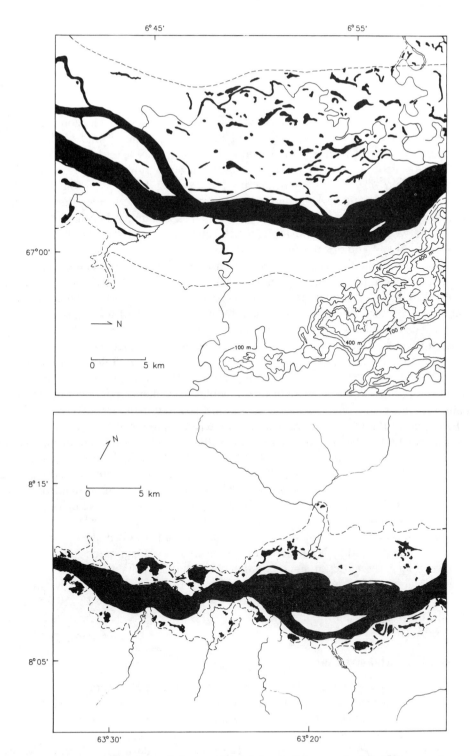

Figure 6.4 Examples of the Orinoco floodplain. Solid areas indicate open-water surfaces, and dashed lines show the floodplain boundaries. The upper panel is a reach between the Apure and Meta rivers, and the lower panel is a reach between the Caura and Caroní rivers. (Adapted from maps published by the Venezuelan Dirección de Cartografía Nacional)

METHODS

Amazon Lake Measurements

Until recently, analysis of the morphometric characteristics of Amazon floodplain lakes was difficult because of the lack of appropriate imagery and maps for the floodplain, much of which is remote and often obscured by cloud cover. However, side-looking airborne radar (SLAR, X-band) imagery, published by the Brazilian *Projeto RadamBrasil* (e.g. Departamento Nacional da Produção Mineral, 1978), covers the entire Brazilian Amazon Basin. This radar imagery provided the opportunity to measure lake sizes and shapes on the Amazon floodplain because of the high contrast between open water surfaces, which cause specular reflection of microwave radiation, and soil and vegetation, which backscatter part of the incident radiation.

We used a combination of prints of semi-controlled radar mosaics and the planimetric maps that were produced from these images (both at 1 : 250 000 scale) to identify and measure lakes. The planimetric maps were used in a few cases for which the print quality of the mosaic was poor, and for areal measurements of the larger lakes.

We measured all lakes ($N = 6510$) on the fringing floodplain of the main stem Amazon River system in Brazil (Figure 6.1). We define the fringing floodplain as area that is inundated seasonally by water from the parent rivers, following Welcomme (1979). Our study area extends from the apex of the Amazon delta up-river to the Peruvian border. Our data include measurements of the shape and length reported by Melack (1984), except that we have excluded lakes of the major tributaries. We have expanded the data set to include lake area as well as shape and length.

Lake measurements included length and width, and area was either measured or estimated. The maximum length (longest straight line across the lake without crossing land) and maximum width (perpendicular to its length) of each lake were measured with a ruler. The areas of the largest 10–15% of the lakes and of all dendritic lakes were measured by electronic planimetry. These lakes appear to be the most irregular in shape. The remaining lakes appear to be elliptic in shape, and their areas could thus be estimated from length and width measurements and the formula for area of an ellipse. Between 5 and 10% of the lakes whose areas were estimated in this manner were irregular, crescentic, or sinuous rather than elliptic in shape; we have probably underestimated their areas. Our measurements were less accurate for the smallest lakes. Lakes with a maximum length $\leqslant 250$ m ($\leqslant 1.0$ mm at a scale of 1 : 250 000) were difficult to distinguish on the images.

Orinoco Lake Measurements

The Orinoco lakes were measured on maps of 1 : 100 000 scale produced from aerial photography by the Venezuelan Dirección de Cartografía Nacional between 1968 and 1975. In the course of field studies on the Orinoco floodplain, we have confirmed that lake and river boundaries on the maps generally correspond to the vegetated shorelines, and hence can be considered the low-water boundaries (i.e. areas that are flooded most of the year).

We measured all lakes ($N = 2294$) on the fringing floodplain of the main stem Orinoco River from 6°40'N latitude down-river to the apex of the delta (Figure 6.3).

Maps were unavailable for a 70 km reach of floodplain that extends from 6°40'N latitude up-river (southward) to the confluence with the Meta River; this reach comprises only about 7% of the total area of fringing floodplain (Hamilton and Lewis, 1990a).

The area and perimeter of each lake were measured with a computerized digitizer. The accuracy of our measurements was lower for lakes with a maximum length ≤ 100 m (≤ 1.0 mm at a scale of 1 : 100 000).

Definition of Floodplain Lakes

We define floodplain lakes as basins that contain water throughout the year. During inundation by the river, the floodplain is typically covered with a continuous sheet of water, within which open-water areas are visible. When the river level falls, water drains from the floodplain, and lakes persist only in depressions that are perched above the river level, or in deeper basins that remain in communication with the river. Floodplain forest and fields of herbaceous plants grow on most of the seasonally exposed land (Junk, 1983). During inundation, the upper portion of the forest canopy generally remains above the water level, and the herbaceous vegetation remains emergent or forms floating mats that rise and fall with the fluctuating water level. For this reason, the boundaries of floodplain lakes appear similar on the radar images regardless of whether the surrounding vegetation was flooded or dry. The shapes of open-water areas that appear on the Orinoco maps, which were drawn from aerial photography, also appear to represent the approximate low-water boundaries of floodplain lakes.

Although in most cases the identification of a discrete floodplain waterbody was unambiguous, there were several situations that required the use of specific criteria for definition of a floodplain lake. Many lakes are interconnected by channels. If the channel is narrow (i.e. the width of a single line on the images or maps) relative to the width of the lakes on either end, the lakes are considered separate waterbodies. Enlarged areas along floodplain channels are not considered lakes unless they are at least three times wider than the channel. Waterbodies that are confluent with the river channel are considered lakes only if the width of the confluence is less than half of the maximum length of the waterbody. Dendritic back-flooded tributary valleys are common along the Amazon floodplain. We consider these to be Amazon floodplain lakes only if they are contiguous with the fringing floodplain of the main stem. Two major tributaries of the Amazon, the Tapajós and Xingú rivers, form large lake-like waterbodies in their lower reaches (Irion, 1984); these are not included in our lake measurements.

Classification of Floodplain Lakes

We subdivided our data sets into three classes of floodplain lakes according to differences in their geomorphological origins. Blocked-valley lakes, which are common on the Amazon but not on the Orinoco, extend inland from the floodplain boundary (escarpment) and are shaped like drainage valleys. All other floodplain lakes located within the boundaries of the fringing floodplain are lateral levee lakes (Hutchinson, 1957). Within this category, we distinguish lakes with dish-shaped basins (dish lakes) from those with channel-shaped basins (channel lakes) (Hamilton and Lewis, 1990b). Dish lakes tend to have lower rates of hydraulic through-flow during inundation, resulting in gently sloping bottoms composed of fine sediment (Blake and Ollier, 1971). Channel lakes tend to have visible currents during inundation, and often have steeper sides and bottoms composed of coarser sediment, or of compacted fine sediment. Channel lakes commonly lie in swales between concentric levees of coarse sediment (scroll bars). We have chosen a length-to-width ratio of 5 as the division between these categories, following the suggestion of Hamilton and Lewis (1990b), who examined the relation between morphometry and limnology in Orinoco floodplain lakes.

RESULTS

The frequency distributions of our measurements of floodplain lake area show strong positive skew. Two common probability distributions with long upper tails are the log-normal and Pareto. The distribution of sizes (expressed as area) for all lakes on the main stem Amazon floodplain is shown in Figure 6.5, which is a log–log plot (base 10) of rank against area (Pareto fit); a Pareto distribution will appear linear on this plot with a slope of $-B$. The parabolic shape in the lower domain suggests a log-normal fit, but the data fit the Pareto distribution better in the upper domain. The transformation from apparent log-normal to Paretian tail behavior occurs at about the 75th percentile, as shown by the box plot. Pareto fits for each lake class on the Amazon floodplain show this apparent transformation from log-normal to Paretian behavior in the upper tail of each distribution (Figure 6.6).

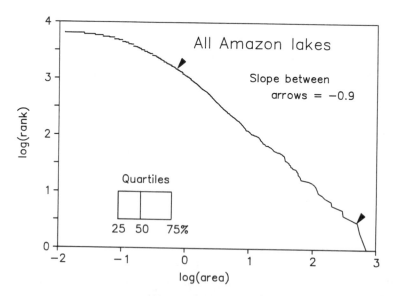

Figure 6.5 Pareto fit for all lakes on the Amazon floodplain. Linearity indicates that the size distribution of the lakes is statistically self-similar. In theory, the fractal dimension D of the surface on which the lakes occur can be estimated from the slope $(-B)$ as $D = 2B$. Units of area are km^2 in this and the following plots. The box plot is aligned along the x-axis to show the quartiles of the distribution of lake areas

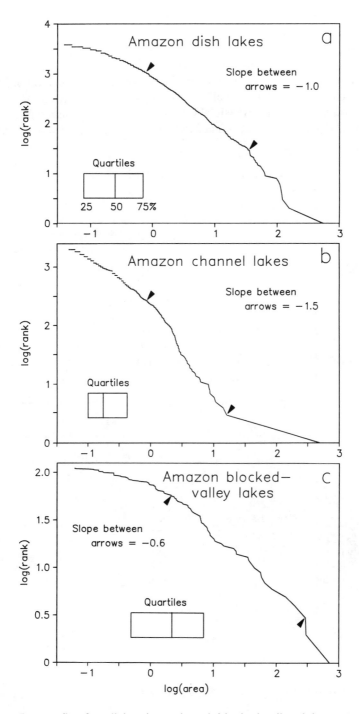

Figure 6.6 Pareto fits for dish, channel and blocked-valley lakes on the Amazon floodplain. See Figure 6.5 for explanation

The three lake classes for the Amazon floodplain show distinct slopes in the linear portion of their Pareto fits (Figure 6.6). Dish lakes, which comprise the majority (68%) of floodplain lakes that we measured along the Amazon main stem, fit the Pareto model above the 75th percentile, except that the largest lakes fall below the extrapolated Pareto line (Figure 6.6a). The slope of the linear portion of the Pareto fit for Amazon dish lakes leads to an estimate for the fractal dimension D of 2.0, which is infeasible. Channel and blocked-valley lakes also fit the Pareto model in the upper domain, although not as well as dish lakes; the values of D estimated from their slopes are 3.0 and 1.2 (Figure 6.6b). Channel lakes thus show the shortest (steepest) Paretian

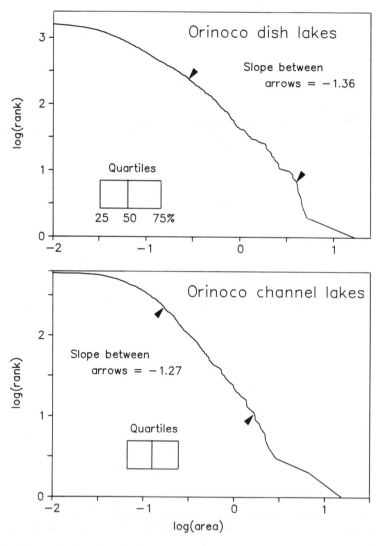

Figure 6.7 Pareto fit for dish and channel lakes on the Orinoco floodplain. See Figure 6.5 for explanation

tail, while blocked-valley lakes show the longest tail. The Pareto fit for blocked-valley lakes is the only one that yields a feasible estimate of D (1.2).

The general shapes of Pareto fits for Orinoco dish and channel lakes resemble those for the Amazon lakes (Figure 6.7). However, the range over which the fit appears linear for the Orinoco lakes lies below that of the Amazon lakes. The slopes of the Pareto fits for Orinoco dish and channel lakes in Figure 6.7 lead to infeasible estimates

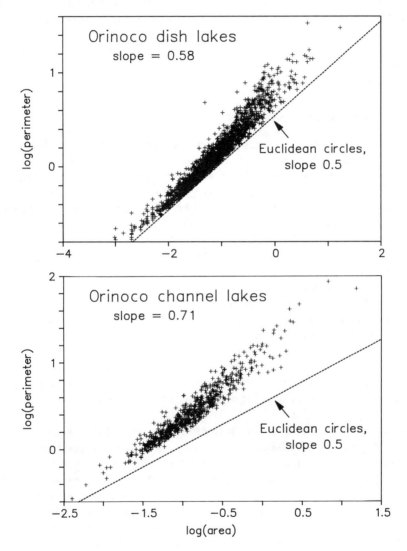

Figure 6.8 Area–perimeter relation for Orinoco dish and channel lakes. The slopes were determined by linear regression; $r^2 = 0.93$ for dish lakes and 0.92 for channel lakes. The line for Euclidean circles (smooth boundaries; $D = 1.0$) is included for comparison. In theory, the fractal dimension D of the surface on which the lakes occur can be estimated from the slope (B) as $D = 2B$; this is an independent method for estimation of the fractal dimension of terrain

of D (2.7 and 2.5), and the slopes of these two lake classes are closer to each other than to the slopes of either dish or channel lakes on the Amazon floodplain. Dish lakes outnumber channel lakes on the Orinoco floodplain by a factor of three.

Our Orinoco data set includes lake perimeters as well as areas, allowing us to examine the area–perimeter relation (Figure 6.8). Simple linear regression was used to fit lines to the empirical data in Figure 6.8. The difference between the slopes of the two regression lines is highly significant (F-test, $P < 0.001$). The slopes of regression lines lead to feasible estimates of D (1.16 for dish lakes and 1.42 for channel lakes). The difference in the area–perimeter estimates of D for the two lake classes is large compared with the similar slopes of the Pareto fits for these lakes.

DISCUSSION

Goodness-of-fit to the Pareto Distribution

Pareto fits for lake areas consistently show linearity in the upper domain, with downward departures of the data in the lower domain and on the uppermost end. This result might be interpreted as evidence for inadequacy of the Pareto distribution (Korčak Empirical Law) to describe our lake area distributions. It is also possible that the data are log-normally distributed in the lower domain, then display a transformation to Paretian behavior in the upper domain; such a transformation is observed commonly in scale-invariant phenomena, and is thought to represent a change in controlling factors (West and Shlesinger, 1990). However, in the following discussion we argue that the departures of the data at the ends can be explained and, indeed, are expected in these data, even if their true distribution is Paretian.

Departure of variates from the extrapolated fit in the tails of a probability distribution can result from truncation or censorship of the variates (Aitchison and Brown, 1963). Truncation refers to the case where values of the variate above or below a certain point either cannot occur or are not observed, whereas censorship refers to the case where, above or below a certain point, the exact values of the variates are unknown but limited knowledge of the variates is available. Both truncation and censorship appear important in the distribution of lake areas that we have presented here.

The apparent log-normal behavior in the lower domain of the distributions probably results because we approached the limits of resolution of lakes on the imagery. In the Amazon lake data (Figure 6.5), the transformation occurs at about the 75th percentile. This point corresponds to a lake area of 0.74 km^2; a circular lake of this area has a diameter of 0.5 km, which is only 2 mm on the 1 : 250 000 radar imagery used for our measurements. Lakes are likely to be increasingly difficult to discern below this size, resulting in increasing censorship of the distribution. The observation that the linear range extended further downward in the Pareto fits for the Orinoco lakes, which were measured on maps of larger scale, supports this interpretation. Coverage of lake surfaces by vegetation probably also obscures some of the smallest lakes because floating vegetation tends to extend outward from the lake shoreline; dense vegetation over water appears as land on X-band radar and in aerial photography. In addition, the smallest basins may not hold water all year simply because of losses by evaporation and seepage. For these reasons, censorship is expected to occur for the smallest lakes even if we could measure them accurately, resulting in a downward departure of the data from the extrapolated Pareto line. Eventually a point of truncation must be reached, below which no lakes can be recognized. Truncation of the lower tail sets the minimum lake area, but unlike censorship it does not affect the slope, because the lakes are ranked from largest to smallest.

Censorship and truncation of the largest lakes also are expected because the floodplain occupies a finite space between the river channel and the escarpment. As lakes become larger, they are increasingly subject to constraints on their area as their boundaries encounter the edges of the floodplain. There are thus fewer large lakes, limited areas of the largest lakes, or both, resulting in increasing censorship of the uppermost end of the distribution. The finite area of floodplain also dictates that there must be a maximum possible lake size, which results in a point of truncation for the distribution.

Censorship and truncation of the upper tail of a Pareto distribution cause the distribution to depart from linearity because the variates are ranked from largest to smallest. A censored or truncated Pareto fit may appear to be approximately linear in the lower domain, but the slope of the 'linear' portion may be altered considerably. To demonstrate the effects of censorship and truncation on the slopes of the Pareto fits, we created an idealized Pareto distribution of lake areas with slope = -0.6 ($D = 1.2$). Figure 6.9 depicts the original distribution (A) and two altered distributions (B and C). In B, we truncated the original distribution by removing the variates of rank 1–5 and re-ranking the remaining variates; this treatment altered the slope considerably across the entire range, although the lower domain still appears approximately linear. Removal of every other variate from ranks 1 to 10, which is a form of censorship, produced same effect on the slope in the lower domain. In C, we reduced the lake areas by an arbitrary proportion that was inversely related to their

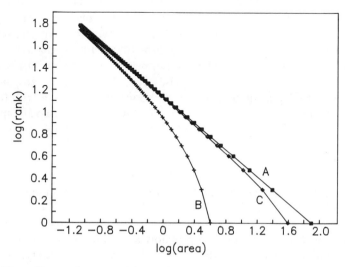

Figure 6.9 The effects of censorship and truncation demonstrated with an idealized Pareto distribution. See text for explanation

ranks; the largest lake was reduced by 50%, the second largest by 25%, the third by 17%, etc. This form of censorship had less effect on the slope. Clearly, censorship and truncation of the upper end can lead to biased estimates of the slopes of Pareto fits and, consequently, of the fractal dimension D.

The variation that we observed in slopes of the Pareto fits among lake classes may be caused simply by differences in the points of censorship and truncation, rather than by differences in the fractal dimension of the terrain. The approximate linearity of the Pareto fits suggests that fractal processes may determine the lake size distributions, but the limited area of floodplain on which the lakes can occur results in steeper slopes that cannot be related to the fractal dimension. The observed differences in Paretian slopes among lake classes seem consistent with this interpretation. On the Amazon floodplain, the slope is smallest for blocked-valley lakes (-0.6), intermediate for dish lakes (-1.0), and largest for channel lakes (-1.5). These differences may reflect different degrees of departure from the slope expected to result from the fractal dimension of the relief. Blocked-valley lakes might be expected to be the least constrained, since they occur outside of the floodplain boundaries. Dish lakes lie in back swamps, which occupy much of the floodplain, while channel lakes are restricted largely to the scroll-bar topography, which covers less area and tends to develop as discrete spatial units compared with the more continuous back swamps where dish lakes occur (Iriondo, 1982). On the Orinoco floodplain, dish and channel lakes appear less segregated (e.g. Figure 6.4); the similar slopes of Pareto fits for these lake classes may reflect their similar spatial constraints.

Comparisons with the FBM Model

Our original objective was to compare our empirical lake size distributions with those produced by the FBM simulation model. However, the parameters of the empirical distributions evidently are altered by censorship and truncation of the upper end. As Goodchild (1988) noted when he presented the Pareto fit for lake areas from the FBM model, the use of a limited area for simulation also results in censorship and truncation of the distribution. The parameters of the Pareto fits are therefore biased in both the simulation model and in the empirical observations, making meaningful comparisons difficult. The FBM model does appear to support Mandelbrot's assertions that fractal relief will have lakes whose areas follow the Pareto distribution. However, beyond that confirmation of fractal theory, its value as a 'null hypothesis' for the analysis of empirical data on lake-rich landscapes remains to be demonstrated.

CONCLUSIONS

Our measurements of the areas of lakes on the Amazon and Orinoco floodplains appear to follow the Pareto distribution, with censorship and truncation on both ends. This indicates that the lakes are statistically self-similar with respect to area over a wide range of observational scales, and suggests that the relief on the floodplain displays fractal characteristics. The observation that all of the lake classes show self-similarity is surprising, because certain lake classes are created by specific geomorphological processes that presumably operate at characteristic scales (e.g. Amazon channel lakes in scroll-bar topography), while others are probably created by more scale-invariant processes (dish lakes).

Statistical self-similarity of lake areas implies that descriptive statistics for the lake populations, such as total abundance per unit land area and the medians and quartiles of lake areas, will vary with the scale of observation. Such statistics therefore cannot be compared between populations unless the same scales are used, or unless a fixed lower limit of lake size is adopted that lies above the resolution limit of the imagery. Statistical self-similarity with respect to area does not imply that the lakes are necessarily self-similar with respect to shape (i.e. that they are isometric rather than allometric (Church and Mark, 1980)).

Censorship and truncation of the upper end of a Pareto distribution are expected to occur for lake areas on floodplains because of the finite space in which the lakes occur. Spatial constraints similarly will affect any analysis of the size distributions of landforms, however, either because of a limited study area or because the landforms occur in a finite space. Consequently, the fractal dimension will be overestimated from size distributions of landforms observed in nature. Goodchild (1981, 1988) found that estimates of the fractal dimension obtained using real and simulated lake size distributions were inconsistent with those obtained by other independent methods, such as the area–perimeter relation. The present study reveals problems with the application of the size-distribution method to empirical data. These problems are likely to explain inconsistency of size-distribution analysis with other methods of estimating the fractal dimension.

ACKNOWLEDGMENTS

We are grateful to S. J. Sippel and L. Meeker for data collection and analysis. This study has been supported by the US National Aeronautics and Space Administration (grant NAS7-918 to J. M. Melack and D. Simonett) and by the US National Science Foundation (grant BSR-86-04655 to W. M. Lewis, Jr. and grant BSR-87-06643 to J. M. Melack and T. R. Fisher). In addition, the senior author was supported by a NASA Global Change Fellowship. L. A. K. Mertes and an anonymous reviewer provided suggestions for improvement of the manuscript.

REFERENCES

Aitchison, J. and J. A. C. Brown (1963). *The Lognormal Distribution, with Special Reference to its Use in Economics*, Cambridge University Press, 194 pp.
Arnold, B. C. (1985). Pareto distribution. In *Encyclopedia of Statistical Sciences*, Vol. 6, Wiley, pp. 568–574.
Badger, W. W. (1980). An entropy–utility model for the size distribution of income. In B. J. West (ed.), *Mathematical Models as a Tool for the Social Sciences*, Gordon and Breach, pp. 87–120.
Blake, D. H. and C. D. Ollier (1971). Alluvial plains of the Fly River, Papua, *Zeitschrift für Geomorphologie, Supplementband*, **12**, 1–17.
Burrough, P. (1985). Fakes, facsimiles and facts: fractal models of geophysical phenomena. In S. Nash (ed.), *Science and Uncertainty*, Proceedings of a Conference Held Under the Auspices of IBM United Kingdom Ltd., London, March 1984, Science Reviews, pp. 151–169.

Church, M. and D. M. Mark (1980). On size and scale in geomorphology. *Progress in Physical Geography*, **4**, 342–390.

Departamento Nacional da Produção Mineral (1978). *Projeto RADAMBRASIL, Folha SA.20, Manaus: Geologia, Geomorfologia, Pedologia, Vegetação e Uso Potencial da Terra*, Departamento Nacional da Produção Mineral, Rio de Janeiro.

Goodchild, M. F. (1981). Randomness at Random Island: stochastic models of the physical landscape. Paper presented to the *Annual Meeting of the Canadian Association of Geographers*, Corner Brook, Newfoundland, August 1981.

Goodchild, M. F. (1988). Lakes on fractal surfaces: a null hypothesis for lake-rich landscapes. *Mathematical Geology*, **20**, 615–630.

Hamilton, S. K. and W. M. Lewis, Jr. (1990a). Physical characteristics of the fringing floodplain of the Orinoco River, Venezuela. *Interciencia*, **15**, 491–500.

Hamilton, S. K. and W. M. Lewis, Jr. (1990b). Basin morphology in relation to chemical and ecological characteristics of lakes on the Orinoco River floodplain, Venezuela. *Archiv für Hydrobiologie*, **119**, 393–425.

Hutchinson, G. E. (1957). *A Treatise on Limnology, 1: Geography, Physics and Chemistry*, Wiley, New York, and Chapman and Hall, London, 1015 pp.

Irion, G. (1984). Sedimentation and sediments of Amazonian rivers and evolution of the Amazonian landscape since Pliocene times. In H. Sioli (ed.), *The Amazon: Limnology and Landscape Ecology of a Mighty Tropical River and its Basin*, Monographiae Biologicae, Vol. 56, Dr. W. Junk, pp. 675–706.

Iriondo, M. H. (1982). Geomorfologia da planície Amazônica. In K. Suguio, M. R. Mousinho de Meis and M. G. Tessler (eds), *Atas do IV Simpósio do Quarternário no Brasil*, Sociedad Brasileira de Geologia, pp. 323–348.

Junk, W. J. (1983). Ecology of swamps on the middle Amazon. In A. J. P. Gore (ed.), *Ecosystems of the World 4B; Mires: Swamp, Bog, Fen and Moor; Regional Studies*, Elsevier, pp. 269–294.

Kaye, B. H. (1989). *A Randomwalk Through Fractal Dimensions*, VCH Verlagsgesellschaft, 421 pp.

Kent, C. and J. Wong (1982). An index of littoral zone complexity and its measurement. *Canadian Journal of Fisheries and Aquatic Sciences*, **39**, 847–853.

Klammer, G. (1984). The relief of the extra-Andean Amazon basin. In H. Sioli (ed.), *The Amazon: Limnology and Landscape Ecology of a Mighty Tropical River and its Basin*, Monographiae Biologicae, Vol. 56, Dr. W. Junk, pp. 47–83.

Korčak, J. (1940). Deux types fondamentaux de distribution statistique. *Bulletin de l'Institut International de Statistique*, **30**, 295–299.

Mandelbrot, B. (1977). *Fractals: Form, Chance and Dimension*, Freeman, 365 pp.

Mandelbrot, B. (1982). *The Fractal Geometry of Nature*, Freeman, 460 pp.

Meade, R. H., C. F. Nordin, Jr., D. P. Hernandez, A. Mejia and J. M. P. Godoy (1983). Sediment and water discharge in río Orinoco, Venezuela and Colombia. In *Proceedings of the Second International Symposium on River Sedimentation*, Water Resources and Electric Power Press, Beijing, pp. 1134–1144.

Meade, R. H., T. Dunne, J. E. Richey, U. de M. Santos and E. Salati (1985). Storage and remobilization of suspended sediment in the lower Amazon River of Brazil, *Science*, **228**, 488–490.

Melack, J. M. (1984). Amazon floodplain lakes: shape, fetch, and stratification. *Proceedings of the International Association of Theoretical and Applied Limnology*, **22**, 1278–1282.

Melack, J. M. and T. R. Fisher (1990). Comparative limnology of tropical floodplain lakes with an emphasis on the central Amazon. *Acta Limnologia Brasiliensia*, **3**, 1–48.

Mertes, L. A. K. (1985). *Floodplain Development and Sediment Transport in the Solimoes–Amazon River, Brazil*. Unpublished MS thesis, University of Washington, Seattle.

Milne, B. T. (1988). Measuring the fractal geometry of landscapes. *Applied Mathematics and Computation*, **27**, 67–79.

Richey, J. E., L. A. K. Mertes, T. Dunne, R. L. Victoria, B. R. Forsberg, A. C. N. S. Tancredi and E. Oliveira (1989). Sources and routing of the Amazon River flood wave. *Global Biogeochemical Cycles*, **3**, 191–204.

Salo, J. (1990). External processes influencing origin and maintenance of inland water–land ecotones. In R. Naiman and H. Décamps (eds), *The Ecology and Management of Aquatic–Terrestrial Ecotones*, Man and the Biosphere Series, Vol. 4, UNESCO and Parthenon, pp. 37–64.

Takayasu, H. (1990). *Fractals in the Physical Sciences*, Manchester University Press, 176 pp.

Tricart, J. (1977). Types de lits fluviaux en Amazonie brésilienne. *Annales de Géographie*, **86**, 1–54.

Welcomme, R. L. (1979). *Fisheries Ecology of Floodplain Rivers*, Longman, 317 pp.

West, B. J. and M. Shlesinger (1990). The noise in natural phenomena, *American Scientist*, **78**, 40–45.

7 Investigating Contemporary Rates of Floodplain Sedimentation

D. E. WALLING, T. A. QUINE and Q. HE
Department of Geography, University of Exeter

INTRODUCTION

The floodplains of most lowland rivers are characterized by extensive deposits of fine sediment resulting from the deposition of suspended sediment during overbank flood events. Where it has proved possible to date these deposits, the results frequently have indicated that deposition has occurred at relatively low rates over extended periods of time. Thus in the case of two UK rivers, Shotton (1978) reports an average rate of deposition of 0.5 cm year^{-1} over the past 3000 years on the floodplain of the Warwickshire Avon, and Brown (1987) refers to typical sedimentation rates of 0.14 cm year^{-1} over the past 10 000 years on the floodplain of the River Severn. Relatively little, however, is known about contemporary rates of overbank deposition, the spatial patterns involved, and the significance of the floodplain as a sediment sink. Such information is needed both to improve our understanding of the contemporary geomorphological development of floodplains and to assess the potential fate of sediment-associated nutrients and contaminants transported through river systems. Furthermore, there is a need to test existing models of floodplain sedimentation against field data in order to verify their assumptions (cf. James, 1985).

The lack of data on contemporary rates of sediment deposition by overbank flows largely reflects the many pratical and conceptual problems associated with any attempt to measure deposition on floodplains. The process inevitably will be characterized by substantial spatial and temporal variability and this introduces major problems for the design of effective sampling strategies. Furthermore, the restriction of the process in many environments to infrequent and essentially random events poses additional problems. Techniques developed for lakes and reservoirs are likely to be inappropriate for floodplains, which are only submerged for short periods of time and where equipment must be deployed in advance of flood inundation.

Existing approaches to documenting contemporary rates of overbank deposition on floodplains may conveniently be subdivided into two groups, largely on the basis of the time-scale involved. The first are essentially event-based and attempt to measure the deposition associated with individual floods. Thus several studies have reported the successful use of sediment traps placed on the floodplain surface in advance of flood events (e.g. Mansikkaniemi, 1985; Gretener and Stromquist, 1987; Lambert and

Lowland Floodplain Rivers: Geomorphological Perspectives. Edited by P.A. Carling and G.E. Petts
© 1992 John Wiley & Sons Ltd

Walling, 1987; Walling and Bradley, 1989a), and the results obtained by these studies provide useful information on the likely magnitude of the rates involved. However, limitations on the sampling densities that can realistically be used, questions of representativeness, difficulties of monitoring very small rates of deposition and the problems of obtaining long-term measurements necessarily restrict the potential of this approach. Attempts also have been made to measure the decrease in suspended sediment load along a floodplain reach and to ascribe the losses to overbank deposition (cf. Lambert and Walling, 1987; Gretener and Stromquist, 1987), but the results obtained can provide information only on the average rate of deposition along the entire reach and provide no indication of the degree of variability or the spatial patterns involved. Furthermore, such results are heavily dependent upon the accuracy and precision of the sediment load measurements, which may frequently be open to question (cf. Walling and Webb, 1989). An alternative approach, which considers individual events, is primarily applicable to rare high-magnitude floods and involves post-event surveys of the resultant sediment deposits aimed at determining their depth and distribution (e.g. McKee, Crosby and Berryhill, 1967; Kesel et al., 1974; Brown, 1983). Other workers, such as Macklin, Rumsby and Newson (1992), also have succeeded in identifying the deposits of known sequences of recent floods within sediment deposits. This general approach is, however, likely to be of limited application under more normal conditions, where deposition rates are low and where it is impossible to distinguish the deposits associated with specific events. For example, in the study reported by Macklin, Rumsby and Newson (1992) vertical accretion rates in the flood deposits investigated in the lower Tyne Valley were noted as being exceptionally high by British standards (2.4 cm year^{-1}) and the authors indicate that the deposits represented 'within-channel sedimentation', rather than 'overbank fines'.

The second group of approaches produces estimates of deposition rates averaged over a number of years and essentially rely on establishing the date of a particular level or horizon and calculating the rate of deposition from the depth of material overlying this level. Various methods have been used to define levels of known date, and these include, firstly, previous benchmark surveys (e.g. Happ, 1968; Leopold, 1973); secondly, the existence of datable surfaces or material (e.g. Costa, 1975); thirdly, relating trace metal concentrations in the sediment profile to the known history of mining activity in the catchment upstream (e.g. Lewin and Macklin, 1987; Popp et al., 1988); and fourthly, the use of fallout radionuclides, such as ^{137}Cs (e.g. Ritchie, Hawks and McHenry, 1975; McHenry, Ritchie and Verdon, 1976; Popp et al., 1988). Although valuable, the first three approaches face limitations in terms of the need for prior surveys, the lack of suitable datable material or objects, and the long time-scales that frequently are involved. In general they have been used to provide estimates of deposition rates at specific points, rather than to study the spatial patterns involved. However, the use of the fallout radionuclide ^{137}Cs would seem to offer particular potential in terms of its global occurrence, the medium-term time-scales involved (i.e. ca. 30 years) and the possibility of assembling data for a large number of points on the floodplain.

Faced with a clear need for increased information on rates and patterns of contemporary floodplain sedimentation and the limitations of existing methods, the authors have explored further the potential for using fallout radionuclides and more particularly ^{137}Cs. Their recent work (e.g. Walling and Bradley, 1989a, b) has

indicated that this approach offers very considerable potential by providing information on average rates of deposition over a clearly defined and spatially consistent
interval of the recent past and a capability for high spatial resolution. Furthermore,
such data can be assembled on the basis of a single site visit and the approach
therefore avoids the need for long-term monitoring and the problems of working on
inundated areas. This contribution outlines the basis for using fallout radionuclides for
investigating contemporary rates of overbank floodplain sedimentation and presents
results from three case studies.

THE APPROACH

Caesium-137 is a man-made radionuclide present in the global fallout of debris
resulting from the testing of nuclear weapons in the atmosphere during the middle
years of the twentieth century. Most of the fallout occurred in the decade between
1956 and 1965, with maximum deposition in 1963, the year of the Nuclear Test Ban
Treaty (Figure 7.1). Since the 1970s, rates of ^{137}Cs fallout have been low, although in
some areas of the world an additional short-term input was received in 1986 as a result
of the Chernobyl accident, which released radiocaesium into the atmosphere.
Chernobyl-derived ^{137}Cs fallout was also accompanied by ^{134}Cs in a near-constant
ratio (cf. Cambray et al., 1987) and it is possible to distinguish bomb-derived and
Chernobyl-derived ^{137}Cs in sediments by measuring the ^{134}Cs content and calculating
the equivalent ^{137}Cs. Caesium-137 has a half-life of 30.17 years and in most environments fallout inputs are strongly adsorbed by clay particles in the surface horizons of
the soil (see Frissel and Pennders, 1983; Livens and Rimmer, 1988). Subsequent to its
deposition as fallout, its redistribution is associated with the erosion, transport and
deposition of sediment particles, and it therefore affords a valuable medium-term
tracer for investigating the erosion, transport and deposition of sediment within the
landscape (e.g. Loughran, Campbell and Elliot, 1982; Ritchie and McHenry, 1990;
Walling and Bradley, 1990).

Caesium-137 has been used widely for dating recent horizons in lake sediment cores
(e.g. Krishnaswami and Lal, 1978). In this application, the chronology commonly is
based either on matching down-profile variations in ^{137}Cs content to the irregular

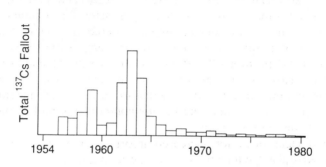

Figure 7.1 The pattern of total annual ^{137}Cs fallout in the Northern Hemisphere. (Based
on Cambray, Playford and Lewis, 1982)

pattern of ^{137}Cs fallout (e.g. Figure 7.1), or on determining the first appearance of the radionuclide in the profile, which will coincide with the onset of significant fallout during the 1950s. In the case of lake cores, the major source of radiocaesium is frequently atmospheric fallout to the lake surface, which subsequently is incorporated into the sediment column. Additional inputs, however, will also be associated with ^{137}Cs fixed to sediment eroded from the drainage basin and deposited in the lake. The latter inputs will tend to blur the temporal record of fallout input, since the radiocaesium content of eroded sediment may reflect more closely the cumulative fallout input to the drainage basin surface rather than the year-to-year variation in fallout amounts.

The use of ^{137}Cs to investigate rates of floodplain deposition may be viewed as a logical development of its application in dating lake cores. The floodplain surface will receive inputs of radiocaesium both directly from atmospheric fallout and in association with deposited sediment eroded from the upstream drainage basin. The latter will frequently dominate the total input. The resultant radiocaesium profile will reflect both the temporal record of these two inputs and any mixing and redistribution that occurs within the subaerially exposed soil profile as a result of biological activity. Floodplain surfaces are commonly occupied by permanent pasture or similar natural vegetation, and such mixing and redistribution is likely to be more pronounced than that occurring in continuously submerged lake sediments. Figure 7.2 compares typical ^{137}Cs profiles (b and c) for two sites under permanent pasture on the floodplain of the River Culm in Devon, UK, with a characteristic profile (a) from permanent pasture in the vicinity of the floodplain, but above the level of inundation. Profile (a) exhibits the typical exponential depth distribution associated with undisturbed sites, which reflects

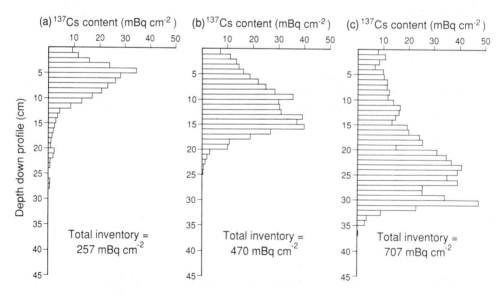

Figure 7.2 Caesium-137 profiles from undisturbed permanent pasture in the Culm Valley, Devon, UK. Profile (a) represents a site above the level of the floodplain, whilst profiles (b) and (c) represent floodplain sites experiencing flood inundation

the effects of biological activity in mixing and redistributing the original radiocaesium input. Because it was collected from an area of flat undisturbed land, the total radiocaesium inventory of this profile (257 mBq cm^{-2}) is indicative of the total local fallout and may be used to estimate a reference value. Profiles (b) and (c) from the two floodplain sites are characterized by higher total ^{137}Cs inventories and by different depth distributions. Both features are consistent with a situation where sediment deposited during overbank deposition has provided an additional input of radiocaesium and has caused an upward 'stretching' of the profile. It is not possible to obtain precise dates for specific levels within the floodplain cores, based on the level at which radiocaesium first appears or the shape of the ^{137}Cs profile, because of the subsequent mixing and redistribution occurring within the soil and demonstrated in Figure 7.2a. Nevertheless, an approximate indication of the depth of deposition during the 30-year period since significant levels of fallout first occurred can be based on the degree of 'stretching' of the profile (see Walling and Bradley, 1989b). In this case the estimates for profiles (b) and (c) are of the order of 10 and 20 cm, and represent average deposition rates of approximately 3 and 6 mm year^{-1}, respectively. At locations where Chernobyl fallout has occurred, the associated short-term input of ^{137}Cs may provide a clearly marked horizon within the sediment profile, which may be used for dating purposes (see Figure 7.10).

In view of the likely spatial variability of floodplain deposition in response to microtopography and local hydraulic conditions, there will frequently be a need to obtain estimates of sedimentation rates at a large number of sites in order to document the patterns involved. The time-consuming nature of gamma spectrometry measurements will, however, frequently preclude the assembly of more than a small number of detailed radiocaesium profiles. An alternative approach to using ^{137}Cs measurements to estimate rates of deposition has therefore been developed (see Walling and Bradley, 1989a, b). In this approach, single whole cores are used to provide values of the total ^{137}Cs inventory (mBq cm^{-2}) at each sampling point. Comparison of these inventories with the reference fallout inventory for the location enables the 'excess' ^{137}Cs to be calculated. This can be ascribed to radiocaesium associated with deposited sediment. If the average ^{137}Cs content of the deposited sediment can be estimated, the value of radiocaesium excess can be converted to an equivalent mass or depth of sediment. In the example provided in Figure 7.2, the average ^{137}Cs content of the deposited sediment has been estimated at 25 mBq g^{-1} using a procedure that is discussed later in this paper. The excess radiocaesium in profiles B and C of 200 and 437 Bq cm^{-2}, respectively, therefore represents approximately 8 and 17 g of sediment and deposition depths of approximately 10 and 21 cm. These results are in close agreement with the estimates based on the degree of stretching of the profile noted previously. A similar procedure can be used for Chernobyl-derived radiocaesium inventories in floodplain sediments. In this case, however, the sediment accumulation will reflect the much shorter period since April 1986.

The limitations of this whole-core approach must, however, be recognized. One potential limitation relates to the use of a single common value for the average ^{137}Cs content of deposited sediment, when converting the excess radiocaesium into an estimate of the depth of deposition. It is well known that radiocaesium is likely to be associated preferentially with the finer fraction of deposited sediment and the use of a common value assumes that the grain size distribution of deposited sediment will be essentially uniform across the floodplain. In many situations this assumption will be reasonable. However, where the grain size distribution of deposited sediment is more variable, the use of a common concentration value, representing an average grain-size distribution, may underestimate deposition depths in areas where coarser sediment is deposited and overestimate deposition at points where finer sediment has been deposited. Furthermore, the low radiocaesium content of sand-sized sediment means that sand deposits will be marked by relatively small increases of ^{137}Cs inventory. Estimates of deposition rates based on ^{137}Cs measurements relate only to fine (clay and silt-sized) sediment. A second potential problem involves the fact that the estimates of accumulation depth or rate obtained are aggregate values for a period extending over about 35 years. These aggregate values may reflect considerable inter-annual variation in deposition depths in response to flood magnitude and frequency and may also conceal longer term trends in the flood and sediment transport regimes such that deposition rates increase or decline during the period. These potential restrictions must, however, be balanced againt the numerous limitations and oper-ational problems associated with any alternative means of collecting equivalent data.

Further illustration and discussion of the use of radiocaesium measurements in documenting rates and patterns of recent overbank deposition on floodplains is provided by considering the results obtained in three case studies. The first introduces work undertaken on the floodplain of the River Culm in Devon, UK; the second reports a study of part of the floodplain of the River Severn near Tewkesbury, UK; and the third reports an investigation of the River Leira in southern Norway.

THE RIVER CULM CASE STUDY

The example presented relates to a detailed investigation of the rates and pattern of recent floodplain sedimentation within a short (ca. 250 m) stretch of the River Culm floodplain near Exeter (Figure 7.3a). Attention focused on the floodplain to the southeast of the river, which was about 200 m wide. The river flows in a meandering gravel-bed channel that is about 16 m wide. The banks are up to 1 m high and are formed largely of fine alluvial material. Overbank flooding is relatively frequent during the winter months and substantial inundation of this area of floodplain generally occurs on about seven occasions each year. Depths of inundation vary with the local topography, but are typically of the order of 40 cm for the mean annual flood and 70 cm for a 50-year flood. The microtopography of the area shown in Figure 7.3a consists of small elevated levee features close to the channel and a broad depression along the outer edge of the floodplain.

More than 150 whole-core samples were collected from this area of floodplain using a grid-based sampling network. Cores were collected using a purpose-built motorized (38 cm^2) percussion corer to depths of approximately 65 cm. The basal section of each core was retained for separate gamma assay, in order to ensure that the core had penetrated the full depth of caesium-bearing sediment. The core samples were returned to the laboratory where they were dried, ground and sieved to < 2 mm prior to measurement of their ^{137}Cs content. These measurements were undertaken using

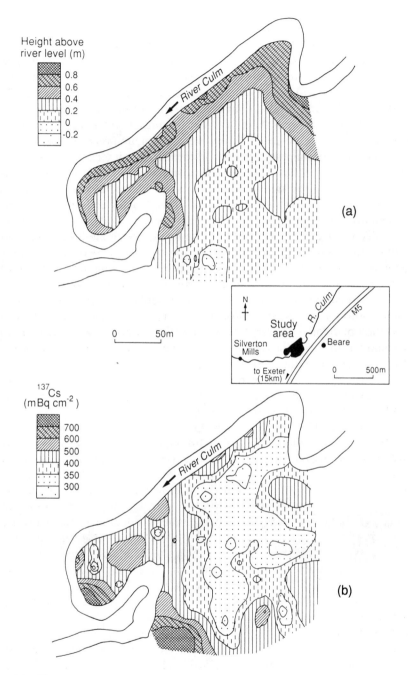

Figure 7.3 The location and microtopography of the study reach of the River Culm floodplain (a) and the pattern of ^{137}Cs inventories demonstrated by the whole-core samples (b)

an Ortec germanium coaxial detector linked to a multi-channel analyser. Count times were typically of the order of 25 000 s, providing a precision of ca. ± 10% at 2 SD.

The resultant ^{137}Cs inventory data have been mapped using a computerized interpolation routine (UNIRAS), and the results presented in Figure 7.3b demonstrate a range between < 300 mBq cm^{-2} and > 700 mBq cm^{-2}. High ^{137}Cs inventory values signify areas with relatively high rates of deposition, whereas low inventory values are associated with areas with relatively low rates of deposition. The core inventories exceeded the local fallout reference inventory of 270 mBq cm^{-2} at all sites, suggesting that scour was unlikely to be a significant process at this location. The pattern of caesium inventories indicates that maximum rates of deposition have occurred in areas bordering the channel and towards the outer edge of the floodplain.

Conversion of Figure 7.3b to a map of deposition rates requires an estimate of the mean ^{137}Cs content of deposited sediment during the period 1955 to the present. This was obtained by detailed analysis of several ^{137}Cs profiles collected from the study area (e.g. Figure 7.4). Using the procedure proposed by Walling and He (1991), it is possible to deconvolve the ^{137}Cs profile for a depositional site into the atmospheric fallout component and the contribution from deposited sediment. The result for one of the profiles is shown in Figure 7.4. By simulating the accumulation of ^{137}Cs fallout within the upstream catchment and its subsequent mobilization by erosion, the deposition of sediment-associated ^{137}Cs on the floodplain, the incorporation of direct fallout inputs into the accreting floodplain surface, and post-depositional mobility of ^{137}Cs with the floodplain sediments, the deconvolution procedure is able to separate the contributions to the total ^{137}Cs inventory made by atmospheric fallout and by deposited sediment over the period 1954 to present. This in turn provides annual estimates of the ^{137}Cs content of deposited sediment during the years 1955–1990 (Figure 7.4b), which can be averaged over the period. Values of this average obtained for several profiles collected from the study area ranged between 19.2 mBq g^{-1} and 30.3 mBq g^{-1} and yielded a mean value of 25 mBq g^{-1}. Since measurements provided a typical dry bulk density for the deposited sediment of 0.8 g cm^{-3}, an inventory to deposition rate conversion factor of 20 mBq cm^{-2} per centimetre of deposition was

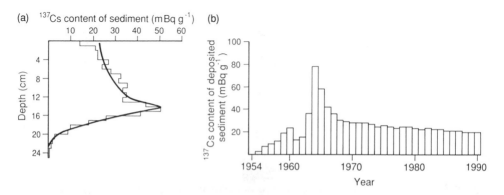

Figure 7.4 Deconvolution of a ^{137}Cs profile from the study site on the River Culm floodplain: (a) indicates the model fit and (b) depicts the temporal trend of the ^{137}Cs content of deposited sediment during the period 1955–1989 estimated using the model

applied to Figure 7.3b in order to produce the map of deposition rates presented in Figure 7.5.

The mean annual rates of deposition depicted in Figure 7.5 assume a constant rate of accretion at each sampling point over the period 1955–1989 and range between < 1 and > 7 mm year^{-1}. From a geomorphological perspective, the pattern shown in Figure 7.5 indicates that maximum rates of deposition are associated with areas close to the channel and may at least in part reflect deposition in areas of active channel migration, such as the inside of the main meander bend. Over the majority of the floodplain, however, rates of general vertical accretion are much lower and of the order of 0–2 mm year^{-1}. Equally, from a sedimentological viewpoint, the pattern indicates that in this location the overbank deposition of fines is not spatially uniform. Rather, deposition is preferentially associated with zones close to the channel and more particularly zones where, during floods, the microtopography promotes migration of sediment from the main channel into the ponded water on the floodplain, where it is deposited. The resultant pattern conforms to the concept of transfer of sediment on to the floodplain by turbulent diffusion proposed by Allen (1985) and James (1985). This concept is further supported by measurements of the grain size distribution of sediment from the floodplain surface, which demonstrate an increased percentage of sand (> 63 μm) in these areas (i.e. ca. 30%) as compared with the inner areas of the floodplain, where values of the order of 10–15% are more typical. A different mechanism must be sought to account for the increased rates of deposition occurring towards the outer margin of the floodplain. These appear to reflect an area of secondary convective flow along this marginal depression. However, the grain size distribution of surface sediment shows no significant differentiation between the

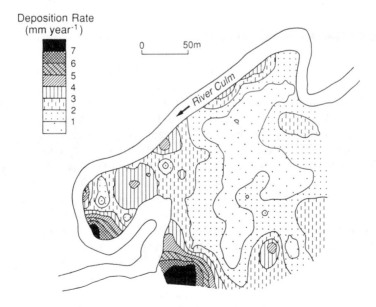

Figure 7.5 The spatial pattern of floodplain deposition rates in the study reach of the River Culm floodplain during the period 1955–1989 estimated from the ^{137}Cs inventories of the whole-core samples

marginal depression and the central area of the floodplain, suggesting that the increased deposition in the former area reflects a slow renewal or exchange of ponded water and associated suspended sediment, rather than the transport of coarser sediment from the main channel. Relatively low rates of deposition are associated with the main body of ponded water occupying the central depression, where more optimal conditions for deposition (i.e. slack water) might be expected to occur. Limited exchange between this body and the channel during overbank floods means that deposition is effectively limited to the sediment contained within the water that initially inundated the floodplain. However, since the highest suspended sediment concentrations are commonly associated with the rising stage of the flood on this river, significant deposition still occurs.

The rates of deposition referred to above represent contemporary rates associated with the past 35 years. It is tempting to extrapolate them back through time and to consider their implications for long-term floodplain development. Any such extrapolation must, however, proceed with caution, since the hydrological and sedimentological regime of the river is likely to have changed in the recent past as a result of landuse change in the catchment upstream and changes within the floodplain itself. Landuse change is likely to have produced an increase in flood magnitude and frequency and in sediment loads and concentrations. A general deterioration in the system of distributary channels and leats that formerly occupied the floodplain also is likely to have reduced the flood capacity of the channel system. Both factors are likely to give rise to increased rates of floodplain sedimentation as the floodplain readjusts. In the medium term, a general raising of the level of the floodplain surface, without concomitant channel aggradation, could be expected to lead to a reduced frequency of overbank flooding and therefore reduced rates of floodplain sedimentation.

THE TEWKESBURY HAM CASE STUDY

This case study involved an investigation of a small portion of the floodplain of the River Severn near Tewkesbury, lying between the main channel of the River Severn and that of the Mill Avon (Figure 7.6a). The 'island' enclosed between the River Severn and the Mill Avon is known locally as Tewkesbury Ham. The microtopography of the study area is marked by elevated areas close to both of the channels, with an extensive intervening lower area. Several small depressions occur within this lower area, and the downstream portion of the eastern bank of the River Severn is bounded by a well-defined ditch of uncertain origin, which was used as a withy bed up to 80 years ago. The entire area of Tewkesbury Ham is inundated during major flood events to depths of the order of 1 m and flooding of at least part of the Ham occurs at least once per year. Because the headwaters of the River Severn had received fallout from the Chernobyl accident (see Walling, Bradley and Rowan, 1989), it was possible to use both bomb-derived and Chernobyl-derived radiocaesium to interpret rates and patterns of floodplain sedimentation at this location. The overall approach followed quite closely that described for the previous case study.

Sixty cores were collected to a depth of 60 cm from the area of the floodplain under investigation, using radial transects. These cores were sectioned into 10 cm increments prior to gamma assay and the total inventories of ^{137}Cs and ^{134}Cs were calculated by

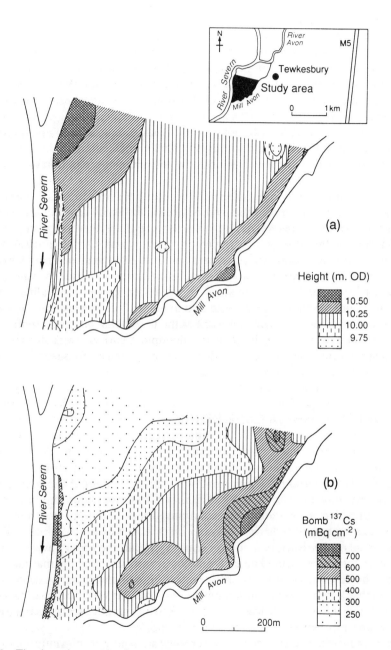

Figure 7.6 The location and microtopography of the study reach of the River Severn floodplain at Tewkesbury Ham (a) and the pattern of bomb-derived ^{137}Cs inventories demonstrated by the whole-core samples (b)

summing the values obtained from the individual sections. Sectioning of the core was undertaken in order to confirm that the cores included the total depth of caesium-bearing sediment and to provide increased precision in the measurements of Chernobyl-derived [134]Cs which was confined to the upper levels of the cores. In all cases, ^{137}Cs was absent from the lower sections of the cores. The measurements of [134]Cs were back-corrected for decay to April 1986 and these values were used to calculate the contribution of Chernobyl-derived ^{137}Cs to the total ^{137}Cs inventory for each core. The ratio of [134]Cs to ^{137}Cs for Chernobyl fallout of 0.6 proposed by Cambray *et al.* (1987) was used in this calculation. By separating the bomb-derived and Chernobyl-derived contributions to the total radiocaesium inventory, it was possible to investigate rates and patterns of floodplain sedimentation over both the period 1954–1989 and the much shorter period since the Chernobyl disaster (i.e. 1986–1989).

Figure 7.6b presents a map of bomb-derived ^{137}Cs inventories for the area of floodplain investigated. In contrast to the River Culm study reported previously, the inventories of several cores were less than the local fallout reference value of 230 mBq cm^{-2}. This suggests that scour has occurred at these points. Over the remainder of the floodplain, ^{137}Cs inventories are significantly in excess of the reference. Conversion of the ^{137}Cs inventory data presented in Figure 7.6b was undertaken using the same procedure used in the River Culm study. An example of the deconvolution of one of the radiocaesium depth profiles is presented in Figure 7.7. The average bomb-derived ^{137}Cs content of deposited sediment was estimated to be 16 mBq g^{-1}, which, coupled with an estimate of average bulk density, provided a conversion factor of 14.5 mBq cm^{-2} per centimetre of deposition. The resultant map of deposition rates over the period 1954–1989 is presented in Figure 7.8.

The pattern of deposition depicted in Figure 7.8 evidences a gradual increase from northwest to southeast. This corresponds in part to the microtopography, which shows a similar trend, but there is also evidence of preferential deposition adjacent to the course of the Mill Avon, where the maximum values of ca. 10 mm $year^{-1}$ occur. Little deposition occurs adjacent to the main channel of the River Severn, except within the deep ditch described previously and on the narrow strip of land between the ditch and the channel. The general lack of deposition along the margin of the main channel of the River Severn and the zone of zero deposition and scour towards the northwest of the area studied can be tentatively accounted for in terms of the relatively high velocities and higher altitudes associated with this area of the floodplain. Where, as along the base of the ditch, conditions are more favourable for deposition, significant depths of accretion do occur. The high rates of deposition along the margin of the Mill Avon may be accounted for in terms of transfer of sediment from the channel, which, during overbank floods, represents a major flow path for water originating from both the Severn and the Avon upstream. Lower velocities undoubtedly occur in this area and these are more conducive to deposition.

Examination of the distribution of Chernobyl-derived [134]Cs inventories across the area of floodplain studied provides a means of assessing the pattern and rate of deposition during the much shorter period 1986–1989. The map of [134]Cs inventories is presented in Figure 7.9a. There was no local fallout of [134]Cs in this area and the existence of substantial quantities of this radionuclide on the floodplain can be accounted for only by deposition of sediment transported from the headwaters of the

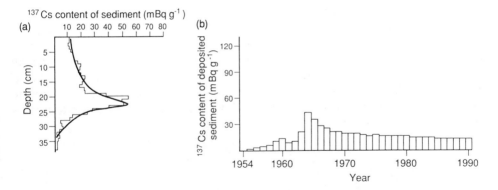

Figure 7.7 Deconvolution of a bomb-derived ^{137}Cs profile from the study site on the River Severn floodplain: (a) indicates the model fit and (b) depicts the temporal trend of the ^{137}Cs content of deposited sediment during the period 1955–1989 estimated using the model

Severn Basin, which received substantial levels of Chernobyl fallout (see Walling, Bradley and Rowan, 1989). The pattern of ^{134}Cs inventories exhibits a similar trend to that shown by the bomb-derived caesium data (Figure 7.6b), although there is more evidence of deposition close to the main channel of the River Severn. Conversion of Figure 7.9a to a map of deposition rates (Figure 7.9b) has been based on a series of direct measurements of the Chernobyl-derived ^{134}Cs content of suspended sediment undertaken on the River Severn during floods at the gauging station at Saxon's Lode, 8 km upstream of Tewkesbury, during the period 1986–1989 (see Walling, Bradley and Rowan, 1989). The mean ^{134}Cs content of suspended sediment during this period,

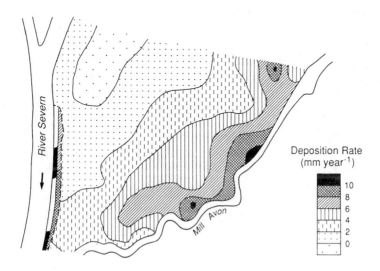

Figure 7.8 The spatial pattern of floodplain deposition rates in the study reach of the River Severn floodplain for the period 1955–1989 estimated from the bomb-derived ^{137}Cs inventories of the core samples

weighted according to the magnitude and timing of the major floods, has been estimated as 27 mBq g^{-1}. Combined with an estimate of the mean bulk density of the surface sediments of the floodplain, this provides a conversion factor of 24.5 mBq cm^{-1}. It is important to recognize that this value is based on the ^{134}Cs content of suspended sediment in transport, rather than deposited sediment, and it is possible that it overestimates the ^{134}Cs content of deposited sediment, since the latter could be expected to be somewhat coarser. The resultant map of sediment deposition rates during the period 1986–1989 is presented in Figure 7.9b.

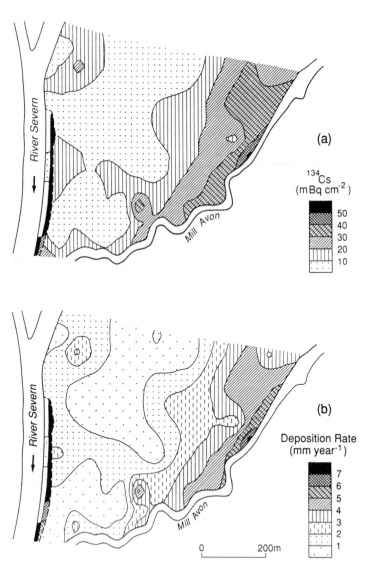

Figure 7.9 The spatial pattern of Chernobyl-derived ^{134}Cs inventories in the study reach of the River Severn floodplain (a) and the associated estimates of floodplain deposition rates during the period 1986–1989 (b)

The rates and pattern of deposition depicted on Figure 7.9b conform closely to those shown on Figure 7.8, although the rates are, in general, only about 60% of those associated with the longer period (1954–1989). This is not unexpected, since rates of floodplain deposition will reflect flood magnitude and the period 1986–1989 included only floods of moderate magnitude. The highest flood during this period had a return period of approximately 5 years. The general concordance between the patterns evidenced by Figures 7.8 and 7.9b nevertheless suggests that at this location the overall pattern of accretion associated with moderate floods is not significantly different from that associated with a period containing higher magnitude events, and that the overall pattern of floodplain sedimentation is a response to a large number of floods, rather than a few high-magnitude events. Some minor differences between the patterns of deposition during the two periods are, however, apparent. More particularly, there is more evidence of deposition adjacent to the main channel of the River Severn and the area of zero deposition depicted on Figure 7.8 is absent from Figure 7.9b. This suggests that scour may be restricted to events of a greater magnitude than those that occurred during the period 1986–1989 and that the high velocity conditions that restrict deposition close to the main channel of the River Severn are associated particularly with high-magnitude events. The existence of high ^{134}Cs inventories in the sediments deposited adjacent to the Mill Avon (Figure 7.9a) provides indirect evidence that the River Severn, rather than the River Avon, is the main source of this sediment. Only very limited Chernobyl fallout occurred over the catchment area of the River Avon (see Walling, Quine and Rowan, 1991) and the ^{134}Cs concentrations found in the deposited sediment are of a similar magnitude to those associated with suspended sediment collected from the gauging station on the River Severn at Saxon's Lode, above the Avon confluence, during the post-Chernobyl period. This further suggests that the course of the Mill Avon provides a secondary route for floodwater from the River Severn during periods of overbank spillage, and that conditions within this zone are more conducive to deposition in response to diffusion-type transfer processes than within the area adjacent to the main channel of the River Severn.

THE RIVER LEIRA CASE STUDY

This case study, which refers to work undertaken by the authors in the catchment of the River Leira in southern Norway, has been included in order to demonstrate the potential for using radiocaesium profiles to obtain information on rates of sediment deposition in environments where accretion is more rapid than in the Rivers Culm and Severn. This area of Norway received significant quantities of Chernobyl fallout and because sedimentation rates are relatively rapid, it has proved possible to use the Chernobyl input to define a datable horizon within the sediment profile.

The lower reaches of the River Leira, which are located about 25 km to the northeast of Oslo, drain an area of marine clays that have been cultivated intensively for cereal production during the past 25 years. In order to develop this area for arable cultivation, large-scale earth moving has been undertaken in order to obliterate many of the gullies that formerly dissected the area and to regrade the fields. Locally these operations are referred to as 'planation'. The River Leira flows into Lake Øyeren and in recent years concern has been expressed over the increased sediment loads of the river, which are detrimental to the water quality of the lake. Some workers have linked these increased sediment loads to the expansion of arable cultivation in the catchment. Studies reported by Bogen and Nordseth (1986) have, for example, indicated that suspended sediment yields from small undisturbed catchments in this area may be as low as 15 t km^{-2} year^{-1}, whereas sediment yields of 530 t km^{-2} year^{-1} have been documented in a small basin where extensive regrading and expansion of cultivation had occurred. Increased sediment loads have been coupled with evidence of increased channel and floodplain sedimentation. Large berms of sediment appear to have accumulated at the margins of the main channel in recent years, but their rates of accretion have not been documented. Radiocaesium measurements were used by the authors to elucidate the rates of deposition involved, and two examples of the results obtained are presented.

The first site investigated involved a small portion of the Leira floodplain near Krokfoss. This was well above normal flood levels, but there was evidence that it would be inundated during severe floods. The radiocaesium profile from this site is presented in Figure 7.10. In this case, both the bomb-derived and the Chernobyl-derived components of the total ^{137}Cs content of the sediment have been plotted. The Chernobyl-derived component was calculated from measurements of the ^{134}Cs content of the sediment. The ^{137}Cs profile sampled in October 1989 shows clear evidence of a buried surface at 9–10 cm depth, which has received substantial amounts of both bomb and Chernobyl fallout. This evidence coincided with field evidence of a buried soil surface provided by organic debris. The overlying sediment must therefore post-date April 1986. The existence of high levels of both bomb and Chernobyl caesium on the buried surface suggests that the surface remained exposed throughout the period from the late 1950s until after the Chernobyl accident. Discharge records for the river indicate that a major flood occurred in 1987 and that this was the greatest flood for several decades. This undoubtedly was responsible for depositing the 9 cm of sediment that now cover the buried surface. At this location, sediment accretion results from high-magnitude infrequent floods that cause inundation of this area of the floodplain.

The second site was located about 15 km downstream at Frogner (Figure 7.10). Here a vertical profile was excavated to a depth of nearly 1.1 m in a berm deposited at the side of the main channel. The bomb-derived and Chernobyl-derived ^{137}Cs content of sediment from this profile has been plotted in Figure 7.10. In this case substantial levels of bomb-derived caesium occur throughout the entire depth of the profile, indicating that the deposits post-date the onset of bomb fallout in the late 1950s. Furthermore, the absence of any significant enhancement of the bomb-derived caesium levels within the profile, which could be related to the period of peak fallout in the late 1950s and early 1960s (see Figure 7.1) suggests that the deposition of sediment at the base of the profile post-dates that period. The bomb-derived caesium data indicate that 1 m of deposition has occurred at this site over a period of something less than 25 years. The existence of a pronounced peak in the Chernobyl-derived ^{137}Cs levels at a depth of about 35 cm indicates that this level represents the surface exposed to fallout in April–May 1986. The 35 cm of overlying sediment have been deposited during the succeeding 3 years prior to sample collection in September 1989. The major flood that occurred in 1987, and which was referred to above, undoubtedly made a major contribution to this rapid accretion. However, the current

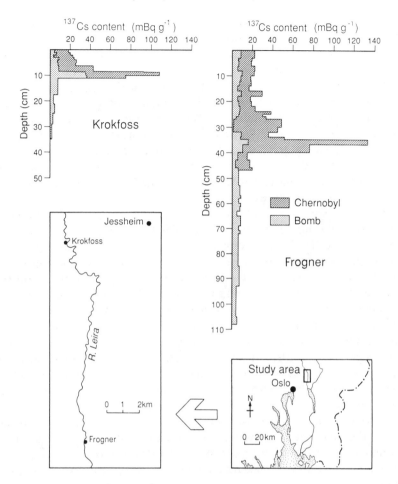

Figure 7.10 The location of the study sites at Krokfoss and Frogner on the River Leira, Norway, and the Chernobyl-derived and bomb-derived ^{137}Cs profiles recorded for the two sites

surface at this site is only about 2.5 m above the normal water stage and the record of sedimentation undoubtedly reflects contributions from a large number of floods, rather than a few high-magnitude events.

PERSPECTIVE

Any attempt to assemble representative information on contemporary rates of over-bank floodplain sedimentation that is capable of meeting the current needs of the geomorphologist must face a number of important problems. These include oper-ational difficulties as well as the fundamental problems of attempting to document processes that are highly variable in both time and space. The use of fallout radionuclides associated with bomb tests and the Chernobyl accident would appear to overcome many of these problems and this contribution reports three case studies where this approach has been applied successfully. In the first, detailed information on rates and patterns of floodplain sedimentation was assembled for a small area of the floodplain of the River Culm in Devon, UK. Rates of accretion averaged over the past 35 years ranged between 0 and 7 mm year^{-1} and the evidence provided on the spatial distribution of deposition rates enabled general conclusions to be advanced as to the main processes involved. Such data also could provide a useful basis for testing and developing existing models of overbank sedimentation and for exploring relationships between sedimentation rates and the microtopography of the floodplain. The second case study afforded similar information for a small portion of the floodplain of the River Severn near Tewkesbury where deposition rates ranged between 0 and 10 mm year^{-1}. However, by exploiting both bomb-derived and Chernobyl-derived fallout radionuclides at this site it was possible to compare the rates and patterns of deposition occurring during the period 1954–1989 with those operating during the shorter period 1986–1989. This comparison confirmed the stability of the general pattern of deposition during the overall period 1954–1989, but also permitted some tentative conclusions regarding the significance of flood magnitude. Again considerable potential clearly exists to use the resultant data to develop and test models of overbank floodplain sedimentation. The third case study focused on the River Leira in southern Norway, where much more rapid rates of deposition had occurred as a result of landuse change. Here, attention focused on the potential for using the record of fallout radionuclide concentrations in the sediment column to date specific levels and therefore to calculate rates of accretion. At one site there was evidence that the 9 cm of deposition since the late 1950s was associated with a single high-magnitude flood event that occurred in 1987. At the other site, accretion was more continuous and averaged about 4 cm year^{-1}, although of the order of 35 cm of deposition had occurred in the short period since 1986. The major flood that occurred in 1987 probably accounted for a large proportion of this deposition.

The use of fallout radionuclides undoubtedly offers very considerable potential for documenting rates and patterns of contemporary floodplain sedimentation. No other currently available technique appears capable of providing equivalent spatial resolution or of yielding information relating to a consistently defined period (35 years) from a single site visit. The temporally aggregated nature of the data obtained might appear a disadvantage in certain contexts, but against this must be set the great difficulty of obtaining temporally representative data from short-term measurements, even if appropriate techniques were available. The possibility of obtaining detailed information on the spatial patterns involved could provide a valuable basis for integrating field measurement and model development, since there have to date been few attempts to test existing models against field data owing to the lack of appropriate measurements.

The approach is, however, not without its problems. These include the time consuming nature of gamma spectrometry measurements (typical count times for individual samples were of the order of 25 000 s), and the possible complications introduced in some areas by the additional inputs of ^{137}Cs associated with Chernobyl fallout. Because of the short half-life of ^{134}Cs (2.06 years) it will soon be impossible to distinguish the bomb-derived and Chernobyl-derived contributions to the total ^{137}Cs inventory, using this indicator. However, in some locations, as in the River Leira example, the existence of an additional input of ^{137}Cs in 1986 provides added possibilities for quantifying recent rates of depositon.

Further work is required to refine the approach and the procedures used. This should include consideration of the effects of the grain size distribution of deposited sediment on the ^{137}Cs inventory. As noted above, the use of a common factor for converting values of 'excess' ^{137}Cs to estimates of deposition depths may underestimate rates of deposition in areas characterized by coarser sediment and overestimate the rates in areas where finer sediment predominates. The main feature of the overall pattern of deposition revealed by ^{137}Cs measurements are, however, unlikely to be changed.

ACKNOWLEDGEMENTS

The authors gratefully acknowledge the support of the Norwegian Water Resources and Energy Administration and of Dr Jim Bogen in the work undertaken on the River Leira, Norway, the cooperation of the landowners in permitting access to the field sites on the River Culm and the River Severn, the valuable contribution of Dr Steve Bradley to early work on the use of ^{137}Cs to monitor floodplain sedimentation on the River Culm, the assistance of Mr Jim Grapes with gamma spectrometry, and the efforts of Mr Rodney Fry in producing the figures. The work on the Rivers Culm and Severn reported here formed part of two larger research projects supported by the Natural Environment Research Council.

REFERENCES

Allen, J. R. L. (1985). *Principles of Physical Sedimentology*, George Allen and Unwin, 272 pp.
Bogen, J. and K. Nordseth (1986). Sediment yields of Norwegian rivers. In *Partikulaert Bundet Stoftransport i Vand og Jorderosjon*, Nordisk Hydrologisk Program Rapport nr. 14, pp. 233–251.
Brown, A. G. (1983). An analysis of overbank deposits of a flood at Blandford-Forum, Dorset, England. *Revue de Geomorphologie Dynamique*, **32**, 95–99.
Brown, A. G. (1987). Holocene floodplain sedimentation and channel response of the Lower River Severn, UK. *Zeitschrift für Geomorphologie*, **31**, 293–310.
Cambray, R. S., K. Playford and G. N. J. Lewis (1982). Radioactive Fallout in Air and Rain: Results to end of 1981. Report no. AERE-R 10485, UK Atomic Energy Authority, 43 pp.
Cambray, R. S., P. A. Cawse, J. A. Garland, J. A. B. Gibson, P. Johnson, G. N. J. Lewis, D. Newton, L. Salmon and B. O. Wade (1987). Observations on radioactivity from the Chernobyl accident. *Nuclear Energy*, **26**, 77–101.
Costa, J. E (1975). The effects of agriculture on erosion and sedimentation in Piedmont Province, Maryland. *Bulletin of the Geological Society of America*, **86**, 1281–1286.
Frissel, M. J. and R. Pennders (1983). Models for the accumulation and migration of ^{90}Sr, ^{137}Cs, $^{239, 240}$Pu and ^{241}Am in the upper layers of soils. In P. J. Coughtrey (ed.), *Ecological Aspects of Radionuclide Release*, Blackwell Scientific, pp. 63–72.
Gretener, B. and L. Stromquist (1987). Overbank sedimentation rates of fine grained sediments. A study of the recent deposition in the Lower River Fyrisan. *Geografiska Annaler*, **69A**, 139–146.
Happ, S. C. (1968). Valley sedimentation in north-central Mississippi. *Proceedings of the Mississippi Water Resources Conference*, Jackson, 1968, pp. 1–8.
James, C. S. (1985). Sediment transfer to overbank sections. *Journal of Hydraulic Research*, **23**, 435–452.
Kesel, R. H., K. C. Dunne, R. C. McDonald, K. R. Allison and B. E. Spicer (1974). Lateral erosion and overbank deposition on the Mississippi River in Louisiana caused by 1973 flooding. *Geology*, **1**, 461–464.
Krishnaswami, S. and D. Lal (1978). Radionuclide limnochronology. In A. Lerman (ed.), *Lakes – Chemistry, Geology, Physics*, Springer-Verlag, pp. 153–177.

Lambert, C. P. and D. E. Walling (1987). Floodplain sedimentation: a preliminary investigation of contemporary deposition within the lower reaches of the River Culm, Devon, UK. *Geografiska Annaler*, **69A**, 393–404.

Leopold, L. B. (1973). River channel change with time: an example. *Geological Society of America Bulletin*, **84**, 1845–1860.

Lewin, J. and M. G. Macklin (1987). Metal mining and floodplain sedimentation in Britain. In V. Gardiner (ed.), *International Geomorphology 1986* I, Wiley, pp. 1009–1027.

Livens, F. R. and D. L. Rimmer (1988). Physico-chemical controls on artificial radionuclides in soils. *Soil Use and Management*, **4**, 63–69.

Loughran, R. J., B. L. Campbell and G. L. Elliott (1982). The identification and quantification of sediment sources using 137-Cs. In *Recent Developments in the Explanation and Prediction of Erosion and Sediment Yield*, Proceedings of the Exeter Symposium, July 1982, International Association of Hydrological Sciences Publication no. 137, pp. 361–369.

Macklin, M. G., B. T. Rumsby and M. D. Newson (1992). Historic floods and vertical accretion of fine-grained alluvium in the Lower Tyne valley, North East England. In P. Billi, R.D. Hey, P. Tacconi and C.R. Thorne (eds.), *Dynamics of Gravel Bed Rivers*, Wiley, (in press).

Mansikkaniemi, H. (1985). Sedimentation and water quality in the flood basin of the River Kyronjoki in Finland. *Fennia*, **163**, 155–194.

McHenry, J. R., J. C. Ritchie and J. Verdon (1976). Sedimentation rates in the Upper Mississippi River. In *Rivers '76*, Vol. II, American Society of Civil Engineers, pp. 1339–1349.

McKee, E. D., E. J. Crosby and H. L. Berryhill (1967). Flood deposits, Bijou Creek, Colorado, June 1965. *Journal of Sedimentary Petrology*, **37**, 829–851.

Popp, C. L., J. W. Hawley, D. W. Love and M. Dehn (1988). Use of radiometric (Cs-137, Pb-210), geomorphic and stratigraphic techniques to date recent oxbow sediments in the Rio Puerco drainage Grants uranium region, New Mexico. *Environmental Geology and Water Science*, **11**, 253–269.

Ritchie, J. C. and J. R. McHenry (1990). Application of radioactive fallout cesium-137 for measuring soil erosion and sediment accumulation rates and patterns: a review. *Journal of Environmental Quality*, **19**, 215–233.

Ritchie, J. C., P. H. Hawks and J. R. McHenry (1975). Deposition rates in valleys determined using fallout cesium-137. *Bulletin of the Geological Society of America*, **86**, 1128–1130.

Shotton, F. W. (1978). Archaeological inferences from the study of alluvium in the lower Severn – Avon valleys. In S. Limbrey and J. G. Evans (eds), *Man's Effect on the Landscape: the Lowland Zone*, Council for British Archaeology Research Report, 21, pp. 27–32.

Walling, D. E. and S. B. Bradley (1989a). Rates and patterns of contemporary floodplain sedimentation: a case study of the River Culm, Devon, UK. *Geojournal*, **19**, 53–62.

Walling, D. E. and S. B. Bradley (1989b). Use of caesium-137 measurements to investigate rates and patterns of recent floodplain sedimentation. *Proceedings of the Fourth International Symposium on River Sedimentation*, Beijing, November 1989, pp. 1451–1458.

Walling, D. E. and S. B. Bradley (1990). Some applications of caesium-137 measurements in the study of fluvial erosion, transport and deposition. In *Erosion, Transport and Deposition Processes, Proceedings of the Jerusalem Workshop*, International Association of Hydrological Sciences Publication no. 189, pp. 179–203.

Walling, D. E. and Q. He (1992). Interpretation of caesium-137 profiles in lacustrine and other sediments: the role of catchment-derived inputs. *Hydrobiologia*, (in press).

Walling, D. E. and B. W. Webb (1989). The reliability of rating curve estimates of sediment yield: some further comments. In *Sediment Budgets, Proceedings of the Porto Alegre Symposium*, International Association of Hydrological Sciences Publication no. 174, pp. 337–350.

Walling, D. E., S. B. Bradley and J. S. Rowan (1989). Sediment-associated transport and redistribution of Chernobyl fallout radionuclides. In *Sediment and the Environment, Proceedings of the Baltimore Symposium*, International Association of Hydrological Sciences Publication no. 184, pp. 37–45.

Walling, D. E., T. A. Quine and J. S. Rowan (1992). Fluvial transport and redistribution of Chernobyl fallout radionuclides. *Hydrobiologia,* (in press).

8 Palaeochannels, Palaeoland-surfaces and the Three-dimensional Reconstruction of Floodplain Environmental Change

A. G. BROWN
Department of Geography and School of Archaeological Studies, University of Leicester

and

M. KEOUGH
Department of Geography, University of Leicester

INTRODUCTION AND BACKGROUND

A major difference between process work on channel change and floodplain construction, and work on the fluvial sedimentary record has been that, while the former has been necessarily three-dimensional, the latter has been restricted generally to two-dimensions using boreholes and exposures. Although multiple exposures do give lateral variation in stratigraphy, including facies change, it is generally at a much larger often regional or macro-reach scale (Brown, 1989). The work described here is an attempt to reconstruct floodplain and channel environments at a reach scale from Lateglacial and Holocene fills of the River Nene in the East Midlands. An additional justification for the reach-scale and three-dimensional approach is that this is the relevant scale for archaeological questions, such as those concerning site location, function and change. At this scale microstratigraphy is often extremely difficult to interpret using standard fluvial sedimentology. This reflects the variety of abiotic and biotic processes in the form of large organic debris, which occur in the river corridor. Indeed the role of biotic processes on stream channels received considerable research in the 1970s and 1980s (Keller and Swanson, 1979; Kellery and Tally, 1979), but its recognition in the sedimentary record remains problematic, as does the recognition of subaerial processes on floodplain surfaces. Vegetation, trees in particular, have, however, been used to investigate floodplain evolution through the stand age pattern created by channel changes (Sigafoos, 1964). Archaeologists have encountered tree trunks and dug-out canoes in floodplain excavations and these have in some cases been used to reconstruct palaeochannel systems and to date channel changes. The best example of this in Britain is at Colwick on the River Trent near Nottingham,

Lowland Floodplain Rivers: Geomorphological Perspectives. Edited by P.A. Carling and G.E. Petts
© 1992 John Wiley & Sons Ltd

where over 50 trunks and the remains of fish weirs indicate considerable channel change since the Medieval period, and, rather uncharacteristically for a large lowland river in the UK, gravel deposition and scroll-bar construction across a significant portion of the floodplain (Salisbury *et al.*, 1984). Large organic debris is far more common in the rivers of central Europe, which have higher stream powers and still flow through forested landscapes. Becker and Schirmer (1977) have used them to provide a dendrochronological framework for floodplain evolution of the Main and upper Danube. The spatial variation of sedimentary and ecological environments on floodplains and the temporal variation of these environments as recorded in multiple palaeosols and alluvial cut-offs have been used often as a model for the interpretation of palaeoecological floodplain studies (Brown, 1982, 1988). Recently this has been formalized by Amoros *et al.*, (1988) and Petts (1989), as synchronic and diachronic analysis; part of a method for applied ecological studies of fluvial hydrosystems. The realization that floodplains are composed of a series of segments of different ages and character that are dependent upon channel behaviour, is applicable not only to ecological studies of floodplains but also to archaeological studies (Needham, 1989) and the modelling of floodplain change (Perez-Trejo and Cincotta, 1990).

THE MIDDLE NENE VALLEY: SITES

The data used here come from three areas within a 23 km long reach of the middle Nene, from Little Houghton near Northampton to Raunds, which lies about midway to Peterborough (Figure 8.1). The Nene has a low regional slope, a non-flashy regime even by UK standards, and it lay outside the area covered by ice during the Devensian glaciation.

Exposures come from suballuvial gravel extraction at Little Houghton, Wollaston, Higham Ferrers and Raunds. At Raunds a large area has been investigated in detail owing to the removal of almost the entire valley floor for aggregate. This area also falls within the Raunds Area Project, a multisite, multiperiod integrated archaeological project, which includes several floodplain sites. These include Bronze Age barrows in the middle of the floodplain, a Neolithic enclosure, a Roman villa complex and a Medieval village.

The next site upstream from the Raunds reach is Higham Ferrers, which is 21 km downstream of Northampton. The floodplain deposits sit on Upper Lias Clay and are between 4.3 and 5.1 m in thickness. Between the basal periglacial gravels and the superficial clay unit is a palaeochannel bed and a reworked sand and gravel unit up to 1.5 m thick. The site at Wollaston also has a very variable, but generally sandy, unit in between the basal gravels and the superficial clay unit. Revealed are two palaeochannels separated by a sand bar that interdigitates with overbank clays. At Little Houghton two sections have been logged, both approximately parallel to the present river and valley axis. Neither revealed a palaeochannel but both had an intervening mixed gravelly, sandy clay (gravelly loam) which contained root material and had formed the subsoil of a palaeoland-surface. On Figure 8.1 there also several other sites investigated by other workers. At the site at Great Billing, which provided a flora and fauna from the periglacial gravels, 0.6–0.9 m of yellow brown Holocene alluvial clay rested on the gravels, apparently abruptly, although further stratigraphic evi-

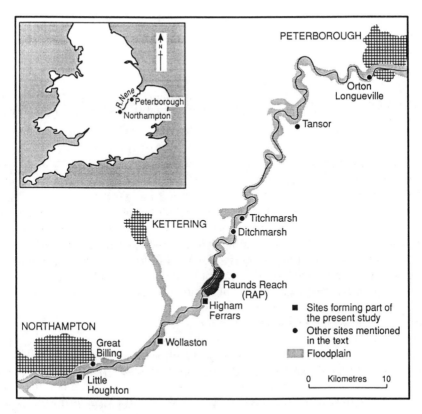

Figure 8.1 Location map of the study reach showing the floodplain extent and sites mentioned in the text. The dense cross-hatching indicates the area covered by the three-dimensional diagram in Figure 8.2. River flow is from south to north

dence is lacking (Morgan, 1969). There also are three sites downstream, between the Raunds reach and Peterborough. At Ditchmarsh abundant large logs of oak have been recovered from a palaeochannel cut into the underlying gravels; the dating of this site is still underway. The sites at Titchmarsh and Orton Longueville were investigated by Holyoak and Seddon (1984). At both they discovered organic sediments of Devensian age within the gravels. The Titchmarsh site also revealed a shallow palaeochannel scar cut into the top of the gravels, filled with detritus mud beneath 2 m of silty clay. Similarly at Orton Longueville a palaeochannel was found covered by 3.4 m of fine sediments. The fine alluvial sediments at Orton Longueville are more complex than elsewhere, with two units, a lower grey-brown silty mud with roots at the base and penetrating the gravels, and an upper blackish-grey detritus mud that is occasionally covered by a grey-brown clayey silt. Largely on pollen evidence, both the palaeochannels and, by implication, the buried land surfaces are dated to the Iron Age. In addition a site at Tansor is at present being investigated by one of the authors (AGB).

As these sites show, the whole Lateglacial and Holocene floodplain stratigraphy is only around 4 m thick and there is no evidence of a higher palaeofloodplain that has subsequently been dissected to the present level. The floodplain is underlain throughout the reach by a planar bedded sand and gravel unit varying in thickness from less than 1 to 3 m.

THE RAUNDS REACH ALLUVIAL ISOPACH AND PALAEOCHANNEL PATTERN

In the Raunds area over 1200 boreholes provided by ARC through English Heritage has allowed the reconstruction of the isopach of the finer sediments sitting on the basal sand and gravel. Each record was interpreted and the depth to the clean sand and gravel of the basal unit used in an attempt to exclude areas and units of reworked sand and gravel. As the historical floodplain is practically flat (under 1m relative relief) this can be regarded as a proxy topographic map of the top of the sand and gravel unit.

The raw pattern is extremely 'noisy' for various reasons, including the lack of surface height control, error in the depth of measurement, and the interaction of the complexity of the suballuvial topography with irregular core cover. Therefore, to investigate pattern in the data, various interpolation procedures were used. Two examples are shown here (Figure 8.2 (a and b)).

Method 1 (Figure 8.2a) uses bilinear interpolation improving the estimate by comparing gradients at the points and using quadratic interpolation. Finally the values for the grid are refined using distance weighting – this effectively produces a rather rough surface.

Method 2 (Figure 8.2b) develops a network of triangles over the surface starting with the pair of points closest together and working outwards from their mid-point. This is followed by fitting a bivariate fifth-order polynomial to each triangle and smoothing. This procedure effectively produces a rather smooth and interconnected surface. Indeed this gave the maximum connectivity of the hollows and depressions. The reason for doing this is to highlight past palaeochannel belts (i.e. their preferred position *sensu* Palmquist (1975)), which exist owing to the nature of Holocene floodplain development in lowland valleys such as the Nene. This is because the dominant process has not been meander migration across the floodplain, but a more complex history of channel stability, bank and bench sedimentation, channel siltation and sudden shifts in channel pattern caused by avulsion (Brown, 1987). The result is a suballuvial topography that reflects the preferred position of channels, which itself is highly temporally autocorrelated. This is because the most important control on channel location is the topographic lows created by past channel activity and through this negative feedback an element of inheritance persists in channel pattern from the early Holocene through to the present. A second result of this process is the preservation of fragments or parcels of landsurface of various ages subsequently buried by overbank sedimentation. Note the coincidence between the excavated palaeochannels, especially the later ones, which are shown on Figure 8.2b (and are tabulated in Figure 8.6), and the bands of thicker and deeper fine alluvium. The low curvature of all the palaeochannels suggests that none of them are neck cut-offs but rather that they are chute channels.

LATEGLACIAL AND EARLY HOLOCENE CHANNEL CHANGE

The basal gravel unit has been regarded as fluvioperiglacial by Castleden (1976, 1977, 1980a, b) largely on the basis of ice-cast wedges and other cryoturbation features. It has been assumed that it was deposited during the mid- or late Devensian, and support for this comes from faunal remains and a radiocarbon date at Great Billing of 28 225 ± 330 years BP (Morgan, 1969). Holyoak and Seddon (1984) also have recovered fossiliferous organics from the basal gravels at Little Houghton, Titchmarsh and Orton Longueville. The Little Houghton organics lie in channels cut into the Upper Lias beneath the gravels and are Middle Devensian in age. The organics at Titchmarsh and Orton Longueville lie about half-way up the gravel unit and are supposedly of Late Devensian age on the basis of pollen and molluscs. One other date from the gravels is available, from Thrapston, which is 2 km upstream of Titchmarsh, where a sample of twigs from 5 m depth gave a radiocarbon date of 25 780 ± 870 years BP (Shotton, Blundell and Williams, 1970). Although this implies that the basal gravels were deposited between the Middle Devensian (ca. 30 000 years BP) and the Late Devensian, the very poor stratigraphic detail from all these studies means that it is impossible to evaluate the possibility of the later dates coming from reworked gravel units cut into, or deposited on, the basal periglacially modified basal gravel unit, which is Castleden's (1976) number 1 or floodplain terrace.

At Raunds (downstream site east section) three units are present, each similar in lithological content and grain size distribution, but separated by an erosional boundary, and at one location a small organic filled depression is present that has given a radiocarbon date of 12 420 ± 60 years BP and an upper date of 10 870 ± 55 years BP (Figure 8.3a). This means that the depression started to accumulate sediment in stage Ic of the Lateglacial and continued to accumulate sediment during the Lateglacial interstadial. From another exposure at the same site (800 m upstream) a basal date on an organic palaeochannel infill cut into the upper surface of the gravels has given a date of 11 395 ± 55 years BP and the date from the top of the organics underlying the silty clay unit is 9375 ± 40 years BP. This indicates reworking of the Devensian gravels, in restricted locations only, and deposition of a fragmentary upper gravel unit during the Lateglacial stadial. The cross-bedding of the middle gravel unit at Raunds, and its restricted occurrence, being cut into the older gravels, suggests that this unit was the result of meander migration, probably during the Lateglacial interstadial. This situation is rather similar to that found under the Gipping floodplain at Sproughton (investigated by Rose *et al.*, 1980). Another rather well-defined palaeochannel at Raunds (RAP. E) has basal organics ^{14}C dated to 9400 years BP. At Wollaston a palaeochannel was abandoned by 9770 years BP and accumulated organics with an increasing inorganic content until around 9000 years BP.

Representative samples have been dated at all the sites and units that contained organic remains within the study area. As can be seen from the cumulative radiocarbon diagram (Figure 8.4) for the reach there is a distinct lack of dates on either debris or *in situ* organics during the early Holocene, especially between 9000 years BP and 5500 years BP. In fact out of 42 dates there is only one that sits within this time window and that is a coarse fraction detrital date from the top of an organic infill of the palaeochannel abandoned before 9770 years BP at Wollaston.

190

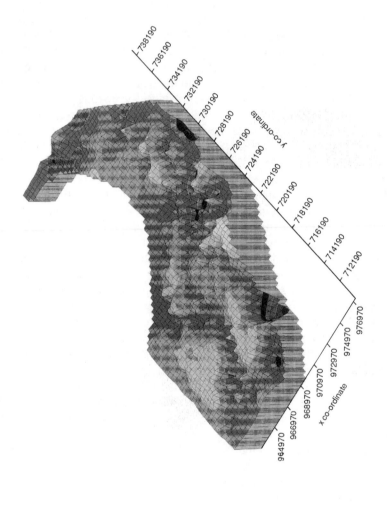

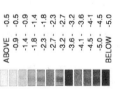

ABOVE	-0.5
-0.9	-0.5
-1.4	-0.9
-1.8	-1.4
-2.3	-1.8
-2.7	-2.3
-3.2	-2.7
-3.6	-3.2
-4.1	-3.6
-4.5	-4.1
-5.0	-4.5
BELOW	-5.0

(a)

191

(b)

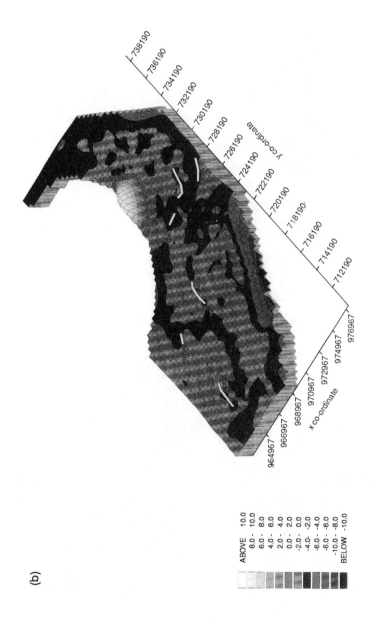

738190
736190
734190
732190
730190
728190
726190
724190
722190
720190
718190
716190
714190
712190

y co-ordinate

964967
966967
968967
970967
972967
974967
976967

x co-ordinate

ABOVE 10.0
8.0 - 10.0
6.0 - 8.0
4.0 - 6.0
2.0 - 4.0
0.0 - 2.0
-2.0 - 0.0
-4.0 - -2.0
-6.0 - -4.0
-8.0 - -6.0
-10.0 - -8.0
BELOW -10.0

Figure 8.2 (a) Isopach three-dimensional diagram of the superficial fine alluvium using Method 1 – see text for details. (b) Isopach three-dimensional diagram of the superficial fine alluvium using Method 2, with palaeochannels and estimated dates of abandonment superimposed – see text for details. Key scales are linear but relative. Refer to Figure 8.1 for the size of the reach covered by these diagrams (dense cross-hatching)

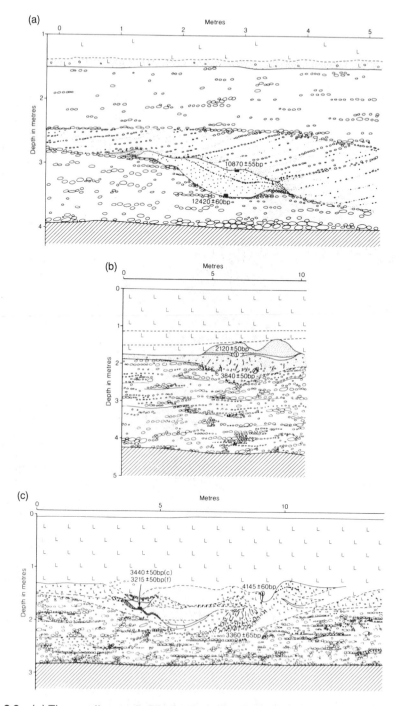

Figure 8.3 (a) The small organic filled depression from the downstream site at Raunds. (b) Raunds: downstream site detail of section showing structure interpreted as tree-throw pit and flood mound. (c) Little Houghton: detail from the southern section showing disturbance of the stratigraphy associated with tree roots

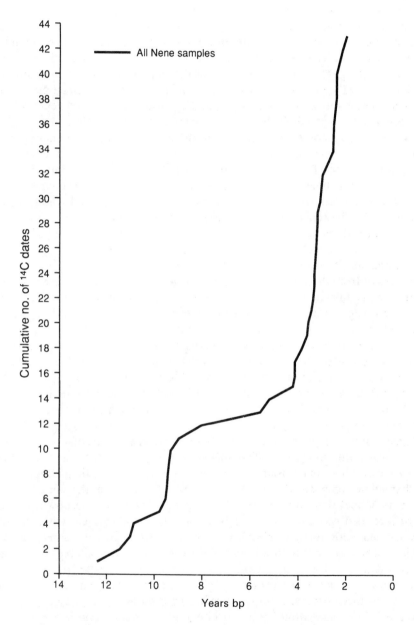

Figure 8.4 Cumulative graph of all 42 radiocarbon dates available for the reach from all alluvial contexts post-14 000 years BP. All dates are uncalibrated

MID-HOLOCENE LAND SURFACES

The removal of the overburden prior to gravel extraction has revealed land surfaces with frequent remains of the past vegetation cover. At the upstream site at Raunds an area some 2200 m^2 was stripped off between two palaeochannels. The palaeochannel at the eastern end of the section is one of the Lateglacial interstadial abandonments, while the other is undated. In between, the land surface had preserved 43 alder root boles all *in situ* (Figure 8.5a). They are all at the same level, are of comparable size and had disturbed the gravels into which they had penetrated to a maximum depth of about 60 cm. The radiocarbon date from one of the roots is 5195 ± 65 years BP, which is an unusual date as it lies in the ^{14}C gap common to East Midland valleys. This probably is because it comes from a land surface, whereas most dated materials from cross-sections come from organic channel fills, of which there are very few from this period. From the size of the boles it appears that the alders were relatively small or shrubby and several boles had multiple stems (2–4+), suggesting coppicing. Coppicing of this age is known from the Somerset Levels, as coppiced wood, including alder, was used in the construction of the Sweet Track, although it is possible both the hazel and alder rods came from naturally coppiced trees resulting from the activity of deer. The ends of the stems have been broken or rotted below the original ground level. The tree density is about 184 trees ha^{-1}. The clustering seen in Figure 8.5 could, in theory, result from there being more than one generation of trees present, although the minimum distance between tree boles is 4 m. It seems likely that the boles come from one or two generations and their preservation implies a rise in the water-table sometime after 5195 ± 60 years BP, when a thin spread of planar bedded sand also was deposited, possibly after some surface erosion. Two other palaeo-alder carr sites are also shown on Figure 8.5: at Little Houghton (Figure 8.5b), which is dated to 3565 ± 65 years BP and Wollston (Figure 8.5c), which is dated to 3605 ± 60 years BP. The alder carr at Raunds can be traced to another land surface 30 m to the northeast and nearer the present channel. On this surface a Neolothic hearth has been found and several tree-throw pits. The palaeosol is thicker and the size and depth of the pits suggests trees much larger and probably of a different species. Artifacts and evidence of burning was found in the pits. This surface is 1.3 m above the alder carr and was either a levee or a floodplain 'island' at the time. If it was a levee then either the small palaeochannel between it and the alder carr and/or a channel in the same location as the present channel was functioning 5000 years ago. About 800 m downstream, the gravel surface also shows evidence of being a palaeoland-surface, with an area of disturbance that closely resembles three-throw disturbance with a pit, roots and asymmetric profile (Figure 8.3b). Two radiocarbon dates are available, one from the lower roots dated at 3840 ± 50 years BP, and another from *Juncus* remains from the organic mud covering the pit, which is 2120 ± 50 years BP. The *Juncus*, which is indicative of surface wetness, seems to have been buried by sand. On stratigraphic and textural ground the sand would appear to be a flood mound deposit rather than the result of a second phase of tree-throw, however, the disturbance is complex and must have involved more than one phase, as is suggested also by the incorporation of third or fourth century pottery, below the 2120 ± 50 years BP date. Any protruberance that increases floodplain roughness will cause localized scour and accelerated overbank sedimentation around it. An example of this is the formation of sandy hemispherical

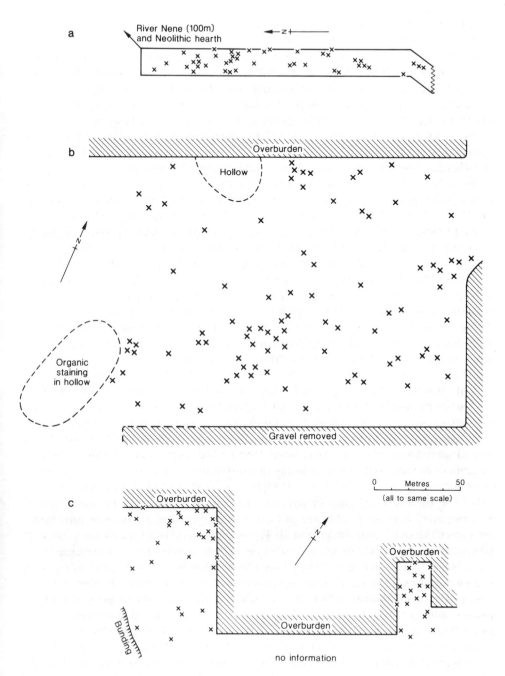

Figure 8.5 Palaeo-alder carrs from (a) upstream site at Raunds, (b) Little Houghton and (c) Wollaston

flood mounds formed around dead tussocks of *Molinia caerulea* during a flood on the River Stour in Dorset, in 1979 (Brown, 1983).

Because of the extensive archaeological interest in the Raunds area it has been possible to trench several of the later phase palaeochannels (Figures 8.2b and 8.6). The earliest of these, which is rather small and poorly defined and undoubtedly was not the main channel, has a basal date of 4200 ± 100 years BP (RAP. C). This was abandoned before the removal of the surrounding alder carr, as is indicated by both pollen analysis and an alder trunk that fell across the infilling cut-off channel. A channel right at the edge of the floodplain at Raunds (RAP. WCT) was abandoned by ca. 2500 years BP, however, it remained open to tributary flow and the youngest abandonment dated is 1970 years BP (RAP. D). The evidence at Raunds, taken together, suggests a stable land surface of some relief that suffered a rise in the water-table and increased overbank deposition from the bed material (probably saltation load) sometime after 5000 years BP, and finally silt and clay deposition by 2200 years BP.

At Little Houghton, above the root horizon, is a pedogenically disturbed gravelly loam that was deposited while the trees were still upstanding. After this the silty clay was deposited. The southern section at Little Houghton (Figure 8.3c) also shows disturbance, with sedimentary structures that cannot be explained in terms of fluvial sedimentology. A large dish-shaped depression cut into the underlying gravels is filled with smaller trough-shaped units with infills ranging from organics to small gravel. There is evidence of pit sedimentation and root penetration. The disturbance is regarded as a tree-throw complex perhaps related to a small channel. Although the dates concentrate around 3440 years BP there are problems:

(1) there is an inversion, with root wood at the top being the oldest at 4145 ± 60 years BP and the lowest being 3360 ± 65 years BP, possibly owing to overturning;
(2) there is some evidence of contamination or mixed origin of the organic-rich sediments in one of the depressions, because the fine organic fraction date is 3440 ± 50 years BP and the coarse 3215 ± 50 years BP, suggesting inwashing of older organic matter from the surrounding soils.

However, the disturbed gravelly unit must have been deposited at the same time as the lower silty clay; this is indicated by the interdigitation of the deposits caused by the erosion of the sandy tree-throw mound. The interpretation of such features as these at both Raunds and Little Houghton, and at other sites, as tree-throw disturbance is not based solely on an absence of alternative causal geomorphological mechanisms, but the similarity of the features with contemporary tree-throw pit and mound stratigraphy. From contemporary studies (Armson and Fessenden, 1973; Schaetzl, 1986,1990) of the effects of tree-throw on soil genesis, and the author's observations, several diagnostic characteristics can be proposed:

(1) asymmetry of the pit in cross-section with the skew being towards the direction of throw;
(2) a mound raised above the pit top formed by the degradation of the uprooted soil disc;

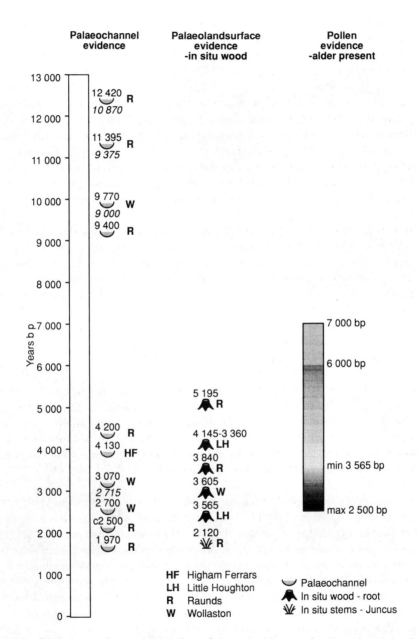

Figure 8.6 Diagrammatic representation of the coincidence of palaeochannel evidence, palaeoland-surface evidence and pollen evidence for this reach of the River Nene during the Lateglacial and Holocene. The minimum and maximum dates by the alder column refer to estimated dates for alder deforestation in the reach

(a)

(b)

Figure 8.7 Tree-throw caused by the October 1987 storms on the floodplain of the River Wandle, South London

(3) organic rich, waterlogged pit sediments often interdigitating with eroded mound
     sediments.

There is little doubt that the importance of tree-throw as a natural process of soil
development and stratigraphic disturbance, has not received due attention outside the
hurricane prone forests of North America. Additionally, although it has proved
difficult to statistically relate the probability of tree-throw to soil type for a single
event (Grayson, 1989), relatively shallow alluvial soils sitting on gravels are probably
more liable to throw than those on deeper soils on weathered bedrock. Indeed Figure
8.7 shows examples from the River Wandle in London, felled during the October 1987
storm. In this case it was clear that failure occurred at the junction of the topsoil with
the gravels below.

At Wollaston, as well as the evidence of an alder carr at 3605 ± 60 years BP growing
over the palaeochannel abandoned in the early Holocene, there is evidence of erosion
into this land surface by a palaeochannel by 3400 years BP, with the deposition of a
sand unit and sand bar that continued to be deposited even after abandonment of the
associated palaeochannel around 2700 years BP. This bar, which became a floodplain
surface feature, continued to be deposited until at least 2300 years BP, by which time
the silt and clay unit had just managed to bury it. This site also illustrated well the
different accumulation rates that coexist within one parcel of the floodplain surface.

At Higham Ferrers there is no evidence of an *in situ* alder carr; instead the site,
which lies adjacent to the southern branch of the present channel, displays evidence of
deposition and some erosion between 4000 and 2300 years BP in a unit 1.5 m thick
that is cut into the basal sand and gravel and underlies the superficial silty clay. There
is evidence of a palaeochannel with an armoured bed and cross-bedded, graded sand
and fine gravel, course grain stringers and mud drapes. There is also evidence of
interdigitating fine sediments within erosional pits. Interestingly none of the wood at
this site is alder or *in situ*. The stratigraphy is interpreted as a channel migration zone
with lateral sedimentation and flood deposition.

DISCUSSION

These sites provide very little support for Castleden's (1976) scheme of floodplain
development for the Nene. Largely on the basis of one radiocarbon date, which seems
anomalously young (Bell, 1969), he suggested continued accumulation of the first
terrace (suballuvial) gravels into the early Boreal, zone V, with erosion 'to floodplain
level' during zones VI and VIIa (8500–5000 years BP), and a well-drained floodplain
in zone VIIb (5000–2500 years BP). The first generation of palaeochannels and their
stratigraphic position on the gravel surface, suggests the body of gravel deposition had
finished prior to the Lateglacial interstadial and that only minor reworking has
occurred since. By zone VIIb the evidence from organics, and of flood erosion and
deposition, suggests that the floodplain had become both wetter and more flood-
prone. The roots and lack of an obvious organic palaeosol underlying the overbank
clay suggests some erosion of the floodplain surface, possibly by as much as 0.5 m in
places. Support for this comes from pits in which organics have been preserved, some

of which reveal contamination by older fine organic matter. While this is not floodplain stripping *sensu* Nanson (1986), it does indicate a changed balance between overbank flows and sediment supply similar to the interpretation by Robinson and Lambrick (1984) of hydrological change on the Thames; initially increased flooding by waters below their transport capacity and later prolonged flooding by sediment-laden waters.

Figure 8.6 illustrates two important aspects of the evolution of the Nene floodplain. Firstly, there is the palaeochannel and palaeoland-surface gap of the early to mid-Holocene and a lack of preservation of Lateglacial land surfaces. The second is that although the palaeoland-surface evidence and the later phase palaeochannel evidence overlap in time, the majority of palaeoland-surface evidence overlaps the early palaeochannels between 4130 and 3070 years BP, with one significantly pre-dating all these abandonments by nearly 1000 years. The palaeoland surfaces therefore provide complementary evidence of floodplain stability and probably are biased (in terms of extent and therefore likelihood of being excavated) towards periods of little channel migration and relatively low rates of overbank deposition followed by periods of an increase in overbank deposition that out-stripped soil development thus causing burial.

CONCLUSIONS

From the detailed study of exposures and palaeochannels from this reach of the middle Nene, five tentative conclusions can be drawn.

(1)   The unconformity/boundary traditionally assumed to separate the common bipartite floodplain fill of the middle Nene is in most cases a palaeoland-surface, which has suffered pedogenic reworking, flood deposition and topsoil erosion but not channel erosion. It is likely that very close attention to the sediments and structures between the major units of other bipartite floodplain sequences also might reveal a polygenetic unit and evidence of a palaeoland-surface.

(2)   There are two major periods of channel change in this reach of the Nene. One began in the Lateglacial interstadial and continued into the very early Holocene. The other occurred between about 4000 and 2000 years BP, after which the channel system has been relatively fixed, and vertical deposition of silt and clay became the dominant mode of floodplain construction along with direct human interference with the channels. The first period of channel change was associated with more reworking of the Devensian gravels than the later. The two periods were separated by a long period of relatively stable channels and land surfaces of greater relief than today's, and soils and vegetation very different to those of the floodplain after the later period of channel change and up to the present.

(3)   The channel change has affected only a relatively small percentage of the floodplain, with avulsion being restricted to the belt that can be identified from the isopach maps. In some places the overbank vertical accretion of the floodplain has only just buried the highest points of the pre-Holocene gravel floodplain surface.

(4)   Accompanying the evidence of channel change from around 4000 years BP, is evidence of a rise in floodplain water-tables and overbank flooding with increased transporting competence. Water-table changes are of critical importance in preserving

floodplain organics and this therefore biases the biotic record. This is the explanation for the observation that pollen evidence indicates without doubt that alder was present in the locality from at least 6500 years BP but that the vast majority of *in situ* macro-organics come from the period 4000–2000 years BP (Figure 8.6). It also explains the pedological disturbance of the frequently encountered gravelly loam middle unit. This rise in water-tables presents more fundamental evidence of the third millenium BP palaeohydrological changes than either flood deposits alone or attempts at regional correlation of the depositional record from valley to valley. Advances in our understanding of the past relationships between environmental change and floodplain formation will come not through regional correlations of previous work with very varying degrees of stratigraphic and chronological sophistication, but through detailed site work reconstructing the changing floodplain environment in three-dimensions and relating observations of floodplain sedimentation and erosion to models of channel behaviour at the reach scale.

(5)   In order to reconstruct Holocene floodplain evolution it is necessary to (i) move out of the channels to the land surfaces and riparian zones between them, and (ii) widen our knowledge of sedimentary structures especially those associated with the floodplain surface, including biological processes.

ACKNOWLEDGEMENTS

This work was done largely while one of us (MK) was in receipt of a NERC research studentship, and radiocarbon dating also was provided by NERC, for which we are grateful. We also would like to thank the gravel pit owners and operators, archaeologists of the RAP, especially C. Halpen, D. Windell and R. J. Rice for help and advice. For her computing and digitizing expertise we thank H. Schneider.

REFERENCES

Amoros, C., A. L. Roux, J. L. Reygrobellet, J. P. Bravard and G. Patou (1987). A method for applied ecological studies of fluvial hydrosystems. *Regulated Rivers*, **1**, 17–36.

Armson, K. A. and R. J. Fessenden (1973). Forest windthrow and their influence on soil morphology. *Soil Science Society of America Proceedings*, **37**, 781–783.

Becker, B. and W. Schirmer (1977). Palaeoecological study on the Holocene valley development of the River Main, Southern Germany. *Boreas*, **6**, 303–321.

Bell, F. C. (1969). *Radiocarbon*, **11**, 265.

Brown, A. G. (1982). Human impact on former floodplain woodlands of the Severn. In M. Bell and S. Limbrey (eds), *Archaeological Aspects of Woodland Ecology*. British Archaeological Reports International Series, 16, pp. 93–105.

Brown, A. G. (1983). An analysis of overbank deposits of a flood at Blandford-Forum, Dorset, England. *Revue de Geomorphologie Dynamique*, **32**, 95–99.

Brown, A. G. (1987). Holocene floodplain sedimentation and channel response of the lower river Severn, United Kingdom. *Zeitschrift für Geomorphologie*, **31**, 293–310.

Brown, A. G. (1988). The palaeoecology of *Alnus* (alder) and the Postglacial history of floodplains: pollen percentage and influx data from the West Midlands. *New Phytologist*, **110**, 425–436.

Brown, A. G. (1989). Holocene floodplain diachronism and inherited downstream variations in fluvial processes: a study of the river Perry, Shropshire, England. *Journal of Quaternary Science*, **5**, 39–51.

Castleden, R. (1976). The floodplain gravels of the river Nene. *Mercian Geologist*, **6**, 33–47.
Castleden, R. (1977). Periglacial sediments in Central and Southern England. *Catena*, **4**, 111–121.
Castleden, R. (1980a). Fluvioperiglacial pedimentation: a general theory of fluvial valley development in cool temperate lands, illustrated from Western and Central Europe. *Catena*, **7**, 135–152.
Castleden, R. (1980b). The second and third terraces of the river Nene. *Mercian Geologist*, **8**, 29–46.
Grayson, A. J. (1989). The 1987 storm impact and responses. *Forestry Commission Bulletin*, **87**.
Holyoak, D. T. and M. B. Seddon (1984). Devensian and Flandrian fossiliferous deposits in the Nene valley, Central England. *Mercian Geologist*, **9**, 127–150.
Keller, E. A. and F. J. Swanson (1979). Effects of large organic debris on channel form and fluvial processes. *Earth Surface Processes and Landforms*, **4**, 361–380.
Keller, E. A. and T. Tally (1979). Effects of large organic debris on channel form and fluvial processes in the coastal redwood environment. In D. D. Rhodes and G. P. Williams (eds), *Adjustments of the Fluvial System*, Kendall Hunt, pp. 169–197.
Morgan, A. (1969). A Pleistocene fauna and flora from Great Billing, Northamptonshire, England. *Opuscula Entomologica*, **34**, 109–129.
Nanson, G. C. (1986). Episodes of vertical accretion and catastrophic stripping: a model of disequilibrium flood-plain development. *Geological Society of America Bulletin*, **97**, 1467–1475.
Needham, S. (1989). River valleys as wetlands: the archaeological prospects. In J. M. Coles and B. J. Coles (eds), *The Archaeology of Rural Wetlands in Britain*, Wetlands Archaeological Research Project/English Heritage, pp. 29–34.
Palmquist, R. C. (1975). Preferred position model and subsurface valleys. *Geological Society of America Bulletin*, **86**, 1392–1398.
Perez-Trejo, F. and R. Cincotta (1990). *The Genesis of Landscapes: the Self-organising Processes of Landscapes*. International Ecotechnology Research Centre (IERC), *News*, **5**, 1–4, Cranfield University of Technology.
Petts, G. E. (1989). Historical analysis of fluvial hydrosystems. In G. E. Petts, H. Moller and A. L. Roux (eds), *Historical Change of Large Alluvial Rivers: Western Europe*, Wiley, pp. 1–18.
Robinson, M. A. and G. H. Lambrick (1984). Holocene alluviation and hydrology in the upper Thames basin. *Nature*, **308**, 809–814.
Rose, J., C. Turner, G. R. Coope and M. D. Bryan (1980). Channel changes in a lowland river catchment over the last 13,000 years. In R. A. Cullingford, D. A. Davidson and J. Lewin (eds), *Timescales in Geomorphology*, Wiley, pp. 159–176.
Salisbury, C. R., P. J. Whitley, C. D. Litton and J. L. Fox (1984). Flandrian courses of the River Trent at Colwick, Nottingham. *Mercian Geologist*, **9**, 189–207.
Schaetzl, R. J. (1986). Complete soil profile inversion by tree uprooting. *Physical Geography*, **7**, 181–189.
Schaetzl, R. J. (1990). Effects of treethrow microtopography on the characteristics and genesis of Spodosols, Michigan, USA. *Catena*, **17**, 111–126.
Shotton, F. W., D. J. Blundell and R. E. Williams (1970). Birmingham radiocarbon dates IV. *Radiocarbon*, **12**, 385–399.
Sigafoos, R. S. (1964). Botanical evidence of floods and floodplain deposition. *U.S. Geological Survey Professional Paper*, **485A**, 35.

9 Influence of Farm Management and Drainage on Leaching of Nitrate from Former Floodlands in a Lowland Clay Catchment

G. L. HARRIS
ADAS Field Drainage Experimental Unit, Cambridge

and

T. PARISH
Ecological Processes Section, NERC Institute of Terrestrial Ecology, Huntingdon

INTRODUCTION

During the mid-1980s, UK agriculture moved from an emphasis on intensive production to systems that are more sympathetic to the environment. As a result there is no longer a perceived need or financial incentive to increase cereal production by draining any remaining floodmeadows or grazing marshes. Effective drainage of clay-based soils is still essential to meet soil management and crop requirements (Trafford and Walpole, 1975) so that in its natural state, agriculture in a floodplain area, often is restricted to low intensity grazing. The impact of the conversion of former floodplains to intensive agricultural production has, therefore, become a topic of interest, since the ecosystems associated with such converted areas are dynamic and representative of a transition phase. In particular the former landuse will have played an important role in the development and stability of structure and organic content in the soil profile.

In recent years, intensive agriculture has been cited widely as being one of the primary causes for the increased presence of nitrate in water systems. However, the effects of conversion of floodplains on the leaching process have not been researched adequately. In particular the effect of ploughing on the organic silty clay floodplain soils needs investigation.

A collaborative experiment was established in 1986 between the ADAS Field Drainage Experimental Unit and the NERC Institute of Terrestrial Ecology to examine the interaction of farm management, hydrology and water quality of a lowland basin. The study area chosen included recently converted floodland in close

Lowland Floodplain Rivers: Geomorphological Perspectives. Edited by P.A. Carling and G.E. Petts
Copyright © 1992 Crown. Published 1992 by John Wiley & Sons Ltd

proximity to an area with a long history of intensive farming. This paper examines the agricultural management and nitrate leaching in these areas and identifies some of the factors that influence nitrate leaching. The work reported covers the two seasons 1987–1988 and 1988–1989 and forms only part of the long-term study.

SITE

Drainage

The site, at Swavesey, Cambridgeshire, consists of a 500 ha basin divided into two key units representative of recently converted floodplain and established traditional intensive agriculture (Figure 9.1). Within the floodplain (Cow Fen), the silty clay soil belongs to the Earith and Mildeney soil series, which frequently are found in river corridors (Seale, 1975). By contrast, in the traditional area (Highfield), the soil is a pelo-stagnogley clay of the Denchworth series (Avery, 1980). The Highfield area was drained by the landowners in the late 1960s and supports intensive mixed agriculture.

Before the installation of the pump drainage in 1985, low input pasture was the only available option in Cow Fen. In the lowest parts, particularly below 4 m OD, even this activity was often restricted by high water-tables or flooding. Some farmers with land above 6 m OD, where there was less risk of flooding, underdrained a few fields and converted them to arable production. An axial flow pump now provides the basic drainage for the agricultural area by transferring water from a storage lagoon within the internal drainage system to an adjacent High Level drain of the arterial river system. The head difference between the ditch water-level in the floodplain and the adjacent main channel governs pump performance. Additionally, in extreme rainfall events, flow occurs into the pumped catchment over a spillway designed to protect the nearby areas of Swavesey village from flooding. In conjunction with the pump

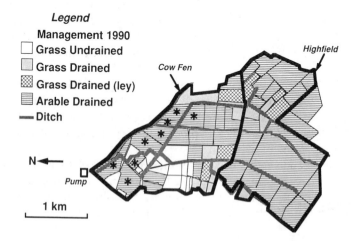

Figure 9.1 Management systems in the catchment in 1990. Fields that changed from grassland to arable following the installation of the pump system are shown by the symbol *

Table 9.1 Cropping patterns in Cow Fen and Highfields, expressed as a percentage of the whole area

	Cow Fen		Highfield	
Crop type	1987–1988	1988–1989	1987–1988	1988–1989
Winter wheat	27	32	55	65
Winter barley	10	8	17	11
Spring wheat	0	4	0	0
Spring barley	2	0	9	0
Spring oats	0˙	3	0	0
Oilseed rape	8	5	6	7
Beans	5	4	0	4
Grass (all)	48	44	13	13

scheme, many fields in Cow Fen that previously were undrained grassland, were drained and then converted to intensive arable cropping (Figure 9.1). Each field has a separate independent subsurface pipe drainage system usually combined with a secondary system, such as mole drainage.

Agriculture

Although the basin supports a variety of arable and grassland regimes the crop most widely grown in Highfield is winter wheat (Table 9.1). Winter barley, oilseed rape and legumes, such as beans, also are grown, with permanent grassland and leys in the eastern section. In Cow Fen the transition towards the lower basin is marked by a gradual increase in grassland, some of which remains undrained or supports non-intensive agriculture (Figure 9.1). Application of nitrogen, predominantly as inorganic fertilizer, varies according to crop type and location. Nitrogen fertilizer applications to winter wheat, in both 1987–1988 and 1988–1989, were similar to the national average of 185 kg ha^{-1} N (Chalmers and Leech, 1988) in Highfield, whereas in Cow Fen the average rate was 60 kg ha^{-1} less. Similarly, applications to the winter barley in Highfield at 159 kg ha^{-1} N were very near the national average of 150 kg ha^{-1} N, but in Cow Fen over 80 kg ha^{-1} N less. The return of nitrogen to the soil after harvest is low as over 80% of the crop residue is removed or burnt. Applications are more variable to grassland, in the range 50–200 kg ha^{-1} N in Highfield, and although up to 200 kg ha^{-1} N in some fields in Cow Fen, many undrained fields received little or no artificial fertilizer. The total applied nitrogen fertilizer to the catchment varied little between the two study years at 54–59 t for cropped land and 7 t to grassland.

EXPERIMENTAL DETAILS

Fields

A record and assessment has been made for each of the 82 individual fields in Cow Fen and Highfield, including details of drainage and management in previous years of the study. Crop seed variety, drilling date and yield is determined from agronomic field surveys and from farm records. By comparison, the record for grass fields includes animal grazing days/stock density and hay/silage cuts.

Water-table Monitoring

The water-table in 16 representative fields within Cow Fen and Highfield, forming 20% of the catchment, was monitored by a series of dipwells, 1 m deep, established in a grid pattern at a nominal 75 m spacing. The manually read dipwells were supplemented by continuous recording water-table meters in selected fields. The operation of the pump was also monitored by eight autographic continuous water-level recorders located in the ditch system, six of which monitored the response to rainfall events in Cow Fen, with two sited in the Highfield area. Rainfall also was recorded at two locations within the basin.

Nitrogen Measurements

Water Samples

Water samples to determine the loss of nitrogen from the catchment were collected from the basin outlet and at points throughout the internal ditch system of Cow Fen and Highfield. With the exception of the lagoon, samples were collected manually by dipping a pre-washed collection vessel into the midstream and subsampling to produce a 30 ml final sample. At the catchment outlet, samples also were collected by a programmable water sampler at a minimum rate of two samples per day. Initially samples were taken at monthly intervals to determine general leaching patterns. In early winter 1988–1989, the sampling was intensified to several times per day and included storm run-off periods. As the nitrite-N and ammonium-N contents in the water generally were found to be low (around 1% of the nitrate-N) subsequent analysis was undertaken for nitrate-N only. Samples taken to the laboratory were examined on a flow injection system where nitrate is reduced quantitatively to nitrite by cadmium metal. The nitrite, along with any originally present in the sample, combines with sulphanilamide and N-1-naphthylethylenediamine to form an azo dye (maximum absorbance 543 nm). The dye then passes through a photometer where the absorbance is measured. The absorbance is proportional to the quantity of nitrate in the sample.

Soil Samples

Soil mineral nitrogen samples were taken from spring 1988–1989 in up to 12 fields in Cow Fen and Highfield. Samples were taken to 900 mm depth, (0–150, 150–300, 300–600, 600–900 mm) and analysed for nitrate-N (NO_3^-) and ammonium-N (NH_4^+) (Ministry of Agriculture, Fisheries and Food, 1986). Soil bulk density values (Seale, 1975) were used to derive the total profile nitrogen in kg ha^{-1} N. Organic matter also was determined for each soil horizon.

Crop Nitrogen Turnover

Nitrogen removed by the crop and animals was assessed for each field within Cow Fen and Highfield. In each arable field, the total N was determined by Kjeldahl digestion (Ministry of Agriculture, Fisheries and Food, 1986) for each crop component (grain, chaff and stalk), from randomized grab samples immediately before harvest, and converted to total removal by utilizing the respective field crop yield. In the grassland area, nitrogen offtake was derived from field stocking rates together with estimates of the effectiveness of each animal group/management system in removing nitrogen. Values used were those suggested by Roberts (1987) who reported highest removal for dairy cows at 120 kg ha^{-1} N per annum for average stocking rates with or without silage/hay cuts. Values for beef cattle were lower at 60 kg ha^{-1} N, whereas sheep at average stocking were considered to remove 25 kg ha^{-1} N.

RESULTS

Hydrology

The variation in rainfall and temperatures in 1988 and 1989 (Table 9.2) was reflected in the depth to the soil water-table observed within the various crop management units in the catchment. In the intensively drained Highfield area, water-tables generally were in the range 400–600 mm for both arable and grassland in winter and spring, although water-tables as shallow as 300 mm below ground-level occurred in wetter periods or where the drainage was poor. As the undrained land remaining in Cow Fen

Table 9.2 Rainfall (*R*, mm) at Swavesey and air temperature (*T*, average of maximum/minimum in °C) from Boxworth Experimental Husbandry Farm compared with the long-term average

	1987		1988		1989		Average	
	R	*T*	*R*	*T*	*R*	*T*	*R*	*T*
January	17.2	0.2	74.4	5.5	26.8	5.8	48	3.0
February	19.3	3.4	21.2	4.7	44.6	5.4	38	3.4
March	43.7	3.7	66.6	6.2	45.4	7.5	38	5.7
April	38.1	10.4	39.7	8.1	79.1	6.2	37	8.4
May	41.5	9.7	41.7	11.8	6.2	13.0	44	11.5
June	92.4	13.1	44.5	13.7	38.1	14.5	47	14.6
July	75.9	16.1	92.6	15.5	40.0	18.8	56	16.5
August	95.1	16.1	26.1	16.3	35.2	17.6	60	16.1
September	22.2	14.5	34.9	13.9	24.2	15.8	49	14.2
October	124.2	10.2	39.5	11.1	40.3	12.5	49	10.7
November	46.9	7.1[a]	30.8	5.1	33.7	6.1	58	6.4
December	20.9	5.6	21.2	6.9	110.8	5.8[a]	50	4.0
Total (*R*)	637.4		533.2		524.4		574	

[a]Estimated owing to loss of data.

Table 9.3   Comparison of depth to water-table (mm)

	1987–1988[a]			1988–1989[a]			
	Winter	Spring	Summer	Autumn	Winter	Spring	Summer
Cow Fen							
Undrained grass	200	400	700	750	450	300	>1000
Established drained arable	500	700	>1000	800	600	600	>1000
Newly drained arable	600	850	>1000	>900	850	800	>1000
Highfield							
Established drained arable	400	650	>1000	750	600	600	>1000
Established drained grass	400	700	800	850	500	600	>1000

[a]Winter, December–March; spring, April–May; summer, June–September, autumn, October–November.

was unsuitable for cropping, the impact of drainage could be determined only in the grassland. Here the depth to the water-table generally was 200 mm less in undrained land than in comparable drained grassland in both winter and spring. However, in the wet spring in 1989, water-tables more typical of winter periods were recorded (Table 9.3). The depth to the water-table in Cow Fen also was 100–200 mm more in fields recently drained and converted to arable cropping compared with nearby long established well-drained arable land. In these new arable fields, high organic matter and good soil structure also was noted.

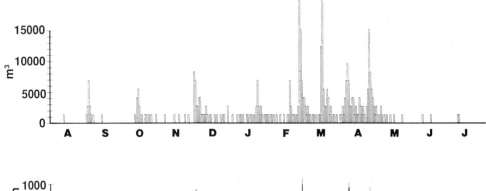

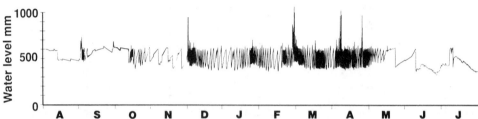

Figure 9.2   Lagoon water-levels in response to inflow following rainfall and operation of the pump, 1988–1989. Conversion to daily pump outflow was made by comparing pump operation with the head difference between the lagoon and the adjacent arterial drain

Table 9.4 Derivation of water-balance (mm) for hydrological year 1988–1989. Evapotranspiration (*Et*) derived from Boxworth based on percentage crop type

Month	Rainfall (R)	Et	R − Et	Effective R − Et	Pump outflow
August	25.8	· 52.0	−26.2	0	0.6
September	34.9	52.3	−17.4	0	3.5
October	39.5	27.2	12.3	12.3	4.6
November	30.8	8.9	21.9	21.9	3.2
December	21.2	8.1	12.3	12.3	9.0
January	26.8	9.7	17.1	17.1	7.8
February	44.6	16.5	28.1	28.1	15.8
March	45.4	31.7	13.7	13.7	18.2
April	79.1	35.0	44.1	44.1	23.7
May	6.2	75.6	−69.4	0	3.5
June	38.1	50.0	−11.9	0	0.6
July	40.0	50.0	−10.0	0	0.6

The operation of the pump system was used to calculate catchment run-off. The response pattern within the catchment for 1988–1989 shown in Figure 9.2 demonstrates that in the drier than average winter, pumping was discontinuous and reflected the individual periods of heavy rainfall. High water-levels in the lagoon, for example in April 1989, indicate that the inflow exceeded the design capacity of a single pump on a number of occasions. By utilizing evapotranspiration data calculated from meteorological records at the Ministry of Agriculture, Fisheries and Food, Boxworth Experimental Husbandry Farm, situated 5 km south of Swavesey, catchment water-balances were determined for 1987 and 1988. In the first year, evapotranspiration estimates were derived for the dominant winter wheat crop only, but in 1988–1989 (Table 9.4) the data were adjusted to reflect the proportions of grass, spring barley and all other cereals at Swavesey. The water-balance of 90% for 1987, including a small spillway input in January 1988 from the village, and of nearly 60% in 1988, indicated either an overestimate of effective rainfall or a loss to groundwater.

Nitrate-N Concentrations

The nitrate-N concentrations measured in the water leaving the catchment are given in Figure 9.3. Although the concentrations vary considerably through both short-term events and over the whole season, the data show values in excess of the EC Directive for potable water supplies (11.3 mg l^{-1} N or 50 mg l^{-1} NO_3^- as nitrate) for long periods in both years. Because the ditch gradient in the lower parts of the catchment was low, water passed relatively slowly to the storage lagoon and some reduction in nitrate loading through denitrification or uptake by aquatic vegetation was likely. In addition some attenuation of either peaks or troughs in the nitrate-N concentration was likely at the catchment outlet owing to the substantial volume of water stored in the lagoon itself prior to each pumping operation. The highest concentration recorded at the catchment outlet was 18 mg l^{-1} N in winter 1987–1988 (monthly sampling) compared with 22 mg l^{-1} N in 1988–1989, see Figure 9.3.

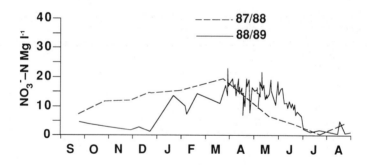

Figure 9.3 Concentrations of nitrate-N in the pump lagoon, autumn 1987 to summer 1989

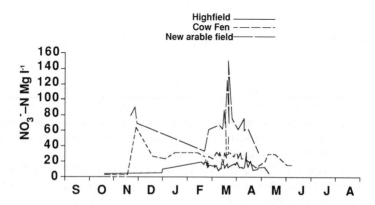

Figure 9.4 Concentrations of nitrate-N in water draining Highfield, Cow Fen (floodplain) and a drainage pipe in a new arable field in Cow Fen

The nitrate-N concentrations in the ditches upstream of the pump lagoon reflected the more immediate run-off from the agricultural land. In the upland Highfield area, representative of long-term intensive mixed agriculture, the concentrations were highest in those areas supporting arable production only. Peak concentrations also exceeded the EC limit for much of the drainage season, with the highest value of 28 mg l^{-1} N in spring 1989 (Figure 9.4). In contrast, samples from ditches either draining from or through the newly converted arable areas of Cow Fen had very much higher concentrations, with peak values between 60 and 70 mg l^{-1} N in winter 1988–1989 (Figure 9.4). Similarly, samples from the field underdrainage system in the new arable fields had the highest concentrations within the study area, with values frequently above 60 mg l^{-1} N and a peak of 160 mg l^{-1} N in spring 1989 (Figure 9.4).

Nitrogen Balance, 1988–1989

Inputs of nitrogen to the catchment were calculated on a field-by-field basis for fertilizers and from estimates for atmospheric deposition of 40 kg ha^{-1} N (Powlson *et al.*, 1986). Total nitrogen applied by the farmers was calculated to be 66 t or an

Table 9.5 Uptake of nitrogen in grain (percentage of dry weight) and crop yield for harvest 1988–1989

	Cow Fen	Highfield
Winter wheat		
Nitrogen (% N)	1.97	2.01
Yield (t ha^{-1})	6.1	6.4
N uptake (kg ha^{-1})	120	129
Winter barley		
Nitrogen (% N)	2.06	2.09
Yield (t ha^{-1})	6.0	6.2
N uptake (kg ha^{-1})	123	130

average of 132 kg ha^{-1} N over the whole area. Total daily removal of nitrate in the soil water leaving the catchment was calculated by correlating pump operation (Figure 9.2), and the nitrate-N concentrations in the lagoon (Figure 9.3). In the drainage period for 1988–1989, when the calculated water-balance was 60%, an estimated 5.5 t of nitrogen was removed through the pump. If the effective rainfall and resultant soil-water movement not recovered through the ditch and pump system had moved to groundwater, and removed nitrate-N at the same rate, then the loss of nitrogen in soil-water would increase to approximately 9 t.

Total uptake of nitrogen by the arable crops was determined for each field and calculated to be 46 t for the whole catchment. As 82% of crop residues were burnt or removed and 80% of the nitrogen offtake was in the grain, the return of nitrogen to the soil from residues was likely to be very small. Little difference in offtake was noted between Cow Fen and Highfield, (Table 9.5). Removal of nitrogen from grassland, assessed from reported data and recorded stocking rates at Swavesey, was estimated to be much lower than from the arable land, at 6 t; predominantly in Cow Fen where substantial areas remained as grassland (Figure 9.1).

Total losses to the atmosphere at Swavesey by denitrification, ammonia volatilization and gaseous loss from growing plants could not be determined in such a large catchment area and were estimated from the literature. Goulding (1988) reviewed rates for losses by denitrification and ammonia volatilization based on season, cropping and whether nitrogen application was as inorganic fertilizer or organic manure (Table 9.6). He suggested that losses could be considerable from organic applications to grassland (up to 30% of applied nitrogen), with losses following inorganic fertilizer applications of up to 13% for arable land and 23% for grassland. Garwood and Morrison (1988) also reported denitrification losses equivalent to 60% of applied fertilizer N for intensively managed and fertilized undrained grassland, reducing to 40% with effective drainage.

At Swavesey, most spring fertilizers were applied in 1989 from late March to the end of April during both wet and mild weather (Table 9.2) resulting in particularly high opportunities for mineralization of organic matter and denitrification. As only 30% of the whole catchment was grassland receiving just 7 t of fertilizer, losses were likely to be dominated by the arable areas, which received 59 t of fertilizer. Losses

Table 9.6 Estimates of atmospheric losses for arable and grassland, after Goulding (1988), expressed as a percentage of applied N

	Fertiliser			Organic manures
	Cut grass	Grazed grass	Arable	Grass
Denitrification	0–10	0–20	0–10 (maximum 30)	< 20
Volatilization	< 3	< 12	< 3	20–30

Table 9.7 Soil mineral nitrogen (0–900 mm) and organic matter (0–150 mm) in early spring and summer 1989

Management unit	Soil mineral spring (kg ha^{-1} N)	Nitrogen summer	Organic matter summer (%)
Cow Fen (floodplain)			
Undrained grass	100–150	50–80	17
Drained grass	100	30–160	14
New arable	150–900	150–750	13
Established arable	125	100	9
Highfield			
Drained grass	50	40–120	5
Drained arable	100–150	50–100	3

Table 9.8 Catchment nitrogen balance 1988–1989

	Tonnes N	Equivalent (kg ha^{-1} N)
Input		
Fertilizer – grass	7	
– arable	59	⎰132
Atmospheric deposit	20	⎱ 40
Total	86	172
Output		
Arable crop/residue	46	92
Grassland	6	12
Leaching (maximum)	9	18
Atmospheric loss	8	16
Total	69	138

near to the upper limit for arable areas (10–15%) were considered likely, an estimate of 12% of the total applied fertilizer to the catchment giving atmospheric losses equivalent to 8 t of nitrogen.

Soil mineral nitrogen and organic matter were determined from spring 1988–1989. Differences were identified between the management units, with the highest organic matter status and soil mineral nitrogen levels in the newly drained arable fields of Cow Fen (Table 9.7). By contrast, the lowest soil mineral nitrogen in the soil profile was found in drained grassland.

An assessment of the nitrogen balance for the whole catchment in 1988–1989 is given in Table 9.8. Although differences in soil mineral nitrogen over the period, including N fixation by legumes, could not be taken into account because of insufficient data, the nitrogen output balance accounts for up to 80% of the estimated N inputs. The catchment nitrogen balance derived suggests that a build-up of soil mineral nitrogen status is occurring owing to nitrogen mineralization of the soil organic matter.

DISCUSSION AND CONCLUSIONS

Nitrate-N concentrations exceeded the EC Drinking Water Directive in both Cow Fen and Highfield throughout most of the drainage season. In both 1987–1988 and 1988–1989, concentrations in the pump lagoon peaked in the spring. In the rest of the catchment concentrations were in agreement with the results of Roberts (1987) and Goss et al. (1988) and were highest in the autumn, then declined over the winter before peaking again in the spring. The spring peak of 22 mg l^{-1} nitrate-N in 1989 coincided with high rainfall and associated run-off (Figures 9.2 and 9.3), resulting in considerable removal of nitrogen from the catchment.

Nitrogen lost by leaching in surface run-off in 1988–1989 from the whole catchment was 11 kg ha^{-1} N, or 8% of the equivalent nitrogen fertilizer applied over the same period. A further 7 kg ha^{-1} N was lost potentially to groundwater. Although this leaching loss agrees favourably with several other catchment studies, such as those by Houston and Brooker (1981) with losses at 8.9 and 10.2 kg ha^{-1} in two predominantly grassland catchments in Wales, the area under grassland is only 30% at Swavesey and fertilizer inputs were low. Comparable large and predominantly arable catchments studied in East Anglia by Edwards (1973), based on limited weekly sampling, recorded annual losses around 12 kg ha^{-1} N. Elsewhere in small isolated plot studies under intensive arable cropping on the Denchworth Series soil (as found in Highfield), Goss et al. (1991) found losses around 20% of the equivalent fertilizer applied. Comparable data to Swavesey also were reported by Wild and Cameron (1980) who found UK losses under arable cropping of 10–15% of the equivalent applied fertilizer nitrogen and around 10% from grassland. Although data from the grassland experiments at North Wyke (Garwood and Morrison, 1988) showed leaching losses of 20 and 55 kg ha^{-1} N from undrained and drained land, respectively, following an application of 200 kg ha^{-1} N, the grassland at Swavesey represented a much lower input system with consequent much lower leachate losses likely.

Analysis of offtake of nitrogen by all arable crops at Swavesey showed that the equivalent of 70% of the applied fertilizer was removed in the grain and residues. As over 80% of the residues at Swavesey were removed, little of the nitrogen taken up by the cropping was returned to the soil. Other UK data provided by Greenwood (1990) for the period 1975–1988 estimated that 51% of the equivalent applied fertilizer is, on average, taken up by the grain. Although uptake of nitrogen in the newly established arable fields in the former floodplain (Cow Fen) was similar to Highfield, fertilizer applications were lower but nitrate leaching was higher. This demonstrated that organic matter mineralization following ploughing and the lowering of the water-tables were important factors in the release of nitrogen. Similar data has been found by Young, Hall and Oakes (1976) for nitrate leaching under ploughed grassland. As organic matter in the 0–150 mm horizon in the undrained grassland at Cow Fen was 14% and in fields converted to drained arable land was at best 13% after 3 years, decreasing to 10% after 15 years, the breakdown process is relatively slow. Together with particularly high soil mineral nitrogen in the soil profile in the new arable fields (Table 9.7), this data suggests that leaching from the floodplain will be substantially above that leaching from established arable land for several decades.

The derived overall catchment nitrogen balance from measured and estimated inputs and outputs was 80%. Of this, the farmer input of 66 t of fertilizer to the catchment in 1988–1989 was calculated predominantly from accurate farm records. Likewise the removal of 46 t in the crop or through the residue was determined from detailed field studies and outweighed any uncertainties in grassland losses, estimated at 6 t or that removed by leaching, calculated as in the range 5.5–9 t. These components together accounted for a balance of approximately 90%. The greatest uncertainties were therefore in atmospheric inputs (estimated as 20 t or 40 kg ha^{-1} N) and atmospheric losses (estimated as 8 t). Of these, the estimate for atmospheric input was derived from data reviewed by Powlson et al. (1986) from a series of experiments over the last 90 years. It is now generally considered from the most recent monitoring that inputs annually are in the range 40–50 kg ha^{-1} N (Goulding, 1989), supporting the estimate used at Swavesey. The estimate for atmospheric losses, mostly by denitrification, at 12% of the equivalent applied nitrogen fertilizer derived from Goulding (1988) and Garwood and Morrison (1988) is supported by Colbourn (1985), who found losses on arable land at 10% of the N applied for each 40 mm rain falling in the 28 days after application. Since most fertilizer was applied to the arable area and was followed by nearly 80 mm rainfall in April 1989 (Table 9.2), losses from the arable land were likely to be high.

Differences over the season in soil mineral nitrogen are therefore most likely to account for the higher inputs than outputs in the year studied, reflecting a build up in soil mineral nitrogen owing to N release from the organic matter following drainage and cultivation. As soil mineral nitrogen has been determined from summer 1989 for representative crop and management systems in up to 25% of the catchment, this aspect will be addressed in future research. However, Harris (1991) reported higher nitrate-N concentrations with commensurate higher total N loadings in this catchment in the following winter, 1989–1990.

The main conclusions from the study were:

(1) Drainage lowered the water-table in the former floodmeadows by at least 200 mm over most of the year. In the newly established arable fields in the

floodplain, good soil structure and high soil organic matter resulted in better drainage and deeper water-tables.

(2) Fertilizer applications to both arable and grass fields were lower in the former floodplain compared with the established arable land.

(3) Nitrate leaching consistently exceeded the EC Drinking Water Directive throughout the whole catchment. The highest concentrations were recorded in drainage water from the former floodplain.

(4) The lowering of the water-tables, together with the mineralization and release of N from the organic matter in the former floodplain, is likely to result in continued high nitrate leaching for several decades.

ACKNOWLEDGEMENTS

We thank ADAS Field Drainage Experimental Unit staff for assistance in the field. Monthly water samples were analysed by Anglian Water and all other laboratory samples in the ADAS analytical laboratories, Cambridge. We are indebted to the farmers of Swavesey who have provided much of the detailed field management information.

Finanacial support for this work from both the Ministry of Agriculture, Fisheries and Food (MAFF) and The Natural Environmental Research Council (NERC) is gratefully acknowledged.

REFERENCES

Avery, B. W. (1980). *Soil classification for England and Wales*, Technical Monograph No. 14, 53. Soil Survey of England and Wales.

Chalmers, A. G. and P. K. Leech (1988). *Survey of Fertiliser Practice, Fertiliser Use on Farm Crops in England and Wales, 1989*, Ministry of Agriculture, Fisheries and Food, London.

Colbourn, P. (1985). Nitrogen losses from the field: Denitrification and leaching in intensive winter cereal production in relation to tillage method of a clay soil. *Soil Use Management*, 1, 117–120.

Edwards, A. M. C. (1973). The variation of dissolved constituents with discharge in some Norfolk rivers. *Journal of Hydrology*, 18(3/4), 219–242.

Garwood, E. A. and J. Morrison (1988). Water deficiency and excess in grassland. The implications for grass production and the efficiency of use of N. In R. J. Wilkins (ed.), *Nitrogen and Water Use by Grassland*, Proceedings of Colloquium 27 March 1987, Institute for Grassland and Animal Production, Hurley, pp. 24–41.

Goss, M. J., K. R. Howse, P. Colbourn and G. L. Harris (1988). Cultivation systems and the leaching of nitrates. *Proceedings of 11th International Conference of the International Soil Tillage Research Organisation*, 2, 679–684.

Goss, M. J., K. R. Howse, P. W. Lane, D. G. Christian and G. L. Harris (1991). *Losses of Nitrate-nitrogen in Water Draining from Under Autumn-sown Crops Established by Direct Drilling or Mouldboard Plough*. Unpublished report, AFRC Institute of Arable Crop Research, Rothamsted, and ADAS Field Drainage Experimental Unit, Cambridge.

Goulding, K. W. T. (1988). Gaseous losses of nitrogen. In *Modelling Nitrogen Losses from Land*, Unpublished report, Ministry of Agriculture, Fisheries and Food, Chief Scientist Group Special Topic Review, London.

Goulding, K .W. T. (1989). Atmospheric deposition. Farming and the quality of natural waters – nitrate, phosphate and pesticides. In *Institute of Arable Crops Research Annual Report 1989*, Section 2, pp. 67.

Greenwood, D. J. (1990). Production or productivity: the nitrate problem? *Annals of Applied Biology*, **117**, 209–231.

Houston, J. A. and M. P. Brooker (1981). A comparison of nutrient sources and behaviour in two lowland sub-catchments of the River Wye. *Water Research*, **15**, 49–57.

Harris, G. L. (1991). *Soil Hydrology and Drainage*, Unpublished Project Report, Ministry of Agriculture, Fisheries and Food, Chief Scientists Group, London.

Ministry of Agriculture, Fisheries and Food (1986). The analysis of agricultural materials. In *Ministry of Agriculture, Fisheries and Food, Agricultural Development and Advisory Service*, Reference Book 427, 3rd edn, London.

Powlson, D. S., G. Pruden, A. E. Johnston and D. S. Jenkinson (1986). The nitrogen cycle in the Broadbalk wheat experiment: recovery and losses of N-labelled fertiliser applied in spring and inputs of nitrogen from the atmosphere. *Journal of Agricultural Science, Cambridge*, **107**, 591–609.

Roberts, G. (1987). Nitrogen inputs and outputs in a small agricultural catchment in the eastern part of the United Kingdom. *Soil Use and Management*, **3**(4), 148–154.

Seale, R. S. (1975). Soils of the Ely district. *Memoirs of the Soil Survey of England and Wales*, Sheet 173, pp. 127–128.

Trafford, B. D. and R. A. Walpole (1975). Drainage design in relation to soil series. In A. J. Thomasson (ed.), *Soils and Field Drainage*, Technical Monograph No. 7, Soil Survey of England and Wales, pp. 49–61.

Wild, A. and K. C. Cameron (1980). Nitrate leaching through soils and environmental considerations. In *Soil Nitrogen as Fertiliser or Pollutant*, Proceedings of a Research Co-ordination Meetings, FAO/IAEA, Brazil 1978, International Atomic Energy Agency; Vienna, pp. 289–306.

Winteringham, F. P. W. (1980). Nitrogen balance and related studies: a global review. In *Soil Nitrogen as Fertiliser or Pollutant*, Proceedings of a Research Co-ordination Meeting, FAO/IAEA, Brazil 1978, International Atomic Energy Agency; Vienna, pp. 307–344.

Young, C. P., E. S. Hall and D. B. Oakes (1976). *Nitrate in Groundwater Studies on the Chalk near Winchester, Hampshire*, Technical Report 31, Water Research Centre, Medmenham, 67 pp.

10 Floodplain Assessment for Restoration and Conservation: Linking Hydrogeomorphology and Ecology

G. E. PETTS, A. R. G. LARGE, M. T. GREENWOOD and
M. A. BICKERTON
Freshwater Environments Group, International Centre of Landscape Ecology, Department of Geography, Loughborough University of Technology

INTRODUCTION

With the world-wide decline in biodiversity and growing concern for the effects of climatic change, international attention is being directed to areas of potentially high conservation value. Following the Sopron Symposium on Land–Inland Water Ecotones in 1988 (Holland, 1988; Naiman and Décamps, 1990) and fostered by the Man and Biosphere (MAB-5) programme of Unesco, attention is being directed to the conservation and restoration of floodplain systems along large, lowland rivers. The potential conservation value of these land–water ecotones has been established (Naiman and Décamps, 1990; Boon, Calow and Petts, 1991).

Geomorphologists consider floodplains as functioning associations of landforms largely produced by alluvial deposition (Lewin, 1978). Ecologists recognize the importance of: (1) the flood pulse in determining many ecological functions both within the floodplain itself and the associated lotic ecosystems (Welcomme, 1979; Décamps *et al.*, 1988; Junk, Bayley and Spinks, 1990); (2) topography and site characteristics in structuring river corridors (Morris *et al.*, 1978; Pautou and Décamps, 1985); and (3) the role of geomorphological dynamics in determining the mosaic of ecological patches that make up natural river corridors (Amoros *et al.*, 1987; Salo *et al.*, 1988; Roux *et al.*, 1989; Salo, 1990).

Throughout the temperate zone – not least within the UK – natural floodplains have been altered dramatically by river regulation over the past 200 years (Petts, Moller and Roux, 1989; Petts, 1990a, b). However, Welcomme (1991) considers that from an ecological viewpoint these floodplains were once homologous to those of tropical regions. For these latter rivers the ecological importance of floodplains has been well established and there is growing evidence to suggest that the floodplains of temperate rivers had similar functional roles. Restoration strategies for temperate rivers are required to enhance conservation, recreation, and educational interests whilst giving due regard to economic returns. However, information is required on the distribution

Lowland Floodplain Rivers: Geomorphological Perspectives. Edited by P.A. Carling and G.E. Petts
© 1992 John Wiley & Sons Ltd

of fauna and flora, on the fundamental hydrological and geomorphological controls on these distributions, and on the biological and chemical interactions between patches so that the functional, as well as structural, characteristics of floodplains are restored.

Over the past decade, geomorphologists have given great attention to uniting fluvial geomorphology and river engineering (e.g. Newson, 1986). Simultaneously, there has been a move away from a focus on impacts to one seeking environmentally sensitive alternatives for river engineering and river restoration (e.g. Newbold, Purseglove and Holmes, 1983; Lewis and Williams, 1984; Gore, 1985; Gore and Petts, 1989; Hemphill and Bramley, 1989; Coppin and Richards, 1990). Geomorphologists have sought to establish links with ecologists (Viles, 1988; Thornes, 1990), but these have focused on the influence of plants on the fundamental earth surface processes of weathering, erosion and deposition. Other important linkages have remained undeveloped.

The research reported herein focuses on the influence of hydrogeomorphology on ecology, specifically on the structuring of the floodplain ecotone. An approach to the assessment of floodplains is summarized and application of the approach is illustrated by reference to ongoing research on the River Trent, UK.

APPROACH

Our interdisciplinary approach to the study of fluvial hydrosystems is based on that developed by researchers at the University of Lyon 1, France (Bravard, Amoros and Pantou, 1986; Amoros et al., 1987; Roux et al., 1989). The approach incorporates fluvial geomorphology, hydrology, hydrobiology, ecology, palaeoecology and historical geography across a range of spatial and temporal scales. Following Lewin's (1978) and, more recently, Salo's (1990) analyses of floodplain systems the focus is on the meso-scale: changes over one to a few hundred years and areas of about 10–100 000 m^2. The approach involves three stages: description and classification, establishment of successional sequences, and assessment of conservation values.

Description and Classification

The first step is to use geomorphological factors, such as channel pattern and stability, and information on human impacts (landuse change and channel modification) to define 'functional sectors', which provide the template for descriptive ecology. At this scale, hydrology and geomorphology are seen to determine habitat diversity and ecological functions.

Secondly, within each functional sector, sets of 'functional units' are defined. Each unit is the biotope and biocoenosis of each elementary landform. A low-level gravel bar and a meander cut-off are examples of functional units. An evolutional (successional) sequence of units forms a 'functional set'. Ecologically, this scale of resolution focuses on communities, their functional characteristics and autogenic successional processes.

Thirdly, 'functional describers' are defined for each unit. These relate to the genesis of the unit and typically include sediment characteristics, such as granulometry (Bravard, 1983) and organic matter (Rostan, Amoros and Juget, 1987), and indicator species (Bournaud and Amoros, 1984; Castella *et al.*, 1984).

Establishment of Successional Sequences

Functional sets are defined by either synchronic or diachronic analyses (Figure 10.1), whenever possible using both. A widely used approach in geomorphology is to substitute spatially sampled data for time series, a form of ergodic reasoning (Paine, 1985). The identification of landforms at various stages of development in the modern landscape allows inferences to be made about changes through time. Synchronic analyses involve comparative studies of the structure and function of units of different age to derive 'successional' models (see Figure 10.5). Particular attention is given to the relative influence of allogenic processes (e.g. water and nutrient fluxes, sediment deposition and channel erosion) and autogenic successional processes (e.g. colonization, competition and eutrophication). These are supported by historical studies designed to establish a chronology of geomorphological events, human impacts and environmental changes. Diachronic analyses, using a range of palaeoecological indica-

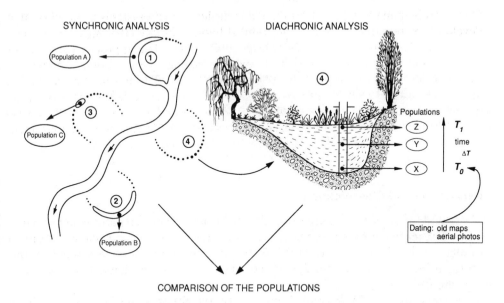

Figure 10.1 A schematic representation of synchronic and diachronic analyses. Functional units, such as cut-off channels (1–4), are dated from historical records (see Table 10.1); they are examined to construct a functional set incorporating a sequence of indicators, such as species or communities (populations A–C). The successional model is then validated by comparison with the results of diachronic analysis. (From Amoros *et al.*, 1987; reproduced by permission of John Wiley and Sons Ltd)

tors (e.g. Amoros and van Urk, 1989) are used to validate the successional models derived from synchronic analyses and historical studies.

Assessment of Conservation Value

Diversity – of communities, species, and habitats – is an important attribute of a functional sector. Descriptions of habitat quality for use in conservation management have involved the use of other criteria, such as rarity and typicalness (Ratcliffe, 1977; Usher, 1980; Eyre *et al.*, 1986; Eyre and Rushton, 1989). Such methodologies use species presence/absence data from primary collections and historical records across habitats from a large area. However, the assessment of floodplain units requires a smaller scale of analysis. Herein, conservation value has been assessed on the basis of species diversity; 'rarity' and 'typicalness', using a modified version of the methodologies of Eyre and Rushton (1989) and Rushton (1987); and a new criterion, 'specialism', which indicates the preference of a species for a particular unit.

Many assessments of conservation value have been based on vegetation. There are two reasons for this. First, ecological patches have intrinsic botanical interest and in the UK this is reinforced by the historical reflection of the inordinate number of botanists who are also conservationists. Secondly, vegetation distribution is assumed to provide a good index of 'habitat' for explaining faunal distributions. However, many components of the fauna relate more directly to other habitat criteria (Nature Conservancy Council, 1989), particularly microclimate and surface characteristics such as granulometry, moisture content, porosity and bulk density – factors that relate to the geomorphological setting. For example, along river margins, gravel bars and south-facing sandy banks may be particularly important.

Beetles (Coleoptera) are one of the most extensively studied groups of invertebrates and have been shown to be useful in assessing environmental quality and changes in standing freshwaters (Eyre and Foster, 1989; Eyre and Rushton, 1989); in assessing human impacts on ecosystems (van Dijk, 1986; Jennings, Houseweart and Dunn, 1986; Luff, 1987; Rushton and Luff, 1988); and for quantifying conservation criteria (Eyre and Rushton, 1989). In this study, Staphylinid beetles have been used to assess the conservation value of each floodplain unit.

Rarity

The method for deriving a rarity score, adapted from Eyre and Rushton (1989), uses species frequencies, defined as the number of traps within which each species was recorded. A geometric scaling (1, 2, 4, 8 and 16+) of the frequencies was then undertaken and an inverse scoring applied (16, 8, 4, 2 and 1, respectively), thereby weighting the rarer species. For each trap the species scores were summed to give a species rarity total (SRT). In order to increase the weighting of traps with several rare species, a rarity association value (RAV) was calculated by totalling the trap species scores but using only scores of two or more, and downweighting any single very rare species to the next highest score. The SRTs and RAVs were summed and this value was then divided by the number of scoring species in each trap to give a rarity quality factor (RQF), which is independent of species richness. Scores for each trap within a unit were totalled to give a unit rarity total.

Typicalness

Using methods similar to Rushton (1987), a typicalness measure for each trap was derived by ordinating the species abundance data using detrended correspondence analysis (within the CANOCO package: Braak, 1988). The centroid of each group of traps, relating to a specific unit, was defined as the mean of the trap scores on the first two axes. The distance of each trap from the centroid in the two axial planes was calculated (see Figure 10.6). Typicalness was defined as the inverse of the distance from the units centroid; traps nearest the centroid being most typical of that unit. The typicalness of each unit was defined as the average of the trap measurements within the unit.

Specialism

Using the subset of species that were frequent in a significant proportion of traps, here the most frequent one-third of species (i.e. those occurring in four or more traps), specialism scores for each unit were generated by dividing the frequency within the unit by the frequency within the sector. Scores of species showing a positive preference (greater than the reciprocal of the number of units) were summed to give a specialism total and this was then divided by the number of scoring species to give a specialism factor.

THE RIVER TRENT FLOODPLAIN

The above approach is being applied to the River Trent (Figure 10.2). The catchment of 10 500 km^2 is underlain primarily by sedimentary rocks of Carboniferous, Permo-Triassic and Jurassic ages, but glacial deposits are widespread – especially in the west and south. The catchment has a population of over 5.5 million, including the major urban centres of Birmingham, Nottingham, Derby and Leicester.

The Flandrian courses of the River Trent at Colwick, near Nottingham, have been described by Salisbury *et al.* (1984). The 2.5 km wide valley floor in this section has been infilled with 2–6 m of large-scale cross-bedded sand and gravel, reflecting lateral point-bar accretion, with a 1–2 m overburden of alluvial silts and clays. That the floodplain deposits at this site had been reworked through their entire depth in historic times is evidenced by the presence of waterfront structures and artifacts dating from Saxon, Norman and Tudor periods within half a metre of the Triassic rocks at the base of the gravel. Active floodplain reworking was clearly common until the seventeenth century. Today, the River Trent is regulated for most of its length, being channelized and with long reaches impounded by weirs and sluices.

Definition of Functional Sectors

The history of the Trent until the eighteenth century is one mainly of conflict between town dwellers, merchants and those engaged in transport on the one hand, and

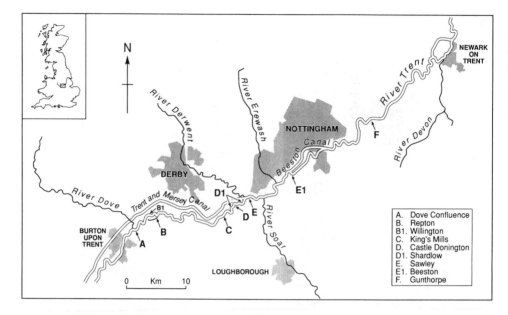

Figure 10.2 The middle Trent locating sites referred to in the text

fishermen and millers, whose trade required the construction of weirs, and the riparian land owners, on the other hand. Consequently, there is a wealth of documentary and cartographic evidence (Table 10.1) describing not only landuse changes and channel works but also the geomorphological character of the river. Using this evidence, together with information from aerial photographs, seven functional sectors have been defined (Figure 10.3).

Historic evidence indicates that the River Trent had a wide, shallow cross-profile with numerous shoals; it flowed in two channels at several points along its course; and several other sectors were characterized by mobile meanders. The corridor of the Middle Trent was characterized by unstable reaches separated by relatively short, stable, control reaches (Figure 10.3). Typically the active floodplain had a width of about 400 m but was much wider at tributary confluences.

Table 10.1 Historic Data Sources: River Trent

1200–1890	Deeds and private legal papers
1590–1850	Local maps
1600–1840	County maps
1620–1890	Estate surveys
1630–1890	Personal correspondence
1699–1922	Government Papers/Acts of Parliament
1750–1850	Enclosure surveys
1800–1820	Topographical reports
1848–1850	Tithe surveys

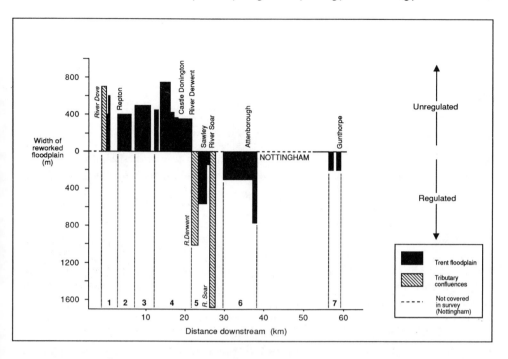

Figure 10.3 Variation of historic floodplain width along the middle Trent from the Dove confluence to Gunthorpe showing the main functional sectors

Systematic works to stabilize the channel, in order to improve navigation, began in the late seventeenth century. Their effect was to divide the river into two groups of sectors (Figure 10.3), upstream and downstream of Shardlow. Channel improvements involved extensive dredging, constructing longitudinal training weirs, and by-passing the worst shoals with side cuts. The riparian zones were cleared of vegetation and snags to create hauling paths.

An Act of 1783 authorized channel improvements from Shardlow downstream to Gainsborough, to maintain a water depth of 24 in. (61 cm) within a navigable channel at least 33 ft (10.06 m) wide, and to provide a hauling path normally 3 yards wide but up to 6 yards (2.74 m) on the 'crocked' parts of the river. The hauling path was completed by 1784. By 1792 it was clear that several of the shoals continued to prevent satisfactory navigation and additional side-cuts with locks and weirs were built between 1793 and 1796, for example at Sawley and Beeston (see Figure 10.2). Thus, sectors below Shardlow have been regulated for 200 years.

The Trent from Shardlow upstream to Burton-on-Trent was controlled following an Act of 1699. However, because of the problems for navigation caused by floods in winter and by the shallows in summer (Wood, 1950) the Grand Trunk Canal was built to link the Mersey and Trent (see Figure 10.2). All navigation on the Trent upstream of Shardlow had ceased by 1805 and since that time the river has been maintained but remained largely unregulated.

Establishment of Functional Sets

Synchronic analyses have been used to establish functional sets and can be illustrated by consideration of the abandoned channels and backwaters. A two-phase successional model is proposed based on two describers: water quality for the aquatic phase and vegetation for the terrestrial phase.

Aquatic Phase

An analysis of backwaters and abandoned channels in the Trent Basin (José, 1988) shows a continuum of forms variably affected by the hydrological regime of the mainstream depending on hydrological connectivity and backwater morphometry. Using the typology of Roux (1982), these range from those dominated by the main river (eupotamon) to those that are not in permanent contact with the mainflow and only mildly influenced by river discharge fluctuations (palaeopotamon). The term 'plesiopotamon' applies to an intermediate category.

For three sites along the River Trent (Shardlow, Willington and Beeston – see Figure 10.2 for locations), water samples were obtained at monthly intervals during 1986 from the main river and at a number of points throughout each backwater. Each sample was analysed for ammoniacal nitrogen, total oxidized nitrogen, orthophosphate, chloride, calcium, sulphate, dissolved oxygen, pH, conductivity and temperature. The typology of sites is well-illustrated by consideration of the orthophosphate data (Figure 10.4).

The site at Beeston is a channel 200 m long that has undergone no morphological change since 1886. It is 4 m wide and up to 1.5 m deep under low-flow conditions. At

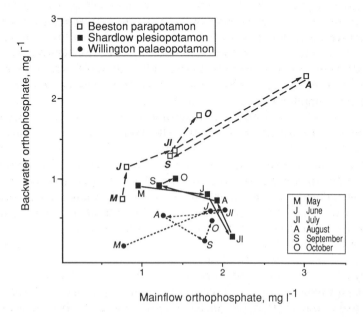

Figure 10.4 Seasonal variation of orthophosphate concentration in three different types of backwater along the middle Trent. (After José, 1988)

Shardlow, the backwater is a former gravel pit that has been connected to the main river since about 1970 to form a boat marina. The backwater has a depth of up to 4 m, an area of 7.5 ha, and a 10 m wide connection with the main river. The cut-off channel at Willington has a length of about 100 m, but the permanent pool extends for no more than 30 m. The 1886 map shows the channel to be linked to the main river at its downstream end at this time.

The Beeston channel is a classic parapotamon, there being only minor differences in water quality between the mainstream and the backwater throughout the year. In contrast, the Shardlow backwater, although according to Roux's typology fitting the parapotamon category, being connected to the main river (effectively) at its downstream end, has water quality characteristics typical of the plesiopotamon type. Whilst having a water quality similar to that of the main river during winter, in summer there are marked differences, especially with reduced orthophosphate and total oxidized nitrogen, reflecting the large volume of the backwater and associated biochemical changes. At Willington marked differences between the water quality in the cut-off channel and the main river were apparent for nutrient, chloride and sulphate concentrations under all flow conditions, although the differences were reduced during high floods. In this typical 'palaeopotamon', substrate siltation and the high levels of organic matter may impede groundwater interactions so that the cut-off channel acts as an isolated floodplain pond.

Semi-aquatic/Terrestrial Phase

Vegetation surveys focused on two sectors of the River Trent, including an unregulated sector (Castle Donnington) and a regulated sector (Gunthorpe: see Figure 10.2). The latter has been described in detail by Large *et al.* (in press). Vegetation patches were distinguished by their seral position and their present-day land management, the main criterion being variation in dominant species present. Patch size was not a definitive character, and varied from 0.1 ha up to 10 ha. The field method adopted was similar to that of Poore (1955a, b), McVean and Ratcliffe (1962) and Birks (1973). Stratified sampling was used (*sensu* Goldsmith, Harrison and Morton, 1986). This involved subjectively dividing the area under consideration into functional units, and then sampling each unit to establish species composition. Uniform areas of vegetation were defined as those that showed little or no obvious variation in the relative abundance, physiognomy or spatial distribution of the most abundant species present.

Characteristics of the vegetation communities are given in Table 10.2. Of the 135 species recorded, 82 occurred in the woodland, 66 in the riparian units and 52 from the seasonal wetland units. The riparian unit, although floristically diverse, was notable in that it had no dominant herbaceous species. The pasture sample comprised 47 species and the margins of the arable fields had only 17 species dominated by ruderals.

Functional Set

By combining the water quality and vegetation classifications described above, the complete successional sequence can be defined (Figure 10.5). Three of the semi-aquatic and terrestrial stages indentified for the River Rhone by Amoros *et al.* (1987) can be identified on the Trent floodplain. However, the successions have been

Table 10.2 Dominant species found in different vegetation patches on the River Trent floodplain at Gunthorpe and Castle Donnington, grouped alphabetically into the five main vegetation patch types

Woodland (82 species)	Riparian (66 species)	Wet area (52 species)	Pasture (47 species)	Arable (17 species)
Anthriscus sylvestris	*Salis fragilis*	*Carex riparia*	*Dactylis glomerata*	*Agrostis gigantea*
Corylus avellana		*Deschampsia cespitosa*	*Holcus lanatus*	*Urtica dioica*
Galium aparine		*Epilobium hirsutum*	*Lolium perenne*	
Prunus spinosa		*Glyceria maxima*	*Poa annua*	
Salix fragilis		*Phalaris arundinacea*	*Poa trivialis*	
Salix viminalis		*Plantago major*		
Urtica dioica		*Urtica dioica*		

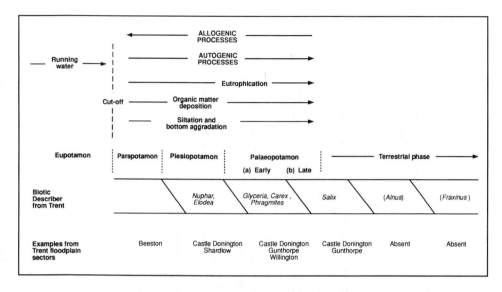

Figure 10.5   Functional set for channel cut-offs along the middle Trent. The diagram shows a schematic representation of the stages of ecological succession, following Amoros *et al.* (1987). Species in parentheses would have been expected to have been present as seral stages but appear to have been prevented by historic management practices

modified and interrupted by various management practices over a historic time-scale. In many places the silted parts of the cut-off channels have been 'improved' and incorporated into the washlands pasture. Woodland generation has been restricted and today is largely confined to the areas of former osier beds. The absence of Alder (*Alnus glutinosa*) is surprising and as yet unexplained. In a study of rivers in eastern England, Mason and MacDonald (1990) found Alder to be one of three common species, with two willows, *Salix alba* and *S. fragilis*. The latter (crack willow) dominates the woody communities of the Trent floodplain and appears to have replaced the osier willow since the turn of the century.

Assessment of Conservation Values

The conservation values of the primary functional units have been assessed for the Gunthorpe sector (see Figure 10.2). Although part of the regulated river, the floodplain here is within the floodbanks and is subjected to major inundation about once every 2 years on average. Seventeen vegetation patches were defined and these were combined into five units, which also have different geomorphological settings (Table 10.3): riparian zone, pasture, arable, wetland and woodland. The last two units form part of a functional set. The wetland and woodland occur in the lower parts of an old cut-off, and the pasture extends over the silted-up, upstream part of this channel. All except the arable unit are subject to regular flooding. The dominant species are as shown in Table 10.2, with the exception that *Carex riparia* is absent. In total, 126 species were identified and 22 species are common to the riparian and wetland units.

Table 10.3 Summary of the characteristics of the functional units within the Gunthorpe sector

	Floodplain			Riparian zone	Terrace
	Woodland	Pasture	Wetland		
Mean elevation (m)[a]	−0.35	+0.286	−0.50	−0.34	+2.22
Soil type	Organic silt–clay	Silt	Organic silt–clay	Silty sand	Sandy silt
Moisture content (%)	30.2	13.6	44.1	9.5	6.0
Number of vegetation species	82	47	43	66	17

[a]Elevation relative to approximate 'bankfull' level.

Table 10.4 Gunthorpe Staphylinids – diversity, rarity, typicalness and specialism indices[b] for the five habitat units

Habitat	Number of species	Mean number of species per trap	Diversity (H)	Rarity total (total of trap RQF scores)	Typicalness (1/distance from centroid)	Specialism Total (SPT)	Average (SPF)	Number of scoring species
Wood	26	4.5	2.77	97	0.4	2.7	0.68	4
Pasture	10	1.3	2.03	66	0.7	0.25	0.25	1
Wet	21	6.8	2.17	70	2.5	4.97	0.62	8
Riparian	31	6.3	2.67	68	0.9	3.14	0.52	6
Arable	6	0.5	1.67	23	0.3	0	0	0

[b]Definitions are given on pp. 220–221.

In May 1990, 12 pitfall traps were located in each unit and beetles collected after 2 weeks. Fifty-five species (Coleoptera: Staphylinidae) were identified. The relatively high diversity within the riparian, woodland and wetland units contrasts with pasture and arable units. Although the total number of pitfall traps was 60, the maximum observed frequency of beetle was 20 (*Tachinus signatus*). Riparian and woodland units have a much higher proportion of very low frequency species compared with the wetland site. Three species *Carpelimus gracilis*, *Neobisnius villosulus* and *Gabrius bishopi* are new records for Nottinghamshire, as is *Ilyobates propinquus*, which also is listed in the Red Data Book (Shirt, 1987) of nationally rare species – all three are found in the wetland unit.

The scores for each unit are given in Table 10.4. The highest rarity score was found for the woodland, but this reflects the large number of species of low frequency present in this unit. The wetland, riparian and pasture units had similar moderately high scores, with arable having the lowest. In terms of typicalness, wetland scored substantially higher than any other unit, indicating that all traps had a similar beetle fauna. Although woodland and riparian units had higher species richness there was less consistency (greater diversity) between traps. The arable typicalness score is unreliable as many traps contained no species. Wetland also had the highest number

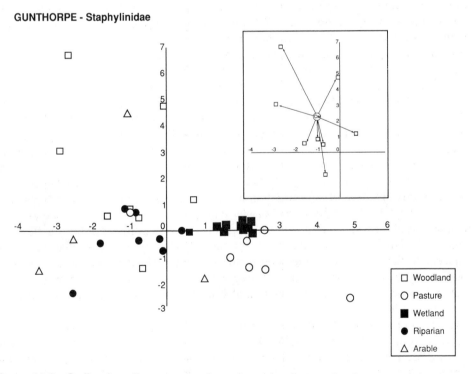

Figure 10.6 Ordination diagram showing plots of axis 1 and axis 2 scores (standard deviation units) from detrended correspondence analysis of species (Coleoptera: Staphylinidae) abundance data for the Gunthorpe functional sector. The inset shows the method for calculating typicalness by measuring the distance of each trap from the unit centroid

of specialist species and highest specialism total. Although having more species, the woodland and riparian units had fewer specialists, either because they shared species with a number of units or had species unique to themselves, but at a low frequency. Pasture had one specialist species only and the arable unit, none.

The characteristics of the units described above are illustrated in the ordination diagram (Figure 10.6). The low scores of the woodland unit are reflected by the wide spread of points in the diagram. Commonly, the high typicalness and specialism scores of the wetland unit are shown by the tight clustering of the traps and the limited overlap with any other unit group. Moreover, the wetland data clearly are different to the riparian data, with which it may have been expected to have similarities.

DISCUSSION

In the European Community, it is estimated that by the year 2000 the area of agricultural land contributing to surplus production will be 12 million hectares (Arnold, Hottges and Rouve, 1989). Corridors may have special roles in conservation (Hobbs and Saunder, 1990) and 'conservation corridors' may become an important theme in landscape ecology. River corridor restoration could make a major contribution to providing for the increasing demands of society for recreation, amenity and nature conservation (Dister et al., 1990). One fundamental problem in river management concerns the width of the 'floodplain' to be incorporated into a restoration programme. Currently, narrow riparian zones, 10–30 m wide on each bank are being proposed for restoration along several European rivers (e.g. Petts, 1990b).

In Great Britain, Sites of Special Scientific Interest (SSSI) can include sections of rivers comprising a core area, incorporating the channel and extending up to a width of 100 m through wide, flat, floodplains, and any adjacent semi-natural, wet habitat – including ponds, grassland, wetland and woodland – that is intimately linked with the river and which is probably dependent on the river for its continued existence (Nature Conservancy Council, 1989). (An SSSI is a category of protected area in Britain. Under the Wildlife and Countryside Act 1981, the Nature Conservancy Council has a duty to notify any area of land which in its opinion is 'of special interest by reason of any of its flora, fauna, or geological or physiographical features'. The conservation of SSSIs is based on cooperation with owners and occupiers in the appropriate management of the land to maintain its special interest.) Furthermore, non-wetland habitats adjacent to the river can be included within the SSSI provided that they contribute significantly to sustaining fauna associated with the river. However, approaches are required to assess the actual and potential values of particular, and assemblages of, floodplain units.

Hydrosystem analysis has been shown to have particular application for post-impact studies and for predictive modelling (Bravard et al., 1986; Amoros et al., 1987; Roux et al., 1989); we suggest also that it has use in evaluating sectors and units for conservation and restoration. On the River Trent two groups of functional sectors have been defined: regulated and unregulated. Functional units have been differentiated according to vegetation type and geomorphological setting, which influences frequency and duration of inundation, and permeability. Within the regulated group of sectors, lack of geomorphological instability has led to succession of units to mature stages with the loss of pioneer and early successional units.

Synchronic analyses have problems associated with all ergodic approaches (Paine, 1985), but when supported by historical evidence and diachronic analyses – the latter are currently being undertaken on the Trent – the functional sets may be reasonably established. Synchronic analysis has been used to define the functional set for cut-off channels using both water quality and vegetation as describers. For the aquatic stages, the preliminary results suggest that artificial backwaters of different size, shape and connectivity with the main channel, can create parapotamonic and plesiopotamonic units. Semi-aquatic and terrestrial stages also may have particular conservation values.

The methods used to assess conservation values are constrained not least by sample size and distribution. None the less, the indices provide a first step towards site assessment. The conservation values of five units within one floodplain sector have been determined using beetles (Coleoptera: Staphylinidae). Three conservation scores (rarity, typicalness and specialism) and ordination were employed. Both pasture and arable units are shown to have low species richness in comparison with the wetland, woodland, and riparian units. Riparian zones are shown to have important faunas, but these are very different from those of the seasonally inundated woodland and wetland units. Although the latter contains a number of nationally rare species, the conservation value is manifested most strongly by the specialism index, which suggests that a diversity of species are dependent on the wetland habitats.

The results indicate that river corridors for conservation must extend beyond the riparian zone to include floodplain wetland units, especially cut-off channels, and include examples of each stage in the functional set. Given variable siltation around cut-off channels (Shields and Abt, 1989) a variety of aquatic and wetland units could be maintained by management or even created artificially within the floodplain. The data show the potential importance in river corridor restoration of seasonally flooded areas, including wetlands and woodlands. Floodplain ecotones containing a mosaic of units appear to have important conservation values, but their effective restoration and management requires that the hydrological and geomorphological dynamics are sustained.

ACKNOWLEDGEMENTS

We wish to acknowledge the contributions of other members of the Freshwater Environments Group to this research: P. M. Wade, C. N. Roberts, G. White, G. E. Revill, P. José, E. J. Darby, S. N. Newbery and S. Ashby. We also would like to thank H. Potter and C. Salisbury for their enthusiastic support; R. J. Marsh for helping with the identification of Staphylinidae; and P. Boon for his constructive comments on an early draft. The research was financed by a grant from the Research Committee of Loughborough University of Technology.

REFERENCES

Amoros, C., A. L. Roux, J. L. Reygrobellet, J. -P. Bravard and G. Pautou (1987). A method for applied ecological studies of fluvial hydrosystems.*Regulated Rivers: Research and Management*, **1**, 17–36.
Amoros, C. and G. van Urk (1989). Palaeoecological analyses of large rivers: some principles

and methods. In G. E. Petts, H. Moller and A. L. Roux (eds), *Historical Change of Large Alluvial Rivers*, Wiley, pp. 143–166.

Arnold, U., J. Hottges and G. Rouve (1989). Removing the straight-jackets from rivers and streams. *German Research*, **1**, 22–24.

Birks, H. J. B. (1973). *Past and Present Vegetation of the Isle of Skye. A Palaeoecological Study*, Cambridge University Press, 415 pp.

Boon, P., P. Calow and G. E. Petts (eds) (1992). *River Conservation and Management*, Wiley, 472 pp.

Bournard, M. and C. Amoros (1984). Des indicateurs biologiques aux descripteurs de fonction-nement: quelques exemples dans un systemè fluvial. *Bulletin Ecologie*, **15**, 57–66.

Braak, C. J. F. ter (1988). *CANOCO – a FORTRAN program for Canonical Community Ordination by (Partial) (Detrended) (Canonical) Correspondence Analysis, Principal Compo-nents Analysis and Redundancy Analysis*, Agricultural Mathematics Group, Wageningen.

Bravard, J. P. (1983). Les sediments fins des plaines d'inondation dans la vallee du-aut-Rhone (approche qualitative et spatiale). *Revue Geographie Alpine*, **71**, 363–379.

Bravard, J. P., C. Amoros and G. Pantou (1986). Impacts of civil engineering works on the succession of communities in a fluvial system: a methodological and predictive approach applied to a section of the Upper Rhone River. *Oikos*, **47**, 92–111.

Castella, E., M. Richardot-Coulet, C. Rowe and P. Richoux (1984). Macroinvertebrates as describers of morphological and hydrological type of aquatic systems abandoned by the Rhône river. *Hydrobiologia*, **119**, 219–225.

Coppin, N. J. and I. G. Richards (1990). *Use of Vegetation in Civil Engineering*, CIRIA Water Engineering Report, Butterworths.

Décamps, H., M. Fortune, F. Gazelle and G. Pantou (1988). Historical influence of man on the riparian dynamics of a fluvial landscape. *Landscape Ecology*, **1**, 163–173.

Dister, E., D. Gomer, P. Obrdlik, P. Petermann and E. Schneider (1990). Water management and ecological perspectives of the Upper Rhine floodplains. *Regulated Rivers: Research and Management*, **5**, 1–15.

Evans, J. T. (1945). *Notes on the History of the Navigation of the River Trent*, Trent Navigation Company, Nottingham.

Eyre, M. D. and G. N. Foster (1989). A comparison of aquatic Heteroptera and Coleoptera communities as a basis for environmental and conservation assessment in static water sites. *Journal of Applied Entomology*, **108**, 355–362.

Eyre, M. D. and S. P. Rushton (1989). Quantification of conservation criteria using inverte-brates. *Journal of Applied Ecology*, **26**, 159–171.

Eyre, M. D., S. P. Rushton, M. L. Luff, S. G. Ball, G. N. Foster and C. J. Topping (1986). *The Use of Invertebrate Community Data in Environmental Assessment*, Agricultural Environ-ment Research Group, University of Newcastle upon Tyne.

Goldsmith, F. B., C. M. Harrison, A. J. Morton (1986). Description and analysis of vegetation. In S. B. Chapman (ed), *Methods in Plant Ecology*, Blackwell Scientific, pp, 437–524.

Gore, J. A. (1985). Development and applications of macroinvertebrate instream flow models for regulated flow management. In J. F. Craig and J. B. Kempar (eds), *Regulated Streams: Advances in Ecology*, Plenum Press, pp. 99–115.

Gore, J. A. and G. E. Petts (eds) (1989). *Alternatives in River Regulation*, CRC Press, 344 pp.

Greenwood, M. T., M. E. Bickerton, E. Castella, A. R. G. Large and G. E. Petts (1991). The use of Coleoptera (Arthropoda: Insecta) for patch characterisation on the floodplain of the River Trent, UK. *Regulated Rivers: Research and Management*, **6**, 4, in press.

Hemphill, R. W. and M. E. Bramley (1989). *Protection of River and Canal Banks. A Guide to Selection and Design*, CIRIA Water Engineering Report, Butterworths, 200 pp.

Hobbs, R. J. and D. A. Saunders (1990). Nature Conservation: the role of corridors. *Ambio*, **19**, 94–95.

Holland, M. M. (1988). *Technical Consultations on Ecotones*, UNESCO-MAB and SCOPE, Paris, pp. 47–106.

Jennings, D. T., M. W. Houseweart and G. A. Dunn (1986). Carabid beetles (Coleoptera: Carabidae) associated with strip-clearcut and dense spruce-fir forests of Maine. *Coleopterists Bulletin*, **40**, 251–263.

José, P. (1988). *The hydrochemistry of backwaters and dead zones*. Unpublished PhD thesis, Loughborough University of Technology.

Junk, W. J., P. B. Bayley and R. E. Spinks (1990). The flood pulse concept in river-floodplain systems. In *Proceedings of the International Large River Symposium. Canadian Journal of Fisheries and Aquatic Sciences*, **106**, 110–127.

Large, A. R. G., K. Prach, N. E. Bickerton and P. M. Wade (in press). Alteration in patch structure and boundary type, and their influence on management and restoration of the floodplain of the River Trent, England. *Freshwater Biology*.

Lewin, J. (1978). Floodplain geomorphology. *Progress in Physical Geography*, **2**. 408–437.

Lewis, G. and G. Williams (eds) (1984). *Rivers and Wildlife Handbook: a Guide to Practices which Further the Conservation of Wildlife on Rivers*, Royal Society for the Protection of Birds, Bedfordshire, and the Royal Society for Nature Conservation, Lincoln.

Luff, M. L. (1987). Biology of polyphagous ground beetles in agriculture. *Agricultural Zoology Reviews*, **2**, 237–278.

Mason, C. F. and S. M. MacDonald (1990). The riparian woody plant community of regulated rivers in eastern England. *Regulated Rivers: Research and Management*, **5**, 159–166.

Morris, L. A., A. V. Mollitor, K. J. Johnson and A. L. Leaf (1978). Forest management of floodplain sites in the northeastern United States. In: R. R. Johnson and J. F. McCormick (eds), *Strategies for Protection and Management of Floodplain Wetlands and other Riparian Ecosystems*, U.S. Department of Agriculture and Forest Service, pp. 236–242.

McVean, D. N. and D. A. Ratcliffe (1962). *Plant Communities of the Scottish Highlands. A Study of the Scottish Mountain, Moorland and Forest Vegetation*, Monograph of the Nature Conservancy, London, 445 pp.

Naiman, R. J. and H. Décamps (eds) (1990). *The Ecology and Management of Aquatic and Terrestrial Ecotones*, Man and the Biosphere series, UNESCO, Paris, 316 pp.

Nature Conservancy Council (1989). *Guidelines for Selection of Biological SSSIs*. Nature Conservancy Council, Peterborough, 228 pp.

Newbold, C., J. Purseglove and N. T. H. Holmes (1983). *Nature Conservation and River Engineering*, Nature Conservancy Council, Peterborough.

Newson, M. D. (1986). River basin engineering – fluvial geomorphology. *Journal of the Institution of Water Engineers and Scientists*, **40**, 307–324.

Paine, A. D. H. (1985). 'Ergodic' reasoning in geomorphology: time for a review of the term? *Progress in Physical Geography*, **9**, 1–15.

Pautou, C. and H. Décamps (1985). Ecological interactions between the alluvial forests and hydrology of the Upper Rhône. *Archives für Hydrobiologie*, **104**, 13–37.

Petts, G. E. (1990a). The role of ecotones in aquatic landscape management. In R. J. Naiman and H. Décamps (eds), *The Ecology and Management of Aquatic and Terrestrial Ecotones*, Parthenon, pp. 227–262.

Petts, G. E. (1990b). Forested river corridors: a new resource. In D. Cosgrove and G. E. Petts (eds), *Water, Engineering and Landscape*, Belhaven, pp. 12–34.

Petts, G. E., H. Moller and A. L. Roux (eds) (1989). *Historical Change of Large Alluvial Rivers*, Wiley, 355 pp.

Poore, M. E. D. (1955a). The use of phytosociological methods in plant investigations. 1. The Braun–Blanquet system. *Journal of Ecology*, **43**, 226–244.

Poore, M. E. D. (1955b). The use of phytosociological methods in plant investigations. 2. Practical issues involved in an attempt to apply the Braun–Blanquet system. *Journal of Ecology*, **43**, 245–269.

Ratcliffe, D. A. (1977). *A Nature Conservation Review*, Cambridge University Press.

Rostan, J. C., C. Amoros and J. Juget (1987). The organic content of the surficial sediments: a tool to study the ecosystem development of abandoned river channels. *Hydrobiologia*, **118**, 45–62.

Roux, A. L. (1982). (ouvrage collectif publié sous la direction de) *Cartographie polythématique appliquée à la gestion ecologique des eux; étude d'un hydrosystème fluvial: le Haut-Rhône français*, CNRS, Lyon, 116 pp.

Roux, A. L., J. -P. Bravard, C. Amoros and G. Pautou (1989). Ecological changes of the French Upper Rhône River since 1750. In G. E. Petts, H. Moller and A. L. Roux (eds), *Historical Change of Large Alluvial Rivers*, Wiley, pp. 323–350.

Rushton, S. P. (1987). A multivariate approach to the assessment of terrestrial sites for conservation. In M. L. Luff (ed.), *The Use of Invertebrates in Site Assessment for Conservation*, Agricultural Environment Research Group, University of Newcastle upon Tyne, pp. 62–75.

Rushton, S. P. and M. L. Luff (1988). The use of multivariate ordination techniques to assess the effects of chlorpyrifos on ground beetle and spider communities in grassland. *Field Methods for the Study of Environmental Effects of Pesticides*, British Crop Protection Council, Monograph No. 40, pp. 171–181.

Salisbury, C. R., P. J. Whitley, C. D. Litton and J. L. Fox (1984). Flandrian courses of the River Trent at Colwick, Nottingham. *Mercian Geologist*, **9**, 189–207.

Salo, J. (1990). External processes influencing origin and maintenance of inland water–land ecotones. In R. J. Naiman and H. Décamps (eds), *The Ecology and Management of Aquatic-Terrestrial Ecotones*, Parthenon, pp. 37–64.

Salo, J., R. Kalliola, I. Häkkinen, Y. Mäkinen, P. Niemalä, M. Puhakka and P. D. Coley (1988). River dynamics and the diversity of Amazon lowland forest. *Nature*, **322**, 254–258.

Shields, F. W. Jr and S. R. Abt (1989). Sediment deposition in cutoff meander bends and implications for effective management. *Regulated Rivers: Research and Management*, **4**, 381–396.

Shirt, D. B. (ed.) (1987). *British Red Data Books 2. Insects*, Nature Conservancy Council, Peterborough, 402 pp.

Thornes, J. B. (ed.) (1990). *Vegetation and Erosion: Processes and Environments*, British Geomorphological Research Group Symposia Series, Wiley, 518 pp.

Usher, M. B. (1980). An assessment of conservation values within a Site of Special Scientific Interest in north Yorkshire. *Field Studies*, **5**, 323–348.

Van Dijk, T. S. (1986). Changes in the Carabid fauna of a previously agricultural field during the first twelve years of impoverishing treatments. *Nederlands Journaal van Zoologie*, **36**, 413–437.

Viles, H. (1988). *Biogeomorphology*, Basil Blackwell.

Welcomme, R. L. (1979). *Fisheries Ecology of Floodplain Rivers*. Longman, 337 pp.

Welcomme, R. L. (1991). River Conservation. In P. Boon, P. Calow and G. E. Petts (eds), *River Conservation and Management*, Wiley, pp. 453–462.

Wood, A. C. (1950). The history of trade and transport on the River Trent. *Transactions of the Thoroton Society*, **54**, 1–44.

11   Channel Changes Related to Low-level Weirs on the River Murray, South Australia

M. C. THOMS and K. F. WALKER
River Murray Laboratory, Department of Zoology, University of Adelaide

INTRODUCTION

Lowland reaches in river systems are distinguished by their relatively low channel gradients and small particle sizes of sediment load and bed material. Long-term sediment storage occurs because flows are insufficient to accommodate the sediment load from the high-energy headwater reaches. The extent of deposition depends on the available storage space as well as the quantity of sediment supply. Storage is limited, for example, in the coastal rivers of eastern Australia, which fall steeply from the Eastern Uplands and deposit sediment over the remaining short distance to the sea (Erskine and Warner, 1987). In contrast, the rivers draining the western slopes have long, low-gradient profiles with extensive floodplains.

Lowland reaches are sediment sinks and potentially subject to continual change (Pickup, 1986). Alterations in the hydrological regime, for example, may cause dramatic changes in channel configuration. Schumm (1968) showed that variable discharges and sediment loads in lowland reaches of the Murrumbidgee River, Australia, caused adjustments in sinuosity, meander wavelength and channel dimensions, although there was little bed degradation. Alterations to the river profile are not as common as changes in configuration, particularly over distances of more than 100 km (Richards, 1982, p. 222). Secular variations in climate also may initiate channel changes, as in the rivers of coastal New South Wales (Erskine and Warner, 1987), but the effects probably have not been significant in the larger catchments ($> 30\,000$ m^2) of southeast Australia (Riley, 1987).

Flow regulation also affects the dynamics of lowland river channels (e.g. Petts, 1984, p. 114). In response, perhaps over several decades, the fluvial system tends towards a new equilibrium. The nature of the changes depends on many factors, including the size and locaton of the regulating structure. Most published studies concern the impact of large dams in headwater regions, and smaller regulators in lowland regions have received scant attention (e.g. Ahmad, 1951). As a result, it is not clear whether the fluvial processes of lowland rivers – and the nature of their response to regulation – are similar to those in headwater regions.

Lowland Floodplain Rivers: Geomorphological Perspectives. Edited by P.A. Carling and G.E. Petts
© 1992 John Wiley & Sons Ltd

This paper outlines morphological changes of the lower River Murray, South Australia, in response to construction of a series of low-level weirs. Supporting information about the nature of the bed and bank sediments and the rate of sediment transport is given by Thoms and Walker (1989, 1991).

THE RIVER MURRAY

The Murray–Darling river system has a catchment of 1.073 million km^2, or about 14% of continental Australia (Figure 11.1a). The basin provides about half of Australia's gross primary production, and 91% of its exploitable water resources are committed to use (Walker, 1985). As the principal river of the system, the Murray rises at 1430 m altitude in the Snowy Mountains and flows northwest to South Australia, gathering flows from the inland slopes of the Eastern Uplands. The average annual discharge of the system is 318 m^3 s^{-1} (range 20–1564; data for Blanchetown, 1950–1980), half of this coming from catchments within 500 km^2 of the Murray's source. This high flow variability, in comparison with world rivers (Finlayson and McMahon, 1987), is a function of low relief and high evaporation potential. A salient feature of the river is its extended long profile (Figure 11.1b). Thus, 89% of the length of the Murray (2560 km) has a channel gradient of less than 0.00017 m km^{-1}. In contrast, slopes in the headwaters region range up to 0.5 m km^{-1}.

The geomorphological history of the Murray is described in a number of intensive studies, notably those of Schumm (1968) and Bowler (1978). The lowland channel cuts into predominantly sand-sized Tertiary alluvium that is up to 60 m thick in South Australia (Twidale, Lindsay and Bourne, 1978). There is a well-developed floodplain (1–20 km) as a legacy of large sediment loads in the past, and a series of floodplain 'palaeochannels' also attests to a long history of channel activity. The largest channels have a meander wavelength up to 3600 m and a width of 1500 m (Bowler, 1978). Their migration rate was about 15 cm year^{-1}, and discharges are estimated to have been 460% greater than those of the modern river (Schumm, 1968).

Regulation has had a profound impact on the flow regime of the Murray (Jacobs, 1989). Flows in the river came under increasing control after about 1929, as new storages were commissioned. In general terms, the effect has been to decrease the frequency of overbank flows and to increase the frequency of flows at or near channel capacity (Bren, 1988). The trap efficiencies of Murray impoundments may be as high as 73% (Thoms and Walker, 1991), indicating a marked reduction in the supply of sediment to downstream reaches. For these reasons, it is likely that regulation has significantly affected the morphology of the river channel.

The 830 km section of the Murray below the Darling junction (the 'lower' Murray) receives no major tributaries. Incoming flows are controlled mainly by large upland reservoirs (Jacobs, 1989), but along the lower Murray there are 10 low-level weirs (Figure 11.1c, Table 11.1). These were constructed in 1922–1935 to ensure safe passage for paddle steamers, but with the decline of the boat industry they assumed a

---

Figure 11.1 The catchment area: (a) the Murray–Darling system, (b) the long profile of the River Murray, and (c) the study area, located below the Murray–Darling confluence

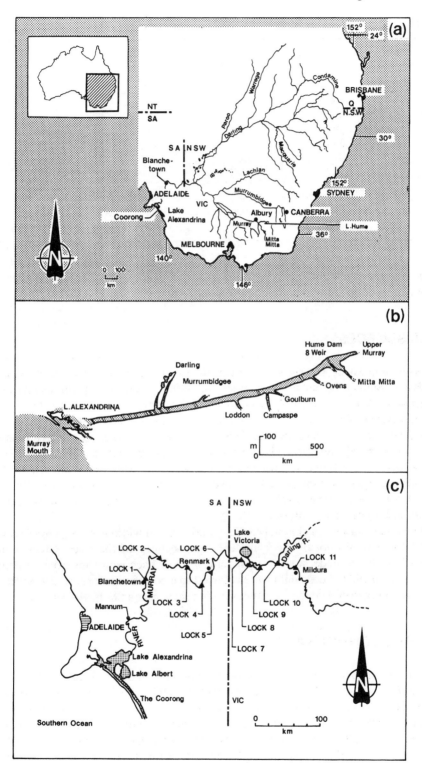

Table 11.1 Data for the regulating structures on the lower River Murray

Lock and weir number	Pool length (km)	Year completed	Storage capacity (Gl)	Upper pool level (m AHD)
1	88	1922	64	3.30
2	69	1928	43	6.10
3	85	1925	52	9.80
4	46	1929	31	13.20
5	57	1927	39	16.30
6	77	1930	35	19.25
7	29	1934	13	22.10
8	39	1935	24	24.60
9	67	1926	32	27.40
10	49	1929	47	34.40

role in flow management. Each weir has an adjacent lock chamber and is assigned a number in a sequence up river (hence Locks 1–10).

DATA SOURCES

Surveys of the lower Murray were undertaken by the South Australian Engineering and Water Supply Department in 1906 and 1988. In processing these data for the following analysis the river was subdivided into 100 km sections and the plans for each section were rescaled and overlain to compare the position and pattern of the channel.

Long profiles were constructed from both surveys. Mean bed elevations were calculated from 50 randomly chosen bathymetric points within 150 m of the actual river distance, at 500 m intervals along the river. The 1906 data were related to the Australian Height Datum (AHD) to allow direct comparisons with the 1988 data. Over 80 000 points from each survey were used to construct the pre- and post-regulation longitudinal profiles.

Analyses of changes in channel capacity and channel width are not possible because the 1988 survey did not obtain complete cross-sections of the river. Cross-sectional surveys have been made at sites immediately below each weir, however, at irregular intervals since 1902. These data were referred to a common reference datum for each weir, the lower pool level, to compensate for inconsistencies in the surveys.

CHANNEL CONFIGURATION

No changes in channel configuration were apparent between 1906 and 1988 (although local changes in channel width were evident in reaches immediately downstream of weirs). The apparent stability of the river planform reflects the low stream energy associated with the low gradient and the cohesive nature of the bank material. Estimates of specific stream power for the Murray range from 0.44 to 5.62 W m^{-2}, considerably lower than those of Ferguson (1981) for high-gradient upland rivers. The cohesiveness of the bank material is reflected in generally high silt–clay contents (12–41%), and in low sinuosities (1.2–2.1) for the individual 100 km survey sections. The silt–clay content of the bed sediment is comparatively low (2–7%), suggesting that erosion of the river bed requires less energy than erosion of the banks. These data compare well with the correlation established by Schumm (1968) for sinuosity and bank silt–clay content in the Murrumbidgee River.

The lower Murray is a 'suspended load channel' according to Schumm's classification (1968, p. 40). This is a simplistic scheme, but it does provide an insight into the changes anticipated in response to flow regulation. In suspended load channels the primary responses to changes in flow and sediment load are changes in bed elevation. As mentioned, flow changes in the Murray may favour bed erosion rather than bank erosion, because of the dissimilarity of the boundary material.

CHANGES IN THE LONGITUDINAL PROFILE

The lower Murray has a low bed gradient (Figure 11.2). Regression analysis of the 1906 survey data indicates an average slope of 0.000052, but with distinct areas of low and high bed elevation (Figure 11.2a). Further examination, using a five-point moving average and spectral analysis (cf. Nordin and Algret, 1965), demonstrated four areas of higher than average bed elevations, separated by areas of low elevations. These regular 'macro-bedforms' have an amplitude of about 5 m and a period of 116 500 m. They are not linked to any obvious changes in the composition of the bed and bank material, channel configuration or morphology.

The 1988 long profile is steeper than for 1906, with an average bed slope of 0.000055 (Figure 11.2b). The regression slopes for the two data sets are significantly different (ANCOVA: $P < 0.05$), but not the intercept values (t test: $P > 0.05$). The increased bed slope appears to be a result of aggradation in the upper region of the profile (Figure 11.2c).

The 1988 profile does not retain the periodic bed elevations of the 1906 data but is 'stepped' in form, particularly in the upper region (Figure 11.2b). The steps are distinguished by relatively flat sections separated from one another by an abrupt local increase in slope. The individual steps are closely associated with weirs. In the upper profile, for example, subreaches 696–725, 725–760 and 760–830 river-kilometres are steps corresponding to Locks 7–8, 8–9 and 9–10, respectively.

Changes are evident in the mean bed elevation over 1906–1988 (Figure 11.2c). The average aggradation is +0.32 m, but site-specific differences range from +10.09 m at 778 km to −11.53 m at 771 km. Despite differences in the nature of the changes, there are four regions with similar changes in bed elevation:

(1)   the upper profile from 835 to 665 km is an area of general aggradation (average +1.5 m);
(2)   over 665–460 km, the bed has degraded by an average −0.5 m – localized deposition is evident, particularly behind Lock 4 (516 km);
(3)   like the upper profile, the middle to lower profile (460–410 km) has aggraded (average +1.5 m), although it is smaller in extent;
(4)   the lower profile (below 410 km) shows no apparent trend.

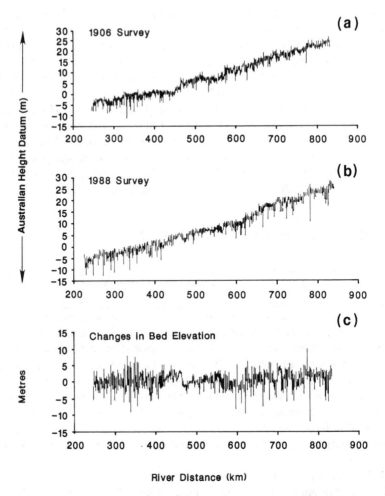

Figure 11.2 The channel profile of the lower reaches of the River Murray. (a) Constructed from the 1906 survey. (b) Constructed from the 1988 survey. (c) Changes in bed elevation between 1906 and 1988

The foregoing data provide for crude estimates of parameters for a simple sediment budget, using average channel widths for each 100 km section. Thus, an estimated 70.75×10^6 m^3 sediment was deposited and 34.58×10^6 m^3 eroded in the lower Murray in 1906–1988, indicating a net accumulation of 96×10^6 t (assuming a sediment density of 2.0 g cm^{-3}). Most of the accumulation (86%) is between 660–830 km, in the upper region of the profile. The average annual sediment input to the lower Murray is 3.2×10^6 t (cf. Woodyer, 1978; Thoms and Walker, 1991). According to these estimates less than 36.5% of the total sediment input between 1906 and 1988 was retained.

The annual sediment yield of the lower Murray, 3.57 t km^{-2}, is low compared with the Australian mean 100 t km^{-2} (Reiger and Olive, 1987). This may be because the low gradient and large temporal variability in flow (Finlayson and

McMahon, 1987) promote an inefficient sediment delivery system (cf. Olive and Reiger, 1986).

CHANGES IN LOCAL GRADIENTS BETWEEN WEIRS

Table 11.2 shows average bed slopes for subreaches between the weirs. Local gradients range from 0.000009 (764–831 km) to 0.000095 (431–516 km), and generally are steeper in the middle reaches. In 1906 the average local gradients between 430–516 km and 619–696 km were 82 and 65% steeper than the overall gradient, whereas in 1988 the bed slope between 619–696 km (Locks 6–7) was 0.000082, some 49% steeper than the average gradient. The two steep areas in the 1906 profile are associated with the macro-bedforms referred to above.

Local gradients in 1988 were lower than those in 1906 for most of the subreaches (Table 11.2), but only six are significantly different from one another (ANCOVA: $P < 0.05$). The bed slopes of the subreaches between 274–362 km (Locks 1–2), 362–431 km (Locks 2–3) and 619–696 km (Locks 6–7) are not significantly different.

A series of river-bed adjustments in the lower Murray has combined to reduce the bed slope between individual weirs:

(1) Degradation has occurred below each weir, the downstream extent ranging from 4 km below Lock 8 to 19 km below Lock 6 (Figure 11.3). However, the maximum degradation does not always occur in this area. For example, 32 km downstream of Lock 2 the river bed was lowered by 8.2 m. The extent of degradation does not appear to be associated with adjustment of the long profile, or with regional or local influences. In general, erosion below dams or weirs may reduce the bed elevation over an entire reach (Galay, 1983), lowering the channel floor but not altering the gradient. It also may promote greater erosion

Table 11.2 Average river bed slopes for reaches located between the individual weirs in 1906 and 1988. Calculated slopes were obtained using the model of Annandale (1987)

Locks	Average bed slope[a–c]		Calculated
	1906	1988	slope
1–2	0.000037	0.000037[a]	0.000032
2–3	0.000021	0.000022[a]	0.000017
3–4	0.000095	0.000053[b]	0.000052
4–5	0.000016	0.000014[b]	0.000016
5–6	0.000064	0.000030[b]	0.000033
6–7	0.000086	0.000082[a]	0.000042
7–8	0.000046	0.000039[b]	0.000041
8–8	0.000052	0.000013[b]	0.000018
9–10	0.000042	0.000009[b]	0.000012

[a] No significant difference between the 1906 and 1988 bed slope at the 5% probabilty level.
[b] Significant difference betwen the 1906 and 1988 bed slope at the 5% probability level.
[c] Differences in bed slopes were tested using an analysis of covariance.

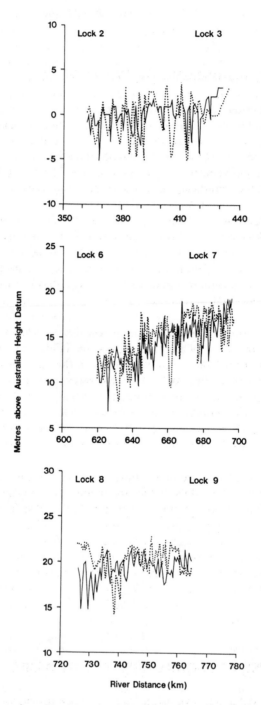

Figure 11.3 Channel profiles between the individual weirs. Three representative weir systems are given. The bold line represents the bed profile in 1906 and the dotted line the 1988 bed profile

below the dam or weir, decreasing with distance downstream (Williams and Wolman, 1984), and thereby reducing the bed slope. In the lower Murray, 62% of the total bed erosion between Locks 5 and 6 occurred within 18 km of the upstream weir. The bed was lowered by 4.2 m in the first kilometre below the weir, but by only 0.8 m in the subsequent 9 km. Unfortunately, the availability of only two surveys precludes a more thorough evaluation of longitudinal degradation over time (but see below).

(2) Aggradation has occurred upstream of each weir, although its longitudinal limits are less distinct than for bed erosion. Figure 11.3 shows the form of the sediment deposits in the various weir pools. In some pools (e.g. Lock 8) there is a well-defined area of aggradation extending up to 50 km from the weir wall. In other pools (e.g. Locks 1–2) the aggradation is much more localized.

At the time of dam closure sediments in the upper reaches of the reservoir may form a broad delta (Bogardi, 1978) that increases in size and migrates downstream toward the dam wall (Garde and Ranga Raju, 1977). Figure 11.3 demonstrates a similar pattern of aggradation behind the weirs on the lower Murray. Locks 3–4 and 8–10 appear to have 'full' reservoir deposits (Annandale, 1987), Locks 5–7 have well-formed deltas where migration is underway and Locks 1–2 show the initial stages of delta formation. Clearly, there are variable rates of general bed aggradation and hence variable adjustments of the local channel gradient between weirs.

The final ('equilibrium') slope of the deposited sediment may be estimated using the average stream power as an indicator of the sediment-carrying capacity of discharge through the reservoir (Annandale, 1987). This is a crude estimate, because it takes no account of changes in the sediment regime, but Table 11.2 shows good general agreement between the estimated and actual slopes for the lower Murray weirs. The estimates for Locks 1–2 (274–362 km), 2–3 (362–431 km) and 6–7 (619–696 km) are lower than the actual slopes, suggesting that further adjustments may occur in these particular subreaches.

There is a general downstream trend in adjustments of the local bed slopes between weirs. The bed slopes of weir pools in the upper area are in equilibrium, whereas those in the lower region are not. In the lower region, local redistributions of sediment between weirs at present are not sufficient for equilibrium and may require the supply of sediment from upstream (cf. Richards, 1982, p. 244). Locks 3–4 and Locks 4–5 are exceptions because the patterns of adjustment there are similar to those in the upper region. This may be the result of a unique set of conditions relating to the time and pattern of weir construction in the lower river, the size of the individual weir systems and the enhanced supply of sediment from local channel sources (the eroding area between Locks 5 and 7: Thoms and Walker, 1989).

CHANGES IN CROSS-SECTIONAL AREA BELOW WEIRS

Figure 11.4 shows repeated surveys of the channel 0.5 km below each weir. Below some weirs there are alternate enlargements and reductions of the cross-sectional area, whilst others show progressive enlargement. Below Lock 4, for example, the area increased by 285% in the 11 years after closure. Enlargements ranged from 171 m^2 (Lock 3) to 384 m^2 (Lock 10), and the time to attain maximal areas varied from 11 to

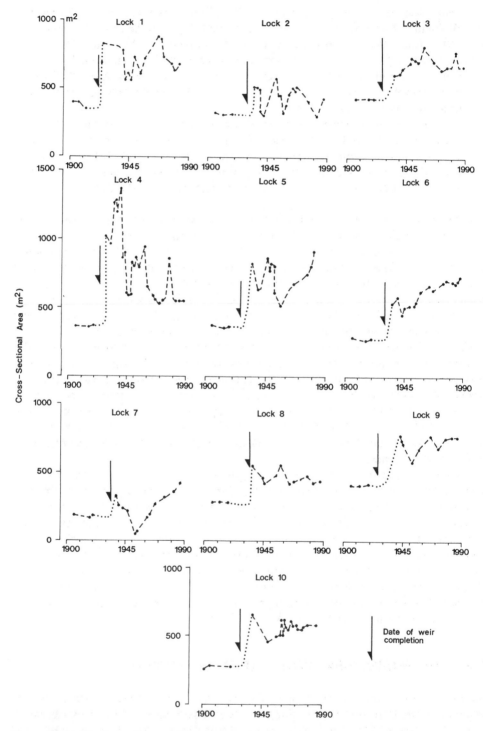

Figure 11.4 The behaviour of cross-sectional areas below the weirs on the lower River Murray. The arrows represent the year when weir construction was completed

55 years. Despite the apparent variability, there are similarities in the response over time; in particular, all sites show an increase in cross-sectional area following construction of the weir.

Enlargement of channels below dams occurs because of high discharge energies and the loss of sediment from upstream (e.g. Petts, 1984). Erosion is first rapid then progressively slower, so that changes in channel dimensions behave as a hyperbolic function (Williams and Wolman, 1984). The data for most of the weirs do not conform to this model because, following the period of initial enlargement, cross-sectional areas decreased below all weirs except Lock 3. From 1937 to 1945, cross-sectional areas were consistently lower than the high initial values, and the channel sections below Lock 2 and Lock 7 retained their pre-regulation dimensions. In this period flows in the river were comparatively low: monthly flows at Lock 6 were an average 42% lower than the means for 1890–1980.

The channel cross-sectional areas below the weirs shows three basic responses:

(1) Stabilizing (Locks 3–4, 8–10). After an initial period of fluctuation the cross-section attains a new dynamic equilibrium, 30–40 years after closure of the weir, where it is 100–200% larger than the pre-regulation value. It is interesting to note the response of these weirs to a major flood in 1976 (peak 1078 $m^3 s^{-1}$). Cross-sections below Locks 3 and 4 increased by 106 and 313 m^2 after the flood, but were restored to pre-1976 values 2 years later. If these cross-sections had not been in equilibrium with the regulated regime the pre-flood values may not have been restored (as happened after a much larger flood in 1956). It is likely that the present cross-sectional areas will be maintained while the regulated regime persists.

(2) Eroding (Locks 5–7). The first stage is similar to the stabilizing response described above in that there is an initial period of fluctuation. Subsequently, erosion and enlargement of the channel have continued since the 1950s.

(3) Fluctuating (Locks 1–2). This response is distinctive because no clear pattern of adjustment is evident and the fluctuations appear to be independent of variations in discharge. There is some synchrony in changes in the cross-sectional area below Locks 1 and 2, and the magnitude of the changes is greatest below Lock 1, the furthest downstream weir.

The responses of cross-sectional area are related to adjustments in the local gradient. The weirs that have reduced their bed slopes significantly apparently have attained a new cross-sectional equilibrium, and those that have not have an eroding or fluctuating response curve.

SCHEMATIC MODEL

Figure 11.5 is a schematic model of the responses of the lower Murray to flow regulation. The model is incomplete because there are no historical data relating to changes in channel dimensions (e.g. width and depth); this is being redressed in work now underway. Nevertheless, the model provides a framework for the changes described in the foregoing discussion. The Murray has a stable, long low profile, reflecting the low relief of the catchment (Figure 11.5a). Regional and local changes in bed slope and elevation have occurred. Thus, there has been aggradation in a 170 km section below the Darling junction, and alternate local areas of bed aggradation and degradation related to the weirs. Sediment from upstream areas has accumulated behind the weirs in the uppermost part of the lower Murray because of increases in local base levels (cf. Leopold and Bull, 1979). Bed degradation has occurred in downstream areas, however, as a response to the reduction in sediment supply despite the increase in local base levels. High local sediment loads have caused aggradation in the mid-section. Given the position of the lower Murray at the terminus of the Murray–Darling system, a long-term supply of sediment is likely to be maintained. Thus, the extent of channel degradation and the short-term increase in the lower river gradient will be reduced over time (Figure 11.5b), with the rate of adjustment being influenced by changes in local slopes and bed elevations.

Local adjustments of the river (Figure 11.5c) are related to regional responses. Equilibrium bed slopes have been attained in aggrading sections, but not in degrading sections because local redistributions of sediment have not been sufficient to reduce bed slopes. In degrading areas more time and a greater redistribution of sediment will be required for equilibrium. Slope adjustment also will be faster in those areas that have a relatively high sediment supply, as in the aggrading sections. Local gradients may attain equilibrium profiles over time but will remain dependent on the state of the upstream gradient and their position in the downstream sequence.

On a smaller scale, the behaviour of the river channel is related to the local bed slope. Active channel erosion is evident below those weirs that have not completed local slope adjustment. Degradation in these areas is part of the overall response between the weirs and not merely a response to clear-water erosion.

The channel profile of the lower Murray has changed since 1906 in response to flow regulation. Bed elevations have changed because of an increase in local base levels, with the actual response being governed by the supply of sediment. There have been changes in the flow regime and supply of sediment (cf. Jacobs, 1989; Thoms and Walker, 1991) and also the size of the sediment being supplied. Sediment cores from behind Locks 2, 3, 6 and 9 indicate that sediment now accumulating in the lower river is finer than that before weir construction: median grain sizes have decreased by 32%. Coarser sediments are trapped in upland impoundments. Changes in gradient must occur in order for a river channel to maintain sediment transport with the available discharge and given channel characteristics (Mackin, 1948).

Channel gradients have changed because of river-bed erosion and sediment accumulation between the locks and weirs. This form of adjustment may have occurred for the following reasons:

(1) the channel banks contain cohesive sediment, as reflected by the high silt–clay content – by comparison, the river-bed sediments have a low silt–clay content and may therefore be less resistant to erosion;
(2) large quantities of sediment (3×10^6 t) supplied to the lower river are available for redistribution between the weirs (hence slope adjustment).

In general a river channel may adjust its gradient in several ways but will do so either to minimize the total amount of work or minimize the expenditure of energy (cf. Richards, 1982).

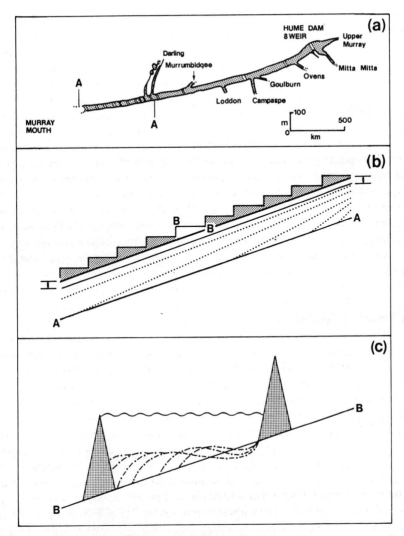

Figure 11.5 Schematic diagrams of the response of the river bed, over time, to flow regulation. (a) The long profile of the River Murray. (b) The channel profile of the lowland region of the River Murray. (c) Adjustment of bed slope between the weirs

The Murray is still adjusting its channel morphology 54 years after the construction of the weirs. A set of interrelated factors contributes to this apparent slow rate of adjustment (cf. Petts, 1984). Low stream energies are characteristic of the Murray, and not conducive to rapid channel changes. As a result, sediment redistribution is an important process; this form of adjustment occurs over longer time periods. In addition, the highly variable sediment transport regime may suppress adjustments (Thoms and Walker, 1991). Given the number of weirs constructed in the lower river, these physical processes have resulted in a complex response to flow regulation.

Adjustments to flow regulation have occurred at a number of different scales and are at various stages of completion. This reflects the time of weir construction (1922–

1935) and the position of the weirs in the downstream sequence. The effects of low-level weirs may be different from those of dams. They are more localized but, in the case of the lower Murray, are extended by the number and close proximity of successive weirs.

CONCLUSION

Lowland river channels may readjust their morphology in response to changes in flow and sediment regimes. This paper indicates that adjustments in lowland channels may differ in character to those experienced in high-energy upland river systems. The River Murray has not experienced major changes in channel configuration or under-gone complete metamorphosis (Schumm, 1968), as reported for high-energy upland rivers. Rather, subtle changes in the channel profile have occurred. This is because of the low-energy conditions, a large supply of sediment and the highly cohesive nature of the boundary sediment. These changes are taking a relatively long time to reach completion.

ACKNOWLEDGEMENTS

We are grateful to the Australian Water Research Advisory Council for financial support (AWRAC Project 86/40), and the Engineering and Water Supply Department (South Australia) for provision of survey data.

REFERENCES

Ahmad, M. (1951). Causes of retrogression below Islam Barrage and remedial measures to cause accretion and push the hydraulic jump to the toe of the glacis. *International Association for Hydraulic Research*, Fourth Meeting, Bombay, A17, pp. 365–393.

Annandale, G. W. (1987). *Reservoir Sedimentation*, Blackwell, 221pp.

Bogardi, J. (1978). *Sediment Transport in Alluvial Streams*. Akademiai Kiado, 328pp.

Bowler, J. M. (1978). Quaternary climate and tectonics in the evolution of the riverine plain, Southeastern Australia. In J. L. Davies and M. A. J. Williams (eds), *Landform Evolution in Australasia*, Australian National University Press, pp. 70–112.

Bren, L. J. (1988). Effects of river regulation on flooding of a riparian red gum forest on the River Murray, Australia. *Regulated Rivers: Research and Management*, 2(2), 65–77.

Erskine, W. D. and R. F. Warner (1987). Geomorphic effects of alternating flood- and drought-dominated regimes on NSW coastal rivers. In R. F. Warner (ed.), *Fluvial Geomorphology of Australia*, Academic Press, pp. 223–244.

Ferguson, R. I. (1981). Channel form and channel changes. In J. Lewin (ed.), *British Rivers*, Allen and Unwin, pp. 90–125.

Finlayson, B. L. and T. A. McMahon (1987). Australia v the world: A comparative analysis of streamflow characteristics. In R. F. Warner (ed.), *Fluvial Geomorphology of Australia*, Academic Press, pp. 17–40.

Galay, V. J. (1983). Causes of river bed degradation. *Water Resources Research*, **19**(5), 1057–1090.

Garde, R. J. and K. G. Ranga Raju (1977). *Mechanics of Sediment Transportation and Alluvial Stream Problems*, Wiley, 483pp.

Jacobs, T. A. (1989). Regulation of the Murray–Darling system. In B. Lawrence (ed.),

Proceedings of the Workshop on Native Fish Management, Murray–Darling Basin Commission, Canberra, pp. 55–96.

Leopold, L. B. and W. B. Bull (1979). Base level, aggradation and grade. *Proceedings of the American Philosophical Society*, **123**, 168–202.

Mackin, J. H. (1948). Concept of the graded river. *Geological Society of America Bulletin*, **59**, 463–512.

Nordin, C. F. and J. H. Algret (1965). Spectral analysis of sand waves. *Journal of the Hydraulics Division, American Society of Civil Engineers*, **92**, 95–114.

Petts, G. E. (1984). *Impounded Rivers: Perspectives for Ecological Management*, Wiley, 326pp.

Pickup, G. (1986). Fluvial landforms. In D. N. Jeans (ed.), *Australia – a Geography*, Vol. 1, Sydney University Press, pp. 148–179.

Olive, L. J. and W. A. Reiger (1986). Low Australian sediment yields – a question of inefficient sediment delivery. *International Association of Hydrological Sciences Publication*, **159**, 355–366.

Reiger, W. A. and L. J. Olive (1987). Channel sediment loads: comparisons and estimation. In R. F. Warner (ed.), *Fluvial Geomorphology of Australia*, Academic Press, pp. 69–85.

Richards, K. (1982). *Rivers: Form and Process in Alluvial Channels*, Methuen, 361 pp.

Riley, S. J. (1987). Secular change in the annual flows of streams in the New South Wales section of the Murray–Darling Basin. In R. F. Warner (ed.), *Fluvial Geomorphology of Australia*, Academic Press, pp. 245–266.

Schumm, S. A. (1968). River adjustment to altered hydrologic regimen – Murrumbidgee River and paleochannels. *U.S. Geological Survey Professional Paper*, **598**, 65pp.

Thoms, M. C. and K. F. Walker (1989). Preliminary observations of the environmental effects of flow regulation on the River Murray, South Australia. *South Australian Geographical Journal*, **89**, 1–14.

Thoms, M. C. and K. F. Walker (1991). Sediment transport in a regulated semi-arid river: the River Murray, Australia. *Special Publications in Canadian Fisheries and Aquatic Sciences*, in press.

Twidale, C. R., J. M. Lindsay and J. A. Bourne (1978). Age and origin of the River Murray and gorge in South Australia. *Proceedings of the Royal Society of Victoria*, **90**, 27–41.

Walker, K. F. (1985). A review of the ecological effects of river regulation in Australia. *Hydrobiologia*, **125**, 111–129.

Williams, G. P. and M. G. Wolman (1984). Downstream effects of dams on alluvial rivers. *U.S. Geological Survey Professional Paper*, **1286**, 83pp.

Woodyer, K. D. (1978). Sediment regime of the Darling River. *Proceedings of the Royal Society of Victoria*, **90**, 139–147.

12 Application of the Instream Flow Incremental Methodology to Assess Ecological Flow Requirements in a British Lowland River

ANDREW BULLOCK and ALAN GUSTARD
Institute of Hydrology, Wallingford·

INTRODUCTION

Water use generally is divided into two primary classes – offstream use and instream use (Lamb and Doersken, 1987). In offstream use, water is withdrawn from the river or aquifer for use beyond its natural flow path, and examples include irrigated agriculture, public and industrial water supply. Each offstream use decreases the volume of water available downstream of the point of diversion and increases availability downstream of the point of return. Instream uses, which by definition do not affect the flow downstream from its point of use, include hydroelectric power generation, navigation, pollution dilution and the environmental requirements of biological, recreational and aesthetic use. The key tool in the management of instream water for the more traditonal economic uses in the UK has been the prescribed flow. Prescribed flows have been set within the framework of Minimum Acceptable River Flows, a strategy for management within the Water Resources Act 1963. Water authorities have faced difficulties in applying the concept of minimum acceptable flows because of practical difficulties of definition of what the principle constitutes and because of a lack of reliable long-term river flows to determine flows accurately (Howarth, 1990). The concept has not been enforced because the Water Resources Act 1963 merely obliged authorities to 'consider' minimum acceptable flows, an obligation later to be replaced in the Water Act 1989 by a power to determine a minimum acceptable flow 'if it thinks it appropriate to do so'. A number of methods have evolved outside of the UK for estimating the increasingly recognized environmental flow requirements, of which the Instream Flow Incremental Methodology (IFIM) developed by the US Fish and Wildlife Service is at present the most sophisticated and widely recognized. The IFIM is a collection of computer models and analytical procedures designed to predict changes in fish and invertebrate habitat arising from flow changes owing to the alteration of physical habitat. A major component of IFIM is the Physical HABitat SIMulation (PHABSIM) system (Bovee,

Lowland Floodplain Rivers: Geomorphological Perspectives. Edited by P.A. Carling and G.E. Petts
© 1992 John Wiley & Sons Ltd

1986), which is a suite of software by which available physical habitat area is obtained as a function of discharge. This chapter describes the first application of IFIM in the UK, to the River Gwash, a lowland river below the Rutland Water impoundment.

Methods for Setting Instream Flow Requirements

Water management in the UK historically has adhered to discharge-based methods in the setting of prescribed flows, being set according to the dry weather flow. The dry weather flow is itself an undefined discharge, but which is indexed by a low flow discharge, typically either the mean annual minimum 7-day flow frequency statistic (Hindley, 1973) or the 95 percentile flow duration statistic. It is only a recent phenomenon in the UK that cognizance is given by resource planners to the ecological value of low river flows; for example, the Yorkshire National Rivers Authority region now employ an environmental weighting scheme, which sets prescribed flows as a proportion of the dry weather flow (DWF) weighted according to a range of environmental characteristics and uses (Drake and Sherriff, 1987). Thus the environmental prescribed flow is set at 1.0 × DWF for the most sensitive rivers and at 0.5 × DWF for the least sensitive, which will determine the amount of water available for offstream uses, pollution dilution and environmental protection.

Recommendations from a review of compensation flows below impounding reservoirs in the UK (Gustard et al., 1987) suggest that a re-evaluation of awards is warranted but that any negotiation of new awards should move away from simply setting prescribed flows as a fixed percentage of the mean flow. The review establishes that many reservoirs provide compensation flows that were determined by industrial and political constraints that no longer apply. Furthermore, the majority of compensation flows were awarded when there were little or no hydrometric data to describe differences in catchment hydrology and little knowledge of the impact of impoundments on downstream aquatic ecology. It is the inheritance of this historical legacy that justifies a reassessment of current compensation flows. Equally, the recognition that aquatic ecosystems have specific flow requirements that perhaps bear little relation to existing compensation awards is a strong argument towards the reassessment of prescribed flows, and towards moving away from discharge-based methods to habitat methods.

However, while quantitative models and design techniques are available for estimating prescribed flow statistics in rivers, for example Low Flow Studies (Institute of Hydrology, 1980), there is a paucity of operational tools for managing aquatic communities in British rivers at a national scale. A notable exception is the development of the RIVPACS (River Invertebrate Prediction And Classification System) technique, developed by the Institute of Freshwater Ecology of the Natural Environment Research Council for predicting macro-invertebrate fauna (Moss et al., 1987) and assessing impacts on macro-invertebrate communities (Armitage et al., 1987). Fish management models tend to be more scheme-specific in nature, for example, the fisheries study downstream of Roadford Reservoir, which commenced in 1984, aimed at developing operating rules to minimize detrimental impacts upon salmonids in the Tamar and Torridge rivers (British Hydrological Society, 1988). The recent development of the HABSCORE technique by the Environmental Appraisal Unit of the Welsh National Rivers Authority region establishes an operational tool for the management of salmonid populations in Welsh rivers. Essentially, both RIVPACS and HABSCORE adopt the same rationale – that the carrying capacities of streams is to a large extent dependent on channel structure and the environmental regime (hydrological, chemical, temperature) experienced within the stream. These characteristics can be measured by a combination of site features (width, depth, substrate, cover, etc.) and catchment features (altitude, gradient, conductivity, etc.). By measuring these features and species populations at a number of pristine sites that have variable habitat, multivariate models can be calibrated that predict species presence and abundance from the environmental variables. The predicted population sets an objective for the river reach based on the habitat that it provides. This type of model may be used to detect anomalies in observed ecological data in relation to the objective population, anomalies which may be attributable to impacting factors. What this type of model is not designed to achieve, however, is the recommendation of hydrological regime or prescribed flow. It would be difficult to apply the RIVPACS or HABSCORE-type model to the recommendation of flow regimes for two reasons; first, because of the inherent problems of using multivariate models based on variations over space for predicting variations in the dependent variable over time; second, because the observed macro-invertebrate and fish data integrate relationships with hydrology, temperature, chemistry and time series effects it is not easy to isolate the effects of flow regime.

Water management in Britain lags a considerable way behind the USA with regard to the development of appropriate management models for recommending flow regime measures that consider ecological demands. In the USA, procedures for evaluating impacts of streamflow changes were first developed and have advanced considerably in the period 1974–1989. Central to these advances has been the concept of instream flow requirements, which recognises that aquatic species have preferred habitat preferences, with habitat defined by physical properties (flow velocity, water depth, substrate and vegetal/channel cover). Because some of these physical properties that determine habitat vary with discharge, so species have different preferences for different discharges. Development of the IFIM by the Aquatic Systems Branch of the US Fish and Wildlife Service, as traced by Nestler, Milhous and Layzer (1989), has allowed the quantification of species preferences for the full range of discharges that may be experienced within a river. This quantification of habitat preferences and the relationship with river flow, notwithstanding the inherent imprecisions, permits the negotiation and setting of optimal flows for ecological management. Setting instream flows in this manner complements purely water-quantity or cost-management objectives by paying cognizance to the physical habitat requirements. By relating ecological demands to discharge, the merit of IFIM lies in providing a quantitative basis that allows river ecologists to negotiate prescribed flows or flow regimes in equivalent terminology to other water resource demands.

Mean annual run-off is below 250 mm in much of central and southeast Britain and regions below 100 mm are extensive. This compares with mean annual run-off in excess of 500 mm in much of upland Britain. Superimposed upon the low values of natural run-off are the high water demands of population centres and arable agriculture throughout lowland Britain, which are met by abstractions of surface and groundwater. There are, for example, well in excess of 5000 current licences issued for abstractions in the Thames and Anglian regions. This coincidence of high water demand and relatively low run-off places increased pressure upon the allocation of instream flow for ecological needs. In addition, the extensive use of structural solutions to river and channel management in lowland Britain, including weirs, channelization schemes and dredging can alter the physical habitat significantly, and IFIM could prove equally appropriate for assessing the impact of channel alterations with regard to flow regime management.

The demand for a scientifically defensible method for both resource allocation and environmental impact assessment in the UK (Petts, 1989) may be satisfied by IFIM when it is considered that the scientific rationale of IFIM has been defended success-fully against legal challenges in the USA. There is therefore scope for the application of IFIM in the UK to yield long-term benefits to instream flow management. By relating ecological requirements to discharge IFIM allows prescribed flows to be determined and set using values that complement quantity-based statistics. The method has received wide international recognition and has been applied extensively to real water resource problems in the USA. The immediate need, and the objective of the project from which this lowland river example is taken, is the assessment of the suitability of the methodology in British rivers.

RATIONALE AND CONCEPTS OF IFIM

The IFIM procedure provides an estimate of habitat loss/gain with changes in discharge. Instream Flow Incremental Methodology itself is a concept or at least a set of ideas and PHABSIM is software (Gore and Nestler, 1988).

The underlying concepts of the IFIM are that:

(1) IFIM is habitat based, with potential usable habitat being simulated for un-observed flow or channel conditions;
(2) evaluation species exhibit a describable preference/avoidance behaviour to one or more of the physical microhabitat variables; velocity, depth, cover or sub-strate;
(3) individuals select the most preferred conditions within a stream, but will use less favourable areas with decreasing frequency/preference;
(4) species populations respond to changes in environmental conditions that consti-tute habitat for the species;
(5) preferred conditions can be represented by a suitability index that has been developed in an unbiased manner.

The purpose of the PHABSIM system is to simulate the relationship between streamflow and available physical habitat defined by the microhabitat variables. The two basic components of PHABSIM are the hydraulic and habitat simulations within a stream reach using defined hydraulic parameters and habitat suitability criteria, as displayed in Figure 12.1. Hydraulic simulation is used to describe the area of a stream having various combinations of depth, velocity and channel index (cover or substrate) as a function of flow. Habitat suitability is based on the preference of species for certain combinations of physical parameters above others. Hydraulic and habitat data are combined to calculate the weighted usable area (WUA) of a stream segment at

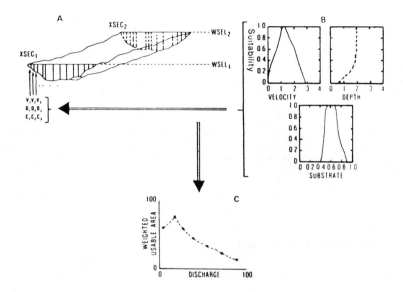

Figure 12.1 Physical habitat simulation (PHABSIM) scientific rationale (reproduced by permission from Gore and Nestler, 1988.)

different discharges based on the preference of selected target species for the simulated combinations of hydraulic parameters.

Physical habitat suitability information for target species, and distinct life stages of those species, can be derived from existing empirical data (including the US Fish and Wildlife Service Curve Library), scientific literature, or direct field sampling.

Calibration of the hydraulic model components is achieved on a transect-defined cell-by-cell basis requiring field observation of channel bed cross-sectional profiles, water surface elevation, mean column velocity, and application of substrate and cover classification schemes. Cells are defined as the boundaries of the data represented by a single survey point, and are defined most commonly in the cross-channel directions as the midpoint between survey points, and in the downstream direction by the inter-transect midpoint.

Observations of these data at calibration flows are necessary to create the data set from which the depth and velocity within cells is simulated at different discharges using the hydraulic programs. Observed channel index values are assumed to be independent of flow.

Cell values of each of the physical parameters are combined with species preference curve information through a selected functional relationship, termed the composite suitability index (CSI), to develop the composite habitat index, termed weighted usable area. The most common CSI functional relationship is multiplicative, but any alternative can be devised. Weighted usable area (WUA), indexed by total surface area of the cell weighted by its relative suitability for a given species, simulates the amount of physical habitat within that cell at different discharges.

Summation of individual cell values within the river reach of interest can be achieved either by a representative reach approach or by habitat mapping and selective identification of field sites. In the representative reach approach, individual transects are assigned a weighting that represents a fraction of the distance to the next downstream transect, according to the distance to the change in habitat type. In the habitat mapping approach, transects are assigned a distance weighting according to the frequency of occurrence of that habitat which the transect represents within the study river as whole.

Once achieved, output comprises a graphical weighted usable area against discharge function for the particular target species under study. Optimal discharges for specific species can be identified from the WUA discharge functions, but must be considered in the context of water availability, water management constraints and ecological objectives. The PHABSIM system comprises a large number of separate programs, which fall into two main categories; hydraulic simulation and habitat simulation.

Hydraulic Simulation in PHABSIM

The hydraulic simulation programs, when calibrated with observed field data, are used to simulate depths and velocities at different discharges selected by the user at transects along a reach of river. To calibrate the hydraulic programs it is necessary to survey the bed profile of the river reach on a transect basis, to measure the distances between transects, and to observe water surface elevation and velocity on a cell-by-cell basis across each transect at a range of different flows. The flows at which the water surface elevation and velocities are measured are termed calibration flows. The discharge for each calibration flow (Q_{CAL}) must be calculated from the observed data. The flows selected by the user when running PHABSIM are termed simulation discharges (Q_{SIM}).

There are three basic hydraulic simulation programs; IFG4, MANSQ and WSP. For the simulation discharges, IFG4 predicts the water surface elevation using a simple stage–discharge relationship and predicts velocities on a cell-by-cell basis using Mannings n and a simple mass balance adjustment using a velocity adjustment factor. Variations in Mannings n with discharge at a point are dealt with by ensuring that the relationship between the velocity adjustment factor and discharge conforms to a requisite shape. In IFG4 and MANSQ each transect is modelled independently. When IFG4 fails to predict water surface elevations sensibly owing to the poor calibration of the stage–discharge relationship, then water surface elevations can be predicted by MANSQ using the solution of Mannings equation. The WSP program is a standard stepbackwater model for the prediction of water surface elevations, which considers transects as dependent and uses an energy balance model to project water levels from one known stage–discharge relationship to all transects upstream. Neither MANSQ or WSP can predict velocities, so once a sensible downstream water surface elevation profile has been predicted for the simulation discharges, then IFG4 is used to predict velocities.

Since none of the hydraulic simulation routines can describe all possible channel conditions it is often necessary to use more than one model to simulate water surface elevations at all discharges at each cross-section. The mixed model approach uses different hydraulic simulation models for the ranges of flow, where each hydraulic model produces the best simulation results for that transect. 'Best' simulation results can be judged on the shape of velocity adjustment factor curves, on checking that water surface slopes are not negative (i.e. that water levels do not increase in a downstream direction), that in IFG4 the exponent of the stage–discharge relationship is between 1.5 and 3 and that the mean error of the discharge against stage regression is low, and preferably less than 5%.

The output from the hydraulic simulation programs is predictions of depth and velocity for each cell for each simulation discharge. Cell values of the channel index (cover or substrate) remain independent of discharge.

Habitat Suitability Curves

The second category is the suite of programs for the simulation of physical habitat space. The input to this suite of programs are habitat suitability curves, which quantify the relative preference of a selected life stage of a target species for depth, velocity and channel index independently. Preference ranges from 0 to 1, with 1 being optimal and 0 being the most unsuitable. The programs, of which the principal is HABTAT, combine the habitat preference values for depth, velocity and channel index for life stages of target species with the predictions of the physical variables from the hydraulic simulations.

The basic form for the expression of suitability is a habitat suitability curve, but there are other categories of curve called utilization curves or preference curves. The distinction between the criteria is the base from which the curves are founded. Essentially, there are three categories (Bovee, 1986):

Category I. The habitat criteria are derived from life history studies in the literature or from professional experience and judgement, and are based on the adjudged suitability of physical habitat variables for target life stages.

Category II. The habitat criteria are based on frequency analysis of microhabitat conditions utilized by different life stages and species as identified by field observations. These criteria are termed 'utilization curves' because they depict the conditions that were being used when the species were observed. Utilization functions may not always accurately describe a species' preference because the preferred physical conditions may be absent or limited at the time of observation.

Category III. These are category II curves in which the criteria are corrected for the bias by factoring out the influence of limited habitat availability. This correction is aimed at increasing the transferability of the criteria to streams that differ from those where the criteria were originally developed, or in the same stream at different flows.

A subsequent category, Category IV, has since been added which consists of conditional curves; essentially Category III curves conditioned for variable factors, such as cover and season.

There are three principal formats in which the microhabitat criteria (i.e. the suitability/utilization/preference for depth, velocity and cover/substrate) can be ex-

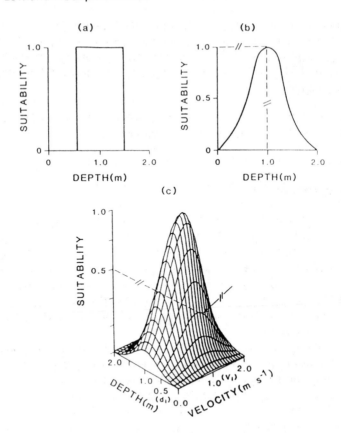

Figure 12.2 Examples of the three formats of habitat criteria: (a) binary, (b) univariate curve, (c) multivariate response surface

pressed; binary criteria, univariate curves, or multivariate response surfaces, as depicted in Figure 12.2.

The binary format establishes a suitable range of conditions for each variable as it pertains to a life stage of a species; within that range the suitability rating is 1.00, beyond it the rating is 0.00. Univariate curves developed from the concept that within the range of conditions considered suitable there is a narrower range that species select as preferred or optimal. The peak of the curve is the optimal value of the physical variable and the tails represent the bounds of suitability. The primary advantage of the multivariate response surfaces is the ability to express interactions among the variables.

Weighted Usable Area and Composite Suitability Indices

Total usable area is defined in terms of the plan area of the water surface with a river reach, expressed in ft^2 per 1000 ft of river length. Total usable area is the summation of the plan surface area of each of the individual cells within the river reach, some of which (principally the near-bank cells) will vary in plan area with discharge. The net suitability of use of a given cell is quantified by the weighted usable area (WUA). The suitability of a cell may be determined by one of four composite suitability indices (CSI), as presented below, where A_i is the plan area of cell i, and $f(v)$, $f(d)$ and $f(c)$ are the habitat suitability indices for velocity, depth and channel index (cover or substrate), respectively:

multiplicative CSI $\text{WUA}_i = A_i \times f(v) \times f(d) \times f(c)$
geometric mean CSI $\text{WUA}_i = A_i \times [f(v) \times f(d) \times f(c)]^{0.333}$
minimum CSI $\text{WUA}_i = A_i \times \min [f(v), f(d), f(c)]$
user-supplied CIS $\text{WUA}_i = A_i \times \text{user selected func} [f(v), f(d), f(c)]$

The multiplicative CSI, in which the gross area of the cell is multiplied by all suitability indices, normally is used and implies a 'cumulative effect' mechanism, a synergistic action whereby optimum habitat availability is achieved only if all variables are optimal (Gan and McMahon, 1990). The geometric mean CSI implies a compensatory mechanism, such that if two of the three variables are in the optimal range then the value of the third variable has little effect unless it is zero. The minimum CSI implies a 'limiting factor mechanism' such that when the cell area is multiplied only by the minimum of the factors the habitat is no better than its worst component. The user-supplied CSI allows the user to define the nature of the CSI function according to the explicit interactions that are sought.

Weighted usable area is calculated cell by cell and summed for the whole reach. Under different flow conditions the values of the physical properties within a cell vary and consequently the habitat suitability indices may alter accordingly to calculate a new weighting factor. At different flows the plan area of certain cells will alter. The variations in these two factors combine to create a weighted usable area relationship with discharge for a river reach.

APPLICATION OF IFIM TO ONE LOWLAND RIVER IN BRITAIN

Selection of sites for IFIM application in two British rivers (Bullock, Gustard and Grainger, 1990) was based on a requirement for the combined availability of pre- and post-reservoir impoundment river flow data and ecological data. Only three reservoirs in the UK were identified as possessing at least 5 years of gauged pre- and post-impoundment flow data by the Institute of Hydrology in 1987 (Gustard et al., 1987), notably Derwent, Blithfield and Brenig Reservoirs.

Acquisition of more recent flow data since 1987 appends Rutland Water to this group. Of these, Blithfield Reservoir and Rutland Water were selected as sites most appropriate for IFIM application, given that the river sites also are biological sampling sites in the River Communities Project (Moss et al., 1987).

The two rivers exhibit contrasting geomorphological features, the River Blithe being an 'upland' pool–riffle river and the Gwash a 'lowland' river with ponded sections. The River Gwash below Rutland Water has been gauged 13 km downstream at Belmesthorpe (Surface Water Archive gauging station number 31006) since 1967. Rutland Water was impounded in February 1975, so the Belmesthorpe record contains a pre-impoundment phase from 1968 to 1974 (7 years) and a post-impoundment phase of 1975–1989 (15 years).

Three reaches were selected on the River Gwash for field measurement of hydraulic data (Figure 12.3): at Empingham (SK954084), immediately downstream of Rutland Water outflow; at Ryhall (TF031109), located 12 km downstream of Rutland Water outflow, immediately upstream of the A6121 road bridge over the Gwash; and at Belmesthorpe (TF041104) located 13 km downstream of Rutland Water outflow, and 1.5 km downstream of the Ryhall site. North Brook contributes river flow between the reservoir outfall and the Ryhall and Belmesthorpe sites, increasing the catchment

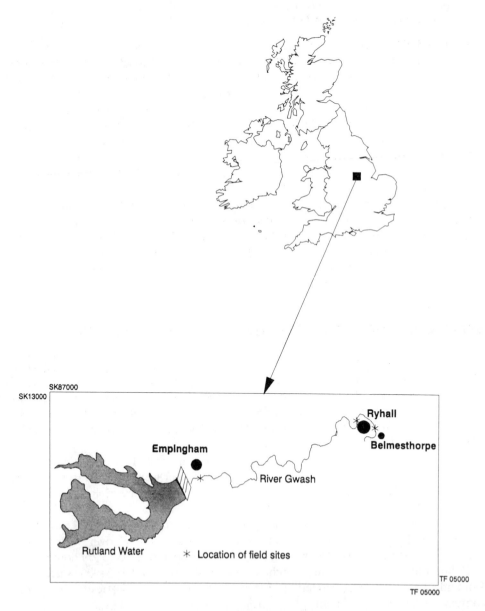

Figure 12.3   Location of sites used in IFIM application

areas by approximately 85 km^2 and 90 km^2, respectively. Empingham represents a well-vegetated ponded reach, Ryhall a pool–riffle sequence and Belmesthorpe a slow-flowing lowland 'drain'.

Actual study sites within these selected reaches are transects across the river channel, with each reach being represented by between 15 and 20 transects. Selection of transect sites was undertaken to represent the microhabitat variability within the reaches, specifically to ensure sampling amongst riffle, pool and straight channel typologies. Each transect is represented by between 23 and 33 cells, the boundaries of which are midpoint between the data observation points across the transect. At each observation point across the transect, data were assembled of bed elevation using levelling. The data are processed to x,y coordinates, with x being the distance away from the transect headpin and y being the bed elevation. Cover and substrate codes were assigned to each cell using the conditional criteria scheme proposed by Trihey and Wegner (1981), presented in Tables 12.1 and 12.2 respectively. In the cover

Table 12.1   Conditional cover classification scheme (from Trihey and Wegner, 1981)

Cover	Description
0	No physical cover
1	0– 25% of the cell affected by object cover
2	25– 50% of the cell affected by object cover
3	50– 75% of the cell affected by object cover
4	75–100% of the cell affected by object cover
5	0– 25% of the cell has overhanging vegetation
6	25– 50% of the cell has overhanging vegetation
7	50– 75% of the cell has overhanging vegetation
8	75–100% of the cell has overhanging vegetation
9	0– 25% of the cell has undercut bank
10	25– 50% of the cell has undercut bank
11	50– 75% of the cell has undercut bank
12	75–100% of the cell has undercut bank
13	0– 25% of the cell affected by object cover combined with overhanging vegetation
14	25– 50% of the cell affected by object cover combined with overhanging vegetation
15	50– 75% of the cell affected by object cover combined with overhanging vegetation
16	75–100% of the cell affected by object cover combined with overhanging vegetation
17	0– 25% of the cell affected by object cover combined with undercut bank
18	25– 50% of the cell affected by object cover combined with undercut bank
19	50– 75% of the cell affected by object cover combined with undercut bank
20	75–100% of the cell affected by object cover combined with undercut bank
21	0– 25% of the cell has a combination of undercut bank and overhanging vegetation
22	25– 50% of the cell has a combination of undercut bank and overhanging vegetation
23	50– 75% of the cell has a combination of undercut bank and overhanging vegetation
24	75–100% of the cell has a combination of undercut bank and overhanging vegetation
25	0– 25% of the cell has a combination of object cover, undercut bank and overhanging vegetation
26	25– 50% of the cell has a combination of object cover, undercut bank and overhanging vegetation
27	50– 75% of the cell has a combination of object cover, undercut bank and overhanging vegetation
28	75–100% of the cell has a combination of object cover, undercut bank and overhanging vegetation

classification scheme, object cover is a combination of features within the channel, such as boulders, tree trunks or vegetation. Under different flows, observations of mean column velocity were made at each observation point, and the mean water surface elevation measured. At each of the three reaches one observation was made per cell of bed elevations and cover/substrate. Two calibration flows were sampled at the Ryhall and Belmesthorpe sites, at which velocities and water surface elevation were re-observed. At each site the lowest calibration flow (CAL1) was of the order of 0.3 $m^3 s^{-1}$ and the highest calibration flow (CAL2) was approximately 0.5 $m^3 s^{-1}$. It is recognized that the magnitude of the calibration flows represents a small range and does not comply with the recommendation for PHABSIM calibration of three flows, each an order of magnitude different. Owing to water resource implications of the 1989 dry summer, releases from Rutland Water were less frequent than normal, with the consequence that no variations of flow from the compensation flow could be measured at Empingham. The Empingham reach could not therefore be calibrated for IFIM application. The range of the physical variables at each transect at Ryhall and Belmesthorpe are presented in Table 12.3.

The development of microhabitat suitability curves for this study was undertaken by the Institute of Freshwater Ecology (Armitage and Ladle, 1989) and by the Institute of Terrestrial Ecology (Mountford and Gomes, 1990). Habitat preference curves for eight species of fish and five species of invertebrate were developed (Table 12.4). The fish curves are based on expert and local knowledge of UK conditions, and fall into the Category 1 type of habitat suitability curves. Curves are developed for four life stages for each species (spawning, fry, juvenile and adult) and express habitat suitability for velocity, depth and substrate. Suitability for cover is expressed for a limited number of adult and juvenile species. The invertebrate curves are derived from analysis of the River Communities Project data base, a large body of sampled information held at the Institute of Freshwater Ecology River Laboratory at Wareham. The curves were based on the frequency of observed occurrence of the five invertebrates in the data set in relation to the physical variables. The invertebrate curves fall into the Category II type and represent habitat utilization curves. Curves are developed for the larval stages of the insects and the adult stage for the pea-mussel, and express habitat utilization for velocity, depth and substrate.

The eight fish species were selected because they are believed to represent the major fish populations of the River Gwash. Invertebrates were selected to represent species with narrow ecological limits, as three insects, the two stoneflies *Leuctra fusca* and *Isoperla grammatica* and the caddis-fly *Rhyacophila dorsalis* are typical of high-

Table 12.2 Substrate classification scheme (from Trihey and Wegner, 1981)

1	Plant
2	Mud
3	Silt (< 0.062 mm)
4	Sand (0.062–2 mm)
5	Gravel (2–64 mm)
6	Rubble (64–250mm)
7	Boulder (250–4000 mm)
8	Bedrock (solid rock)

Table 12.3   Ranges of physical habitat variables observed at calibration discharges

Transect	Substrate Min.	Substrate Max.	Velocity (m s⁻¹) Cal1 Min.	Cal1 Max.	Cal2 Min.	Cal2 Max.	Depth (m) Cal1 Min.	Cal1 Max.	Cal2 Min.	Cal2 Max.
River Gwash, Ryhall										
A1	1.9	6.1	0.00	0.61	0.00	0.84	0.00	0.25	0.00	0.27
B2	1.0	6.1	0.00	0.73	0.00	0.93	0.00	0.22	0.00	0.27
C3	1.0	6.5	0.00	0.95	−0.03	1.06	0.00	0.24	0.00	0.26
D4	1.0	6.7	−0.10	0.84	0.00	0.83	0.00	0.39	0.00	0.46
E5	1.0	3.5	0.00	0.22	−0.05	0.50	0.00	0.74	0.00	0.80
F6	1.0	6.6	0.00	0.30	0.00	0.38	0.00	0.36	0.00	0.43
G7	1.0	5.3	−0.03	0.45	0.00	0.54	0.00	0.38	0.00	0.44
H8	1.0	4.8	−0.04	0.35	0.00	0.48	0.00	0.38	0.00	0.42
I9	1.0	6.3	0.00	0.35	0.00	0.53	0.00	0.26	0.00	0.31
J10	1.0	5.8	0.00	0.63	0.00	0.60	0.00	0.20	0.00	0.26
K11	1.0	6.7	−0.04	0.64	0.00	0.59	0.00	0.34	0.00	0.42
L12	1.0	6.6	0.00	0.63	0.00	0.60	0.00	0.33	0.00	0.40
M13	1.0	5.4	0.00	0.61	0.00	0.92	0.00	0.34	0.00	0.42
N14	1.0	7.0	0.00	0.50	0.00	0.65	0.00	0.18	0.00	0.27
O15	1.0	6.6	0.00	0.51	0.00	0.58	0.00	0.33	0.00	0.43
P16	1.0	5.8	0.00	0.50	0.00	0.60	0.00	0.28	0.00	0.38
Q17	1.0	5.8	0.00	0.47	0.02	0.62	0.00	0.34	0.00	0.43
River Gwash, Belmsethorpe										
A1	1.1	4.0	−0.04	0.46	0.00	0.39	0.00	0.55	0.00	0.60
D4	1.0	4.2	0.00	0.20	0.00	0.27	0.00	0.50	0.00	0.57
E5	1.0	3.6	0.00	0.18	0.00	0.29	0.00	0.68	0.00	0.72
G7	1.0	5.0	0.00	0.34	0.00	0.39	0.00	0.40	0.00	0.44
H8	1.1	2.0	0.00	0.21	0.00	0.27	0.00	0.63	0.00	0.69
I9	1.0	2.4	0.00	0.18	0.00	0.20	0.00	0.70	0.00	0.74
J10	1.0	3.6	0.00	0.10	0.00	0.23	0.00	0.78	0.00	0.86
K11	1.0	2.5	0.00	0.20	0.00	0.24	0.00	0.55	0.00	0.89
L12	1.0	4.0	0.00	0.15	0.00	0.31	0.00	0.47	0.00	0.23
M13	1.0	7.0	0.00	0.27	0.00	0.35	0.00	0.35	0.00	0.40
N14	1.1	7.3	0.00	0.26	0.00	0.36	0.00	0.37	0.00	0.43

Table 12.4   Summary of target species used in the application of IFIM in the UK

Fish (in each case adult, juvenile, fry and spawning life stages are considered)	Macro-invertebrates	Macrophyte
Brown trout	*Isoperla grammatica* (stonefly)	*Ranunculus fluitans*
Grayling	*Leuctra fusca* (stonefly)	(river water-crowfoot)
Dace	*Rhyacophila dorsalis* (caddis-fly)	
Chub	*Polycentropus flavomaculatus*	
Roach	(caddis-fly)	
Bream	*Sphaerium corneum* (pea-mussel)	
Pike		
Perch		

velocity reaches, while, in contrast, the caddis-fly *Polycentropus flavomaculatus* and the pea-mussel *Sphaerium corneum* are characteristic of low-velocity reaches.

Habitat preference curves were developed for the aquatic macrophyte *Ranunculus fluitans* Lam., commonly known as river water-crowfoot. This species was selected because of its abundance in both rivers. Category II utilization curves were constructed as univariate curves from the observed abundance of the species in the River Blithe. No smoothing functions were applied to the observed data. Example preference curves are presented in Figure 12.4.

Model simulations were run using IFG4 on the Ryhall and Belmesthorpe sites, using MANSQ to simulate water surface elevations on transects where the velocity adjustment factor curves were theoretically unrealistic. The simulated longitudinal water surface profiles are presented in Figure 12.5. Despite the inconsistent prediction of water surface elevations in a downstream direction at certain reaches no attempt was made to improve the simulation output. In all simulations, substrate was used as the channel index and the channel cover classification values were not used in this study. A multiplicative composite suitability index function was adopted throughout.

Results

Physical habitat simulations (PHABSIM) have been used to generate estimates of weighted usable area for each of the simulation flows for each life stage of each target species. Output data are presented in three ways:

(1) weighted usable area against discharge functions for each reach for selected species;
(2) weighted usable area duration curves for the four life stages of brown trout and the stonefly *L. fusca* at Ryhall;
(3) Pre- and post-impoundment weighted usable area duration curves for adult brown trout and *L. fusca* at Ryhall.

Weighted Usable Area Against Discharge Functions for Each Week

Weighted usable area versus discharge functions for each life stage of each target species are presented in Bullock, Gustard and Grainger (1990). The River Gwash in the vicinity of Rutland Water is characterized by fish populations of brown trout, grayling, chub, dace, roach and bream (Armitage and Ladle, 1989), and functions for these and the macro-invertebrate and macropphyte species are presented in Figure 12.6. Habitat area, both total and weighted, is expressed in units of ft^2 per 1000 ft of river length. In the case of total habitat area, these units represent the total plan area of water surface within the reach, and in the case of the WUA the total habitat area has been multiplied by the composite suitability index. Total available habitat area is similar at the two sites (12.6a, b), being of the order of 20 000–24 000 ft^2 per 1000 ft. Total available habitat area increases with discharge as the plan surface area extends with rising stage over the sloping channel banks. The increase in plan area is small and constant for incremental increases in discharge owing to the rectangular nature of the channel forms.

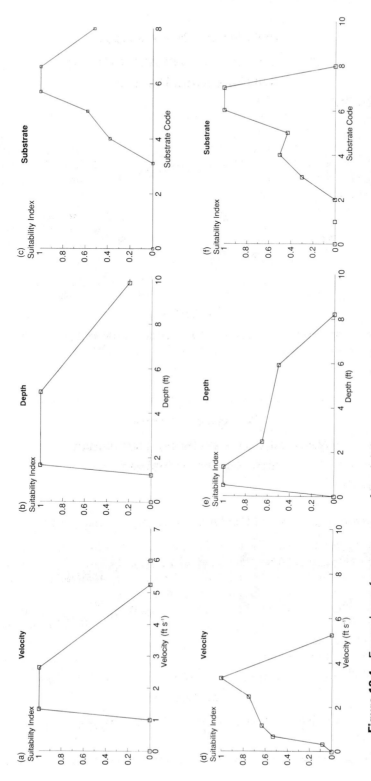

Figure 12.4 Example preference curves for adult brown trout (a, b, c) and a macro-invertebrate, *Leuctra fusca* (d, e, f)

(a)

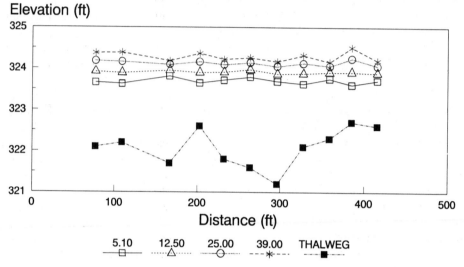

(b)

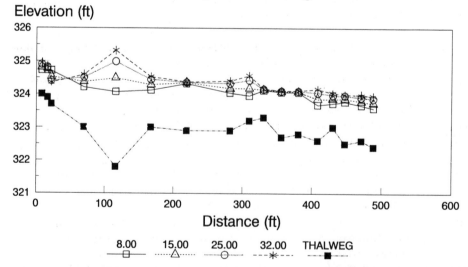

Figure 12.5 Simulated longitudinal water surface profiles for example discharges

The weighted usable area versus discharge relationships exhibit different forms and absolute values amongst different life stages of the same fish species, between species and between sites. Whilst the relationships can be similar for different life stages, for example bream at Belmesthorpe, others, such as brown trout, exhibit diversity amongst the life stages. In general, there is most physical habitat available for the juvenile stage of the salmonids at both Ryhall and Belmesthorpe and more available for the fry stages of the cyprinids, especially at Ryhall.

There are striking differences in amounts of habitat for different species; for example, a total absence of available habitat for adult bream compared with up to 50% of the total habitat being available for adult brown trout, which is not surprising given the pool–riffle characteristics of the Ryhall site. Belmesthorpe, on the other hand, offers considerably greater available habitat for chub (especially fry) and bream.

The invertebrate relationships exhibit a much greater consistency in form and absolute values than the fish species. At Ryhall, macro-invertebrate species that prefer lower velocities, *P. flavomaculatus* and *S. corneum*, possess greater amounts of available habitat than the species with preference for higher velocities. Indeed, habitat availability curves for these two species appear to flatten, whereas for the other species habitat increases at higher flows, and therefore higher velocities. However, at Belmesthorpe, where velocities are generally lower, all curves continue to increase with discharge. At Belmesthorpe, habitat availability for *R. fluitans* remains constant and insensitive to discharge but is very much steeper at Ryhall.

The weighted usable area versus discharge relationships displayed by the life stages can be said to accord to seven basic forms, as summarized in Figure 12.7. In interpreting the format of the curves one must be aware of the tendency of the PHABSIM system to extrapolate the WUA versus discharge relationship below the lowest simulated flow to zero flow. This can introduce sharp inflections into the curve shapes, so identification of the seven basic forms is founded upon the simulated flow range alone. The characteristics of each form, and an example from the relationships in Figure 12.6, are summarized as follows:

Type A – no physical habitat available throughout the simulated range of flows. Example: adult bream at Belmesthorpe.
Type B – no physical habitat available in the lower range of simulated flows but above a certain threshold the amount of physical habitat increases with discharges. Example: adult brown trout at Belmesthorpe.
Type C – physical habitat is available across the full range of simulated flows and the amount available continues to increase with higher discharges. Example: spawning chub at Ryhall.
Type D – physical habitat is available across the full range of simulated flows and the amount available continues to decrease with higher discharges. Example: fry brown trout at Belmesthorpe.
Type E – essentially a type C curve, that attains a peak of physical habitat availability followed by continuous declining habitat. Example: fry chub at Belmesthorpe.
Type F – essentially a type D curve that attains a trough of physical habitat availability followed by continuous increasing habitat. (No example at Ryhall or

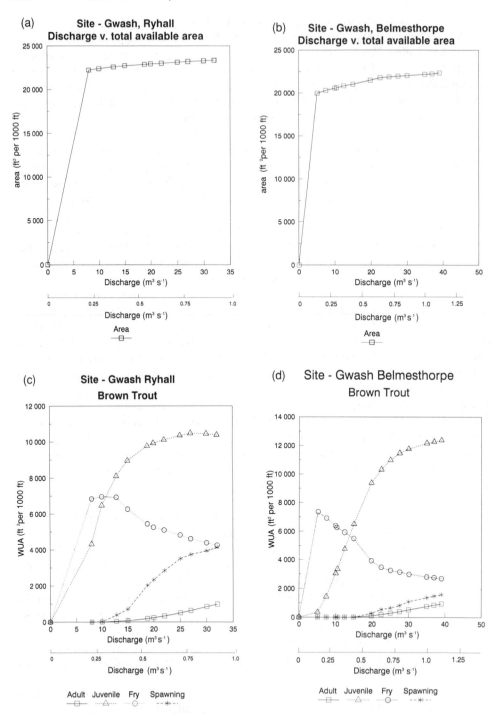

Figure 12.6 Weighted usable area against discharge relationships for target species in the River Gwash

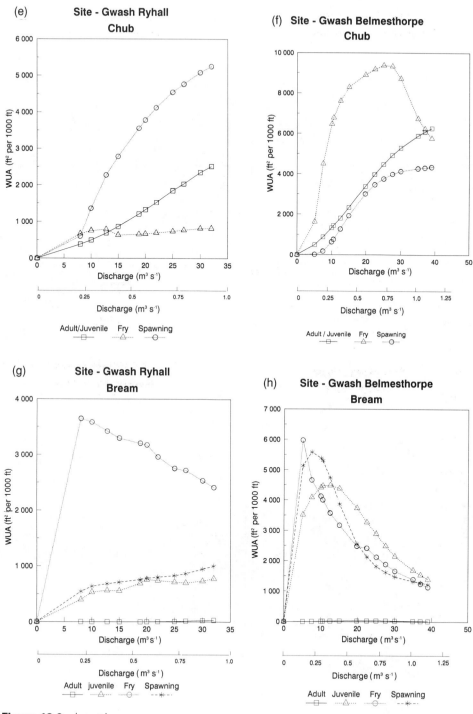

Figure 12.6 *(cont.)*

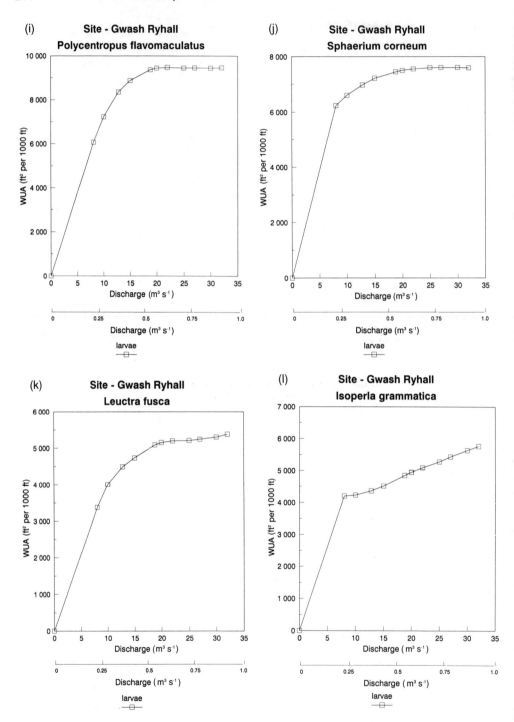

Figure 12.6 (*cont.*)

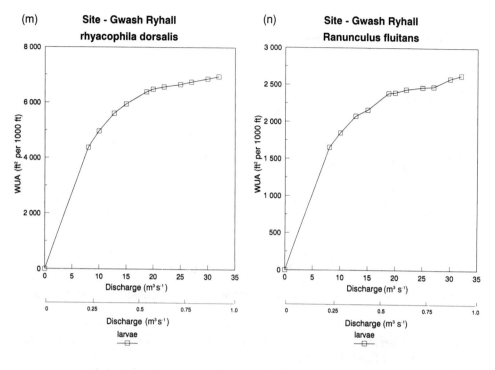

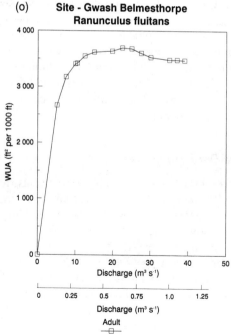

Figure 12.6 (*cont.*)

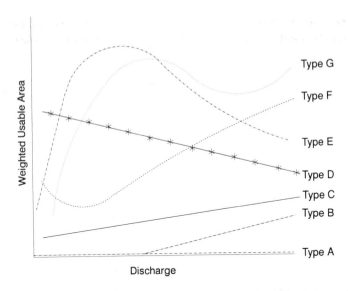

Figure 12.7   Basic forms of weighted usable area against discharge relationships

Belmesthorpe but refer to spawning bream at Hamstall Ridware (Bullock, Gustard and Grainger, 1990).)

Type G – bimodal type E or type F. Example: fry roach at Ryhall.

It will be obvious that some of these type curves are not in fact unique and that some could represent the sampled limb of another type curve were the simulatd flow range to be extended (for example type C is merely the rising limb of type E). However, the distinction is maintained because the form of the curve within the full flow range of a river reach is the critical aspect, even though this may, for example, represent only a single limb of a more complex curve.

Weighted Usable Area Duration Curves

The previous section presented weighted usable area versus discharge relationships without reference to the frequency with which discharges occur in the river reaches. Any consideration of instream flows, and habitat availability, must consider the distribution of river flows over time to enable sensible interpretation and, in the longer term, feasible water management. This section represents a selection of the weighted usable area versus discharge relationships superimposed upon a gauged natural flow duration curve at the Ryhall site (Figure 12.8). The flow duration curve represents the percentage of time that a given discharge (expressed as a percentage of the natural mean flow) is exceeded – for example the 95 percentile flow is the discharge that is exceeded 95% of the time – on all but 18 days per year on average. Superimposed upon the flow duration curve is the total weighted usable area versus discharge curves for selected species, in which the simulation discharges have been assigned an exceedance percentile. It is stressed that the habitat availability curves do

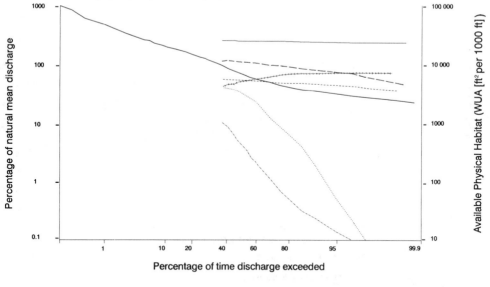

Figure 12.8 Weighted usable area against discharge relationships superimposed upon flow duration curves at Ryhall

not represent true habitat duration curves, for which the time series of daily flow data would have to be converted to a time series of daily habitat data.

The merit of superimposing weighted usable area versus discharge curves upon the natural flow duration curve is the capability to associate key thresholds, inflections or peaks and troughs of the habitat–discharge relationship with the probability of their occurrence. For example, it becomes evident that there is no physical habitat available for adult brown trout for 5% of the time on average per year for brown trout at Ryhall. The diagrams also identify that the discharge that provides the optimal physical habitat for the spawning, juvenile and adult life stages of brown trout is higher than the 40 percentile flow (equivalent to the mean flow). Higher discharges would, however, reduce the available habitat for the fry stage.

Comparison of Pre- and Post-impoundment Weighted Usable Area Duration Curves

Figure 12.9 presents pre- and post-impoundment flow duration curves for the Ryhall site, in which flow is expressed as a percentage of the natural (i.e. pre-impoundment) mean flow. Under the two different flow conditions, the simulation discharges (and

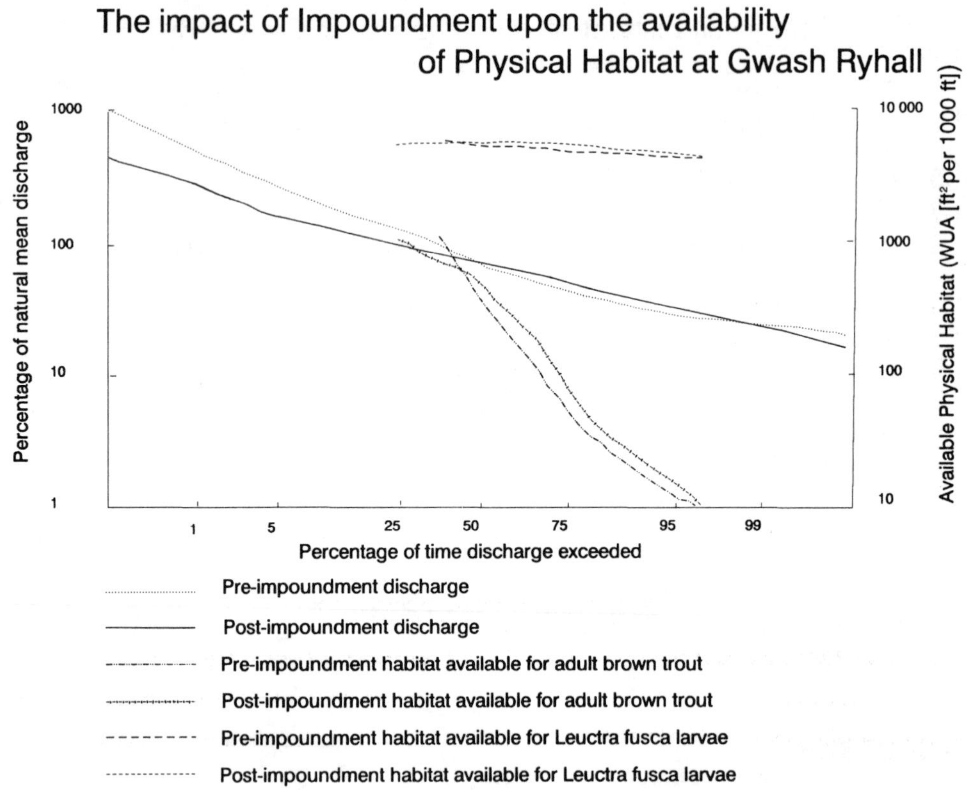

The impact of Impoundment upon the availability of Physical Habitat at Gwash Ryhall

Figure 12.9 Pre- and post-impoundment flow duration curves and superimposed weighted usable area against discharge relationships below Rutland Water

hence WUA) are assigned different exceedance percentiles, which distinguishes the pre- and post-impoundment habitat availability curves.

In the case of the impoundment of Rutland Water, the effect upon the flow regime is complex. Flows in excess of the median (50 percentile flow) are reduced by impoundment, but the low flows are higher, with the exception of the extreme low-flow events. Consequently, at lower flows (in the 50–95 percentile range) there is increased available physical habitat for adult brown trout at a given flow percentile after impoundment than previously. The gain in habitat, however, in absolute terms is of a small degree, the curves being displaced by 5 percentiles at a maximum. At flows in excess of the median, the available physical habitat is lower after impoundment and the disparity appears likely to widen at high flows given the divergence of the flow duration curves. *Leuctra fusca* do not gain or lose habitat significantly, owing to the flatness of their curves at these flows.

Use of Physical Habitat Availability Results

Prescribed flows can be extracted from the habitat versus flow relationships, and it is seemingly most logical to argue for the peak of the function or significant thresholds or inflections. In doing so, however, there are several concepts that must be borne in mind (Bovee, 1982):

(1) a flow change that is beneficial to one life stage may be detrimental to another life stage;
(2) a flow change that is beneficial to one species may be detrimental to another species;
(3) various life stages and species may require different amounts of water at different times of the year;
(4) a flow that maximizes usable habitat in one part of the stream may not maximize, and may indeed decrease, habitat in another part of the same stream;
(5) more water does not necessarily mean more habitat.

In addition, sensible interpretation and subsequent negotiations must consider, amongst other factors:

(1) water availability, and its seasonal distribution, within existing water management and institutional constraints;
(2) that other biological factors, such as food supply, temperature and water quality may be more important than physical habitat;
(3) how the biological population levels will respond.

DISCUSSION

It is most definitely not an objective of this project to recommend flow regimes for the River Gwash or to reassess the compensation flows from Rutland Water. Indeed any interpretation of the presented results within a management context would be grossly misrepresentative. It is emphasized that these results are of a preliminary nature, being the result of the first application of IFIM in the UK and should not be used on any account as an indication of habitat availability for design purposes.

Although the results are preliminary, the first application of IFIM in the UK generated valuable experience and lessons. The IFIM is based upon the assumption that habitat is important to ecological communities, and that populations respond to availability of physical habitat. Greater research emphasis has been placed upon the prediction of how hydraulic factors vary with flow than how ecological populations vary with flow. Population models for specific species and life stages have not yet developed to the extent of enabling prediction to incorporate change in velocity and depth. As a result, habitat modelling remains the best approach for assessing instream flow requirements.

Before accepting the methodology for application in the UK, there is a need to identify the sensitivity of the model to river regimes where water levels vary with the seasonal growth of macrophytes and can be controlled by variable sluices and lock gates. Similarly, there is a need to develop a channel index for rivers where channel vegetation is likely to exert a greater impact upon variability of velocities with transects. Furthermore, it must be recognized that seasonally varying cover codes and species-specific and life-stage specific channel indices must be adopted where possible.

The identification of well-defined channel controls facilitates the application of the different hydraulic simulation routines in PHABSIM. However, channel controls in lowland rivers are not always identified easily, and the control upon a stage–discharge relationship within a reach is more likely to vary with discharge than in steeper upland channels. Variable backwater effects will occur with greater frequency in lowland rivers, and a greater dependency can be envisaged upon WSP, which is the only simulation routine currently within PHABSIM for coping with backwaters.

Best results are achieved from PHABSIM simulations when wide stage–discharge relationships have been established. However, the variability of flows in some lowland rivers is small, especially in rivers draining permeable chalk aquifers and, in particular, within a moderate time-scale. Furthermore, the small ranges of discharges are typically associated with small ranges of stage, to the detriment of PHABSIM simulations.

There is a need to develop habitat preference curves for target species for which curves do not exist, and to refine and check the applicability of those that do exist. Existing data bases of fish information are not converted to habitat preference criteria easily, while the River Communities Project will aid the development of criteria for macro-invertebrates, despite the observed data being based on macrohabitat rather than microhabitat variables.

The habitat versus flow relationships developed for the River Gwash exhibit many facets of interest, including the different habitat relationships for different life stages of one species, for different species, variations between the two reaches on the same river and the different forms to which the relationships accord. The potential of habitat versus discharge relationships from IFIM for setting prescribed minimum flows, reviewing compensation releases and other instream flow demands in lowland rivers is demonstrated clearly by this investigation and valuable experience has been gained in its accomplishment.

ACKNOWLEDGEMENTS

The project was carried out by a collaborative research group, headed by the Institute of Hydrology and involving the Institute of Freshwater Ecology, the Institute of Terrestrial Ecology and Loughborough University. The application described is part of work commissioned by the Department of the Environment. The opinions expressed are those of the authors and are not necessarily those of the Department.

REFERENCES

Armitage, P. D. and Ladle, M (1989). *Habitat Preferences of Target Species for Application in PHABSIM Testing*, Institute of Freshwater Ecology contribution to Bullock, Gustard and Grainger (1990) report to the Department of the Environment, London.

Armitage, P. D., R. J. M. Gunn, M. T. Furse, J. F. Wright and D. Moss (1987). The use of prediction to assess macroinvertebrate response to river regulation. *Hydrobiologia*, **144**, 25–32.

Bovee, K. D. (1982). A guide to stream habitat analysis using the instream flow incremental methodology. *U.S. Fish and Wildlife Service, Instream Flow Information Paper No. 12*, FWS/OBS – 82/26, 248 pp.

Bovee, K. D. (1986). Development and evaluation of habitat suitability criteria for use in the instream flow incremental methodology. *U.S. Fish and Wildlife Service, Instream Flow Information Paper No. 21*, US Fish and Wildlife Service Biologic Report 86(7), 235 pp.

British Hydrological Society (1988). *Aspects of Hydrological Studies in S.W. England: Monitoring and Resources*, BHS Circulation No. 17 (January), Wallingford, pp. 4–6.

Bullock, A., A. Gustard and E. S. Grainger (1990). *Instream Flow Requirements of Aquatic Ecology in Two British Rivers – Application and Assessment of the Instream Flow Incremental Methodology*, Institute of Hydrology Report No. 115, Wallingford, 149 pp.

Drake, P. J. and J. D. F. Sherriff (1987). A method for managing river abstractions and protecting the environment. *Journal of the Institution of Water and Environmental Management*, **1**(1), 27–38.

Gan, K. and J. McMahon (1990). Variability of results from the use of PHABSIM estimating habitat area. *Regulated Rivers: Research and Management*, **5**, 233–239.

Gore, J. A. and J. M. Nestler (1988). Instream Flow Studies in perspective. *Regulated Rivers: Research and Management*, **2**, 93–101.

Gustard, A., G. Cole, D. C. W. Marshall and A. C. Bayliss (1987). *A Study of Compensation Flows in the United Kingdom*, Institute of Hydrology Report 99, Wallingford, 198 pp.

Hindley, D. R. (1973). Definition of dry-weather flow in river flow measurement. *Journal of the Institution of Water and Environmental Management*, **27**(8), 438–440.

Howarth, W. (1990). *The Law of the National Rivers Authority*, University College of Wales, Aberystwyth.

Institute of Hydrology (1980). *Low Flow Studies*. Institute of Hydrology, Wallingford, 43 pp.

Lamb, B. L. and H. R. Doersken (1987). Instream water use in the United States – water laws and methods for determining flow requirements. *National Water Summary 1987 – Water Supply and Use: Instream Water Use*, USGS Water Supply paper, pp. 109–116.

Moss, D., M. T. Furse, J. F. Wright and P. D. Armitage (1987). The prediction of the macroinvertebrate fauna of unpolluted, running water sites in Great Britain using environmental data. *Freshwater Biology*, **17**, 41–52.

Mountford, O. and N. Gomes (1990). *Habitat preference of river water-crowfoot (Ranunculus fluitans) for application in PHABSIM testing*, Institute of Terrestrial Ecology contribution to Bullock, Gustard and Grainger (1990) report to the Department of the Environment.

Nestler, J. M., R. T. Milhous and J. B. Layzer (1989). Instream habitat modeling techniques. In J. A. Gore and G. E. Petts (eds), *Alternatives in Regulated River Management*, CRC Press, pp. 295–315.

Petts, G. E. (1989). Methods for assessing minimum ecological flows in British rivers. *British Hydrological Society National Hydrology Meeting – Prescribed Flows, Abstraction and Environmental Protection*, London, 16 June 1989, Institute of Civil Engineers, pp. 17–25.

Trihey, E. W. and D. L. Wegner (1981). *Field Data Collection Procedures for Use with the Physical Habitat Simulation System of the Instream Flow Group*, Cooperative Instream Flow Service Group, Fort Collins, Colorado, 11 pp.

13 Restoration of Lowland Rivers: the German Experience

KLAUS KERN

Institut für Wasserbau und Kulturtechnik, Universität Karlsruhe

INTRODUCTION

Lowland rivers especially have been subject to various regulation works, such as straightening for flood control, water power plants and navigation purposes (Petts, Möller and Roux, 1989). In Germany, all of the major rivers, e.g. the Rhine, Elbe, Moselle, Main and Danube, have been regulated. Unlike other large European rivers, such as the River Loire, not a single section of the larger German rivers has remained untouched. In most cases large-scale regulation works began in the middle of the nineteenth century. Today river regulation is still being implemented, e.g. on the Danube to complete the missing link of the Rhine–Main–Danube Canal, which will connect the North Sea with the Black Sea.

There has been a considerable loss of river-bed habitats resulting from the construction of these uniform cross-sections. Former wetland habitats located in many floodplains have been turned into agricultural or forest land. Only a few square kilometres of the vast original floodplain wetlands still remain in the upper Rhine Valley (Dilger and Späth, 1988).

In the 1980s, when nature conservation interest began to focus on stream rehabilitation, concepts for the ecological restoration of larger rivers were established. There are two projects currently investigating the River Rhine: the 'Integrated Rhine River Programme', which covers a river section of about 220 km from Basel, Switzerland to Mannheim, Germany; and an international programme, 'Salmon 2000', which covers the entire river downstream from Lake Constance. There is another programme that is working towards the restoration of a 160 km section of the Danube in the State of Baden-Württemberg. None of these programmes have resulted in any real action yet, although some old technical projects have been stopped.

OBJECTIVES AND LIMITATIONS OF RIVER RESTORATION

The main objective of river restoration is the conservation of nature. Successful reintroduction of plant and animal species typically found within a river ecosystem

Lowland Floodplain Rivers: Geomorphological Perspectives. Edited by P.A. Carling and G.E. Petts
© 1992 John Wiley & Sons Ltd

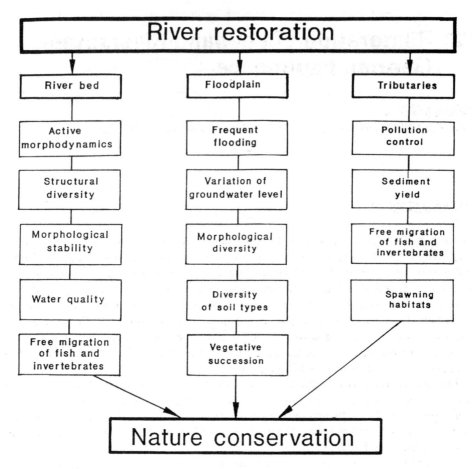

Figure 13.1   Concept of river restoration leading to the conservancy of nature

requires comprehensive studies and management of the channel, floodplain and tributaries. Figure 13.1 shows some aspects of these three components.

River Bed

Active Morphodynamics/Structural Diversity

The structural diversity of the river-bed habitat is a result of continuous erosion and deposition of sediments. For example, disturbances in sediment transport by backwater effects of weir construction in many cases cause irreversible changes in the aquatic habitat.

Morphological Stability

Morphological dynamics of the river bed are essential for structural diversity. Continuous erosion or accumulation with a considerable change of the average river-bed level within a few decades is often artificial and usually causes a loss of species. Rivers should be free to meander but not to erode.

Sufficient Water Quality/Pollution Control

Restoration of river-bed structures cannot ignore water quality problems. Structural restoration must never be regarded as a substitute for pollution-control efforts along a river and its tributaries despite inherent self-purification effects. Pollution control can be quite successful. The water quality of the River Rhine and River Neckar has improved considerably within the last 20 years. Many limnologists regard class II (less polluted) of the water quality classification system commonly used in Germany as sufficient for the restoration of watercourses (Braukmann, 1991).

Migration of Fish and Invertebrates

Most weirs and drop structures interrupt the necessary migration of many aquatic species. Long, backwater reaches with a total change of stream-bed pattern, velocity, temperature and nutrients also present insurmountable barriers. Nature conservation is concerned with all migrating species, not only those of economic value. In this respect conventional fish ladders have proven to be insufficient (Gebler, 1989).

Floodplain

Frequent Flooding

The ecological value of the floodplain depends on the frequency, duration and height of inundations (Dister, 1985b). If the hydraulic capacity of the channel has been increased considerably, a secondary development of wetlands in the floodplain is impossible.

Variation of Groundwater Level

Groundwater level variations are just as important as frequent flooding. In many cases, river regulation can lead to severe bed erosion causing lower groundwater levels. Sealing the bed of the channel also might prevent corresponding variations of surface and groundwater levels.

Morphological and Soil Diversity

Depending on a river's history, the floodplain can be characterized by a pattern of depressions, oxbows, ridges and plains. According to variable sedimentation processes, soil properties vary across the floodplain. Owing to increased transport capacity along the channel, coarser grains are deposited near the river banks leading to rather dry habitats, while silting occurs on the floodplain. With the shifting of the river bed the succession of different layers will result in heterogeneous profiles. Floodplain formation must be regarded as a geological process, which cannot be developed artificially within a short amount of time.

Vegetation

When physical habitat properties have been restored, there is no need to accelerate vegetation growth. Many wetland plants can adjust to rapid morphological changes.

Tributaries

Sediment Yield

Within many watersheds, landuse has increased the suspended load while channelization has decreased the bedload typically found in these watercourses. The effects on river restoration projects may be significant. There can be considerable siltation of floodplains as well as erosion problems in the channel. Reduction of field erosion might be possible while the necessity for bank protection ought to be re-examined.

Habitats for Spawning

River restoration has to include the construction of spawning habitats within all tributaries. Weirs and drop structures must allow for fish migration.

PLANNING PRINCIPLES

Feasibility

The feasibility of any river restoration project has to be evaluated by first answering the following questions.
(1)   What are the objectives of the project?
(2)   Which sections need to be restored?
(3)   Where are pollution sources and what can be done to improve water quality?
(4)   Is there an equilibrium in sediment transport?
(5)   What are the prospects for restoration of floodplain habitats (existing landuse, development plans)?
(6)   Are there any legal constraints by water and fishing rights?
(7)   Are there plans to use water resources in the future?
(8)   What are the demands for flood protection?
Some of these questions need further study, others can be answered by the appropriate water authorities. In larger projects, studies that include field observations have to be carried out covering the historical aspects of river regulation and landuse, physical and chemical limnology, channel morphology, riparian vegetation, floodplain habitats, etc. Infra-red aerial photographs may be helpful in floodplain mapping (Konold, Pfeilsticker and Jöst, 1991). A survey of the fauna should be included in every project (e.g. fish, aquatic invertebrates for water quality control, special groups of insects, birds).

The 'Leitbild' Concept

After analysis of the collected data, a 'guiding image' or 'Leitbild' has to be developed as a guide for the final restoration concept (Figure 13.2). The 'Leitbild' gives a

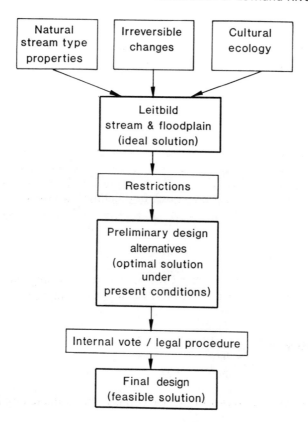

Figure 13.2 The guiding image ('Leitbild') concept as a planning instrument and steps to the final design

description of the desirable stream properties with regard to the natural potential only. Economic and political aspects are not considered. It is based on three elements, which are:

(1) natural stream-type properties (e.g. stream pattern, morphodynamics, floodplain morphology, natural flow dynamics and flooding, potential vegetation, etc.);

(2) irreversible changes of abiotic and biotic factors (i.e. the run-off regime or sediment transport, changes in the pattern of alluvial sediment deposition on the floodplain, extinction of plant or animal species, etc.);

(3) aspects of cultural ecology (historical landuses, such as, grassland irrigation, cultivation of willows for basket making, etc., favoured species that are currently endangered by modern agriculture).

The 'Leitbild' of the stream system represents the *ideal solution*. Existing conditions, such as current landuse, water rights, flood protection requirements, etc., are not considered. Only in a very few cases is it possible to carry out the 'Leitbild' draft without major concessions. Normally numerous restrictions are imposed on a project, thereby preventing an 'ideal' solution. Therefore it is up to the planners to find the *optimal solution* with regard to the essential ecological rehabilitation of existing conditions. This preliminary design may be based on several alternative drafts, with different ecological and economic effects. The optimal solution will have to be discussed with all participants involved, including the public. Usually only minor alterations are imposed by legal procedures. The final project design may be called the *feasible solution*.

CASE STUDIES

Restoration of the Upper Danube

Two branches contribute to the source of the Danube (Figure 13.3). Beyond this confluence, the young Danube flows past the Swabian Alb within a deep, narrow valley. A considerable part of its water is lost to the Rhine river system owing to the karst geology. The river then follows the edge of the Swabian Alb reaching the Bavarian border at the city of Ulm.

Geology

The geomorphology of this watercourse is rather complicated, because different glacial periods have left their marks on the landscape. The ancient fluvial systems have eroded and aggraded their valleys with the change between warm and cold periods (Geyer and Gwinner, 1986). The Danube itself used to have a much larger catchment area, with a significantly higher sediment yield than it does today. After leaving the Jurassic mountains of the Swabian Alb, the river enters an area with soft layers of marine sediments, called the Molasse. At the edge of the Swabian Alb, it again encounters resistant Jurassic formations that hinder further erosion, with average gradients of 0.08–0.11%. In this section the river's morphology is similar to that of lowland rivers with respect to valley width and floodplain extent.

Nineteenth Century Rectification

Figure 13.4 shows a small section of the Danube with the old 1840 channel and the regulated river of 1990. Numerous oxbows were adjacent to the channel indicating rapid movement of the meanders in the soft Molasse sediments. Regulation works, which started in the mid-nineteenth century, cut through all the meanders at the edge of the Swabian Alb and reduced the river length by 20% on some sections. The resulting higher gradient caused the channel bed to erode up to 2.5 m within 90 years (Figure 13.5). This led to a corresponding reduction of water levels. Since the groundwater level was also affected, the few remaining uncultivated wetland habitats also were endangered.

Restoration Programme

The water authorities addressed these problems in 1988 by implementing a restoration programme on the Danube. The entire river section located within the State of Baden-Württemberg has been investigated with regard to:

Figure 13.3 The Danube in the State of Baden-Württemberg, Germany

(1) historical river morphology;
(2) riparian vegetation and ecology;
(3) physical and chemical parameters of water;
(4) insect larvae and fish species;
(5) weirs and drop structures with respect to accessibility for fish migration;
(6) hydrogeology
(7) vegetation and landuse of the ancient floodplain using aerial infra-red colour photographs;
(8) special groups of insects and birds.

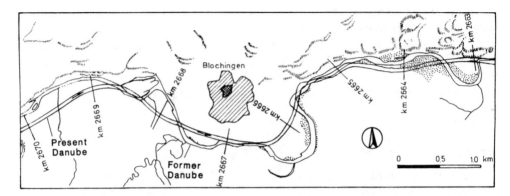

Figure 13.4 Historical and present course of the Danube near the village of Blochingen, Germany

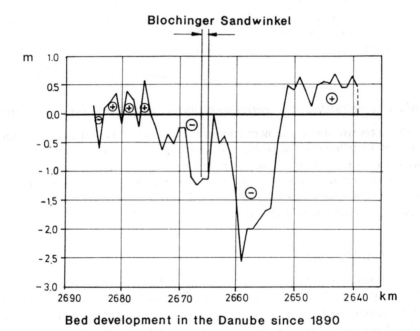

Bed development in the Danube since 1890

Figure 13.5 Bed level development of a Danube section in the State of Baden-Württemberg. The 0-level corresponds to the river bed in 1890 compared with the bed level of 1980

After the devastating flood of February 1990 (a 200 year event), hydrological and hydraulic studies for flood protection have been added to the research programme.

Meander Restoration

One project proposal near the village of Blochingen (Figure 13.4), has been studied in detail as a restoration example. The map shows the shortening of a meander, which was done in 1874. The old meander is still visible, but instead of free flowing surface water, trees are growing in the former channel. Only higher floods, such as the one that occurred in February 1990 reveal the depressions in the floodplain marking former channels. The area between the old and the new channel has been cultivated with very little success. In fact the soil is very poor, consisting of coarse gravel and sand. It is actually one of the driest soil types, based on a classification scale for soil humidity.

What Would the 'Leitbild' of this Danube Reach Look Like?

As far as the morphology is concerned, the soil would pose very little resistance to riparian erosion. We know that the old channel moved as much as 100 m in only 30 years. Catastrophic events could alter the channel leaving oxbows exposed to eventual siltation. The river bed would become broader with gravel bars and possibly could be split by islands. The channel would carry little more than the annual flood. The groundwater level would vary a few feet below the floodplain. The habitats adjacent to the river, with the exception of former oxbows, would stay very dry owing to the properties of the soil. The floodplain and the river banks would be vegetated with forest species.

What Must be Regarded as Irreversible Damage Done to the River System?

Very little is known about sediment transport. Obviously the banks of the Danube and many of its tributaries have been protected against erosion. In addition, weir construction for upstream water power plants may retain sediments. Some endemic fish species are extinct, other species have been introduced. The water quality, classified as 'less polluted' is quite satisfactory. Flood protection for the nearby village of Blochingen has to be kept in mind because higher flood levels will not be tolerated.

Hydraulic Model Results

Figure 13.6 shows a river diversion with a new meander on the right side, and a further meander on the left bank enclosing two large islands. Two gently sloping drop structures – permitting migration – raise the average water level in the channels as well as in the floodplain. The floodplain will be reshaped with oxbows, ridges and depressions. There will be no bank protection whatsoever within the new meander channels.

Based on hydraulic performance, optimal drop structure positions and heights, as well as the gradient and width of the new channel, were investigated by a model test in the Theodor–Rehbock Laboratory at the University of Karlsruhe (Dittrich et al., 1989). The drop structures allowed the average water level (corresponding to $24 \text{ m}^3 \text{ s}^{-1}$) to be raised by almost 1 m compared with 1.3–1.5 m of bed erosion (Figure 13.7). On the other hand, there is better protection from higher flood levels for the village of Blochingen. As a result of this diversion, the 100-year flood levels (amounting to about $400 \text{ m}^3 \text{ s}^{-1}$), in the vicinity of the village will be 30–40 cm lower than before. The old channel will be used mainly for flood flow. Approximately three quarters of

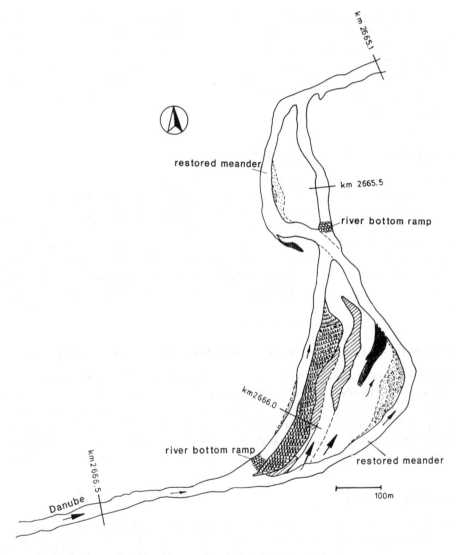

Figure 13.6 Project site 'Blochinger Sandwinkel' of the Danube showing the current channel and the meanders that will be restored

the 100-year discharge will pass through the present channel. During average flow conditions most of the discharge will flow through the new meanders.

What has been Achieved?

(1) Restoration of the former average water level and free flow conditions in the new channel.
(2) Raised water levels up to a discharge of 160 m^3 s^{-1}, which corresponds to the average annual flood.
(3) Increased protection from floods higher than 160 m^3 s^{-1} in the vicinity of the village.

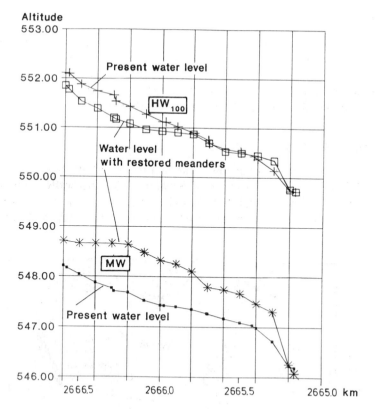

Figure 13.7 Danube water levels near Blochingen: average flow (24 m³ s⁻¹) and flood flow (400 m³ s⁻¹) under present conditions compared with restored meanders. Results of a model test in the Theodor-Rehbock Laboratory at the University of Karlsruhe

(4) Unrestricted bank erosion will be allowed in the new meander (a 100 m strip has been purchased along the outer banks).
(5) Control of bed erosion (which could be calculated from the model test).
(6) Improvement of floodplain habitats.

What are the Concessions to be Made?

(1) Drop structures are not typical for the Danube in this section.
(2) The old channel habitat will correspond to neither the still-water conditions of oxbows nor to the free flow conditions of the channel.
(3) Recolonization of vegetation and fauna will be according to present-day conditions.
(4) High costs have to be accepted for the restoration of rivers, especially those with erosion problems. The calculated costs for the above Danube project near the village of Blochingen are 1.5 million DM. The restoration of the entire section of the Danube within the State of Baden-Württemberg will cost more than 100 million DM. The project will be regarded as a model test in a scale of 1 : 1 with an intensive survey of morphological and ecological development.

Upper Rhine River

The morphology of the upper River Rhine between Basel, Switzerland and Mannheim, Germany has been used by numerous authors as an illustration of a braided and meandering channel pattern (Figure 13.8). The natural channel morphology of the upper Rhine was first replaced by a regulated river system during the early nineteenth century, as described by Kunz (1975) and Lehle (1985).

The Historical Situation

Devastating floods, recorded in historical reports dating back to 1306, destroyed villages, and killed people and livestock (Schäfer, 1974). All settlements in the postglacial floodplain of the upper Rhine Valley were more or less endangered by flooding. In addition, the continually changing course of the River Rhine prevented

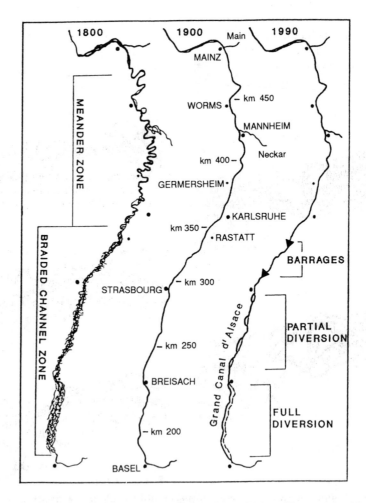

Figure 13.8 Historical development of the upper River Rhine. (Reproduced by permission from Dister, 1985a)

the establishment of a well-defined border between France and the then sovereign State of Baden.

Nineteenth Century Rectification

In 1815 Johann Gottlieb Tulla started planning the 'total rectification' of the upper River Rhine. His general idea was to concentrate water flow into one channel in order to lower water levels through bed erosion (Tulla, in Schäfer, 1974). Navigation or water power usage was not a consideration at this time. Figure 13.9 shows the different steps of the regulation works over a period of 40 years within the braided channel section. Between Basel, Switzerland and Strasbourg, France a total volume of about 9 10^6 m^3 of earth had to be transported for the construction of some 440 km of levees and dams – an incredible achievement for those times, which was accomplished by two generations of village people (Kunz, 1975). The river naturally eroded an additional 4 10^9 m^3 to form its new channel bed. These figures point out the significance of Tulla's concept of a river's natural ability to regulate itself. Figure 13.9 also shows that Tulla did not completely cut off the floodplains from the Rhine, as his successors did. In fact, he was well familiar with the principles of flow dynamics. Subsequently, civil engineers have tended to ignore downstream problems caused by upstream construction works.

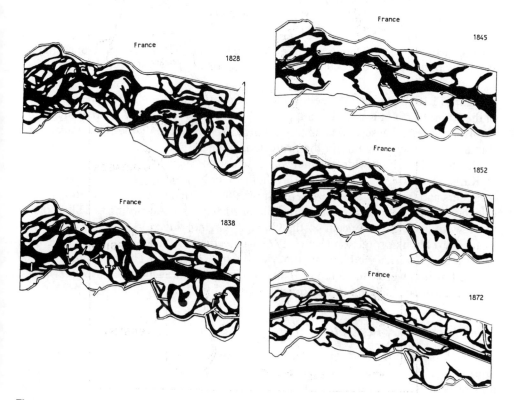

Figure 13.9 Historical river training works at the upper River Rhine in the braided channel section using small dams to guide the current. (Reproduced by permission from Schäfer, 1974, © Senckenberg Institut, Frankfurt)

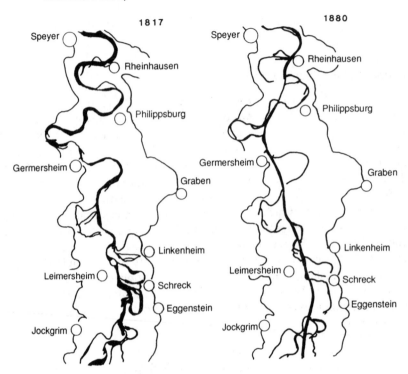

Figure 13.10   River regulation at the upper River Rhine in the meander zone. (Reproduced by permission from Schäfer, 1974, © Senckenberg Institut, Frankfurt)

Figure 13.10 shows the nineteenth century regulation works in a meandering section of the Rhine north of Karlsruhe. Supported by diversion structures, the river naturally cut its meanders. These new channels initiated erosion and only the river banks were stabilized in order to maintain the projected channel.

The new 200–250 m wide channel was designed to carry 2000 m³ s⁻¹ (average flow at Karlsruhe 1200 m³ s⁻¹). At a threshold of only 1000 m³ s⁻¹, the pre-existing meanders and furcation channels were hypothetically flooded while the floodplain was calculated to accept overflow at a threshold of 2000 m³ s⁻¹. According to Tulla's plans there were no levees adjacent to the channel; engineering works being confined to surface-level bank protection (Kunz, 1975). Tulla died in 1828, 10 years after the first project had been carried out in the vicinity of Karlsruhe. The regulation works planned by him were completed in 1880 and the central drawing in Figure 13.8 shows the degree of rectification at this time.

Twentieth Century Regulation

Although protection from severe floods was provided by nineteenth century regulation works, the channel bed still contained irregular gravel bars and scour holes, making the Rhine unsuitable for navigation. Between 1907 and 1939 a low-flow channel was established by installing groynes. This reduced the free flow channel width to 75–150 m. A system of levees was also implemented to improve flood protection, as requested by the growing population within the upper Rhine Valley. Consequently many of the diversion meanders and furcation channels were taken out of the flood regime.

After World War I, France was given the rights to all the waterpower along the French border. They built the 'Grand Canal d'Alsace', which was completed in 1959 (Figure 13.8). At this time, four power stations were taking 1200 $m^3 s^{-1}$ off the River Rhine and leaving only 15–20 $m^3 s^{-1}$. After the German–French reconciliation, a less damaging solution was found for using the waterpower. Upstream of the barrages, built in a new channel adjacent to the river, short sections of the main river bed only affected by backwater were included ('partial diversion' in Figure 13.8). Downstream of Strasbourg, two more dams were built in conjunction with levees in the riparian zone. Downstream of the last dam south of Karlsruhe (built in 1974) erosion problems have been controlled successfully by artificial addition of sediment. The amount of sediment added to the river is calculated according to the actual tested transport capacity of the flood wave.

Hydrological and Ecological Consequences

Bed Erosion. Tulla's rectification plans were based on calculated river-bed erosion amounts in order to obtain lower flood- and groundwater levels. Figure 13.11 shows the development of the average water level since 1828 within the Basel–Strasbourg section (Raabe, 1968). The nineteenth century regulation works led to almost 4 m bed erosion at the beginning of the channelized section. By 1950 the bed levels had cut down to 7 m. The eroded upstream material was transported through the downstream sections; thus erosion processes were partially compensated for and accumulation was registered around 260 km. It is not known at which level erosion would have ceased without the construction of these dams.

Groundwater Levels. Within the section where the river water was completely diverted into the Grand Canal d'Alsace, the decreasing groundwater level has caused considerable damage to the surrounding environment. The former wetlands com-

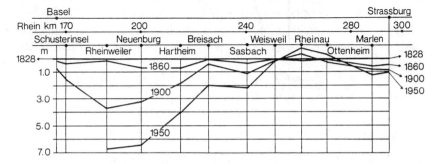

Figure 13.11 Bed level development of the upper River Rhine between Basel, Switzerland and Strasbourg, France, since the beginning of regulation up to the construction of the river diversions. (Reproduced by permission from Raabe, 1968, © Deutscher Rat für Landespflege, Bonn)

pletely lost their natural character. Some of these areas have turned into upland habitats with no connection to the river at all. Downstream of Breisach (Figure 13.8) the groundwater level, influenced by the head and backwater of the dams, has increased to constant levels.

Flood Levels. The regulaton works after 1950 decreased the active floodplain area by 130 km^2 (60% of the former area). Consequently the flood waves accelerated downstream showing higher peaks. The flood protection level for the cities of Karlsruhe and Mannheim changed from a 200-year to a 50-year event. Under present conditions the historical flood of 1882/1883 would have had a 25% peak discharge increase at Worms occurring 48 h earlier (Figure 13.12). It was calculated that an additional storage volume of 200 10^6 m^3 would be necessary to restore these former conditions (Lehle, 1985).

Flow Dynamics. The regulation works (especially after 1950) considerably altered the flow dynamics that are essential to habitat structure and organism survival. The diversion channel sections and oxbows suffered not only from altered water levels but also from changes in temperature, oxygen, nutrients and suspended load regime. These are the main abiotic factors within a floodplain ecosystem (Dister, 1985a, b).

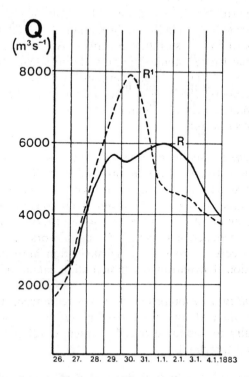

Figure 13.12 Calculated peak discharge (R') under present conditions compared with the historical Rhine flood peak (R) of 1882/1883. (Reproduced by permission from Dister, 1985a)

Technical Solution and River Restoration

Faced by the increased danger of flooding, the water authorities started building detention reservoirs adjacent to the river. These were supposed to store about 55 10^6 m^3 (at reservoir levels up to 9 m of water). The additional storage volume was to be obtained by two weir constructions in the old Tulla channel (about 100 10^6 m^3) and by special management of the power station dams. These solutions did not consider any ecological aspects. In the 1980s, environmental impact studies started changing the objectives of the programme, and in present discussions solutions that integrate ecological and technical aspects are being considered (Dister, 1985a, b; Dilger and Späth, 1988). With reference to Figure 13.1, river restoration must cover the river bed, the floodplain and tributaries.

River Bed

Water Quality. In general, water quality has improved considerably in Germany in the last 20 years. In the state of Baden-Württemberg, 92% of all residents are connected to a biological waste-water treatment plant. As far as organic pollution is concerned, the River Rhine has reached values of II and III on the five-step classification scale (Landesanstalt für Umweltschutz Baden-Württemberg, 1990). Chemical parameters (e.g. chloride, nitrate, phosphate, Cd, Pb, Hg, organic carbon substances) have been stabilized. In addition, catastrophic events, such as the November 1986 fire at Sandoz industries in Basel, Switzerland, still endanger organisms. This poses new questions to river restoration practices. After the release of toxic (fire-fighting) water into the Rhine, monitoring proved that aquatic fauna in the adjacent wetlands was damaged as severely as the river fauna itself (Landesanstalt für Umweltschutz Baden-Württemberg, 1987). However, the recovery of the ecosystem actually took less than the 3 years forecast by limnologists. The sediments were not affected as severely as originally suspected.

Active Morphodynamics. Sediment transport and river-bed changes are restricted to those sections where free-flow conditions still exist. This is true for the reach downstream of the last dam at Iffezheim (345 km) as well as for small upstream sections in the old channel with low water levels. Weir constructions in these sections are used to sustain groundwater levels. Restoration of morphodynamics means that the channel must be widened and bank protection works removed. This can be accomplished only in certain areas, because navigation must be guaranteed by an international convention. Uncontrolled sedimentation cannot be tolerated.

Morphological Stability. Erosion problems downstream from Iffezheim might possibly be terminated by the addition of oversized grains instead of a natural grain size distribution. Eventually an armoured river bed might develop hindering further bed erosion up to a certain threshold.

Structural Diversity. Structural diversity is related closely to active morphodynamics of the river bed. Gravel bars, vegetated islands, deep water zones with high flow velocities over coarse gravel, and shallow riparian reaches with muddy layers represent a variety of habitats typical of larger rivers. The restoration of channel patterns is limited to free-flow river sections. Downstream of the last weir construction the requirements for navigation will confine restoration works more or less to ecological improvements at the river banks. Upstream of Strasbourg only short reaches of the old channel bed still have free-flow conditions but with very little water.

Migration of Species.

Weir constructions with backwater reaches are a significant interruption of flowing water ecosystems. Conventional fish ladders only help a few species in migration. Parallel flow systems around the backwater reaches of weirs might be a solution.

Floodplain

Frequent Flooding. The flow regime of the Rhine system is characterized by high flood levels during the summer vegetation growth period. In 1987 lower floodplain areas were flooded for 175 days (Späth, 1988), in 1978 for 210 days and in 1966 as long as 300 days (Dister, 1983). Examination of the vitality of single forest species after flood periods (Späth, 1988) proved that not only are the duration and level of floods critical but the flow velocity of water carrying oxygen is just as important. Downstream of Karlsruhe former floodplain areas may be restored simply by breaking adjacent levees. Upstream of Karlsruhe the operation of detention reservoirs has to consider natural flooding, i.e. natural frequency, storage levels according to natural flood levels and flowing water inside the reservoir. Higher storage levels would be allowed only during severe flood events. Since the intake and outlet structures of the existing reservoirs were designed to manage only the extreme flood events, completely new technical solutions have to be developed.

Variation of Groundwater Level. In those sections where the water levels are fixed and flood flow is confined to the channel, only a new system using the floodplain for flood flow could restore the variations of the groundwater level.

Morphological Diversity and Diversity of Soil Types. The natural floodplain morphology still exists in large areas. In some regions dredging of gravel pits has caused irreversible damage.

Tributaries

Sediment Yield. All tributary rivers are used intensively for water power. Supposedly most of the bedload is retained by weir constructions. Nevertheless higher floods, such as the one that occurred in February 1990, cause considerable damage to river banks and protection channel works, rendering a certain amount of sediment.

Migration of Species and Spawning Habitats. Programmes for the restoration of tributaries as spawning habitats are in consideration.

Prospects for Restoration

Restoration of the upper Rhine is limited to a few ecological improvements unless major concessions are made with regard to current use of land and water resources.

The chances for restoration are more favourable downstream of Karlsruhe, where free-flow conditions still exist.

ACKNOWLEDGEMENTS

I am grateful to P. Larsen for valuable suggestions and for presenting the lecture at the conference as I was unable to attend. I thank A. Dittrich for carrying out the Danube model tests and P. Carling for reviewing the manuscript.

REFERENCES

Braukmann, U. (1991). *Limnologische Untersuchungen bei naturgemäßer Gewässergestaltung-Bedeutung für Planung und Erfolgskontrolle*, Institut für Wasserbau und Kulturtechnik, Publication 180, Universität Karlsruhe, pp. 177–195.

Dilger, R. and V. Späth (1988). *Rheinauenschutzgebietskonzeption im Regierungsbezirk Karlsruhe*, Ministerium für Umwelt Baden-Württemberg Bd. 1, 188 pp., 2 maps.

Dister, E. (1983). Zur Hochwassertoleranz von Auenwaldbäumen an lehmigen Standorten. *Verhandlungen der Gesellschaft für Ökologie*, **X**, 325–335.

Dister, E. (1985a). Taschenpolder als Hochwasserschutzmaßnahme am Oberrhein. *Geografische Rundschau*, **37**(5), 241–247.

Dister, E. (1985b). Auelebensträume und Retentionsfunktion. In *Die Zukunft der ostbayerischen Donaulandschaft*, Seminar, Akademie für Naturschutz und Landschaftspflege, Laufen, pp. 74–90.

Dittrich, A., K. Kern, P. Larsen and R.-J. Gebler (1989). *Modellversuche zur Umgestaltung der Donau auf der Gemarkung Blochingen*, Institut für Wasserbau und Kulturtechnik, Universität Karlsruhe, unpublished report, 47 pp.

Gebler, R. -J. (1989). Fischaufstiege, derzeitige Situation und zukünftige Konzeption. *Wasserwirtschaft*, **79**(2), 64–68.

Geyer, O. and M. Gwinner (1986). *Geologie von Baden-Württemberg*, Schweizerbart, 472 pp.

Konold, W., R. Pfeilsticker and M. Jöst (1991). *Vegetationskartierung der Donauaue bei Riedlingen mit Hilfe von Infrarot-Luftbildern*, Institut für Wasserbau und Kulturtechnik, Publication 180, Universität Karlsruhe, pp. 158–176.

Kunz, E. (1975). Von der Tulla'schen Rheinkorrektion bis zum Oberrheinausbau. In '*Naturschutz und Gewässerausbau*', *Jahrbuch für Naturschutz und Landschaftspflege*, **24**, 59–78.

Lehle, M. (1985). Hochwasserschutz am Rhein für den Raum Mannheim. *Wasserwirtschaft*, **1**, 11–14.

Landesanstalt für Umweltschutz Baden-Württemberg (1987). *Der ökologische Zustand des Rheins und seiner Nebengewässer nach dem Sandoz-Unfall*, Appendix to Gütezustand der Gewässer in Baden-Württemberg 4, Ministerium für Ernährung, Landwirtschaft, Umwelt und Forsten Baden-Württemberg, 12 pp.

Landesanstalt für Umweltschutz Baden-Württemberg (1990). *Gütezustand der Gewässer in Baden-Württemberg 5*, Ministerium für Umwelt Baden-Württemberg, 6 pp., 3 maps.

Petts, G. E., H. Möller and A. L. Roux (1989). *Historical changes of large alluvial rivers: Western Europe*, Wiley, 355 pp.

Raabe. W. (1968). Wasserbau und Landschaftspflege am Oberrhein. *Proceedings of Deutscher Rat für Landschaftspflege*, **10**, 24–31.

Schäfer, W. (1974). Der Oberrhein, sterbende Landschaft? *Natur und Museum*, **104**(11), 334–337.

Späth, V. (1988). Zur Hochwassertoleranz von Auenwaldbäumen. *Natur und Landschaft*, **63**(7/8), 312–315.

Index